T0406342

# Politics of Citizenship and Migration

Series Editor
Leila Simona Talani
Department of European and International Studies
King's College London
London, UK

The *Politics of Citizenship and Migration* series publishes exciting new research in all areas of migration and citizenship studies. Open to multiple approaches, the series considers interdisciplinary as well political, economic, legal, comparative, empirical, historical, methodological, and theoretical works. Broad in its coverage, the series promotes research on the politics and economics of migration, globalization and migration, citizenship and migration laws and policies, voluntary and forced migration, rights and obligations, demographic change, diasporas, political membership or behavior, public policy, minorities, border and security studies, statelessness, naturalization, integration and citizen-making, and subnational, supranational, global, corporate, or multilevel citizenship. Versatile, the series publishes single and multi-authored monographs, short-form Pivot books, and edited volumes.

For an informal discussion for a book in the series, please contact the series editor Leila Simona Talani (leila.talani@kcl.ac.uk), or Palgrave editor Isobel Cowper-Coles (isobel.cowpercoles@palgrave.com).

This series is indexed in Scopus.

Belachew Gebrewold

# Postcolonial African Migration to the West

## A Mimetic Desire for Being

Belachew Gebrewold
Social Work and Social Policy
Management Center Innsbruck
Innsbruck, Austria

ISSN 2520-8896 ISSN 2520-890X (electronic)
Politics of Citizenship and Migration
ISBN 978-3-031-58567-8 ISBN 978-3-031-58568-5 (eBook)
https://doi.org/10.1007/978-3-031-58568-5



This Palgrave Macmillan imprint is published by the registered company Springer Nature Switzerland AG.
The registered company address is: Gewerbestrasse 11, 6330 Cham, Switzerland

Paper in this product is recyclable.

# Acknowledgment

This book is the product of many years of research, and discussion with researchers, decision-makers, colleagues, friends, relatives, etc. The exchange of thoughts and the discussions I had with different people have helped me a lot to reflect on my thoughts and refocus my research and arguments. I thank all those who contributed to the successful completion of this book. Without the financial support of the Tiroler Wissenschaftsfond this research would have not been possible. I thank my university, Management Center Innsbruck (MCI), for the support and facilities it offered me, which greatly helped me complete the research and publication process.

# Praise for *Politics of Citizenship and Migration*

"Belachew Gebrewold takes us on an African journey of critical self-reflection, creatively using Girard's mimetic theory to parse some of the deepest challenges facing the continent. Field research on migration brought him to the realization that many who are well-off want to migrate to the West. "Why?" he asked, and this led to a profound realization that they are on a quest for recognition and being that is driven by an internalized sense of inferiority. They have to prove that they can make it in the West, the model of a colonized mimetic desire. As Girard points out, mimetic desire for an object morphs into ontological mimetic desire—a desire to Be the Other. With this realization Gebrewold puts forth a powerful challenge to Africans—to own their own problems of corruption, social inequality, poor governance, and relative poverty. Rather than blame the former colonizers from the West, he claims that Africans can transform their continent by claiming their own ontological strength as Africans. For those of us who are not Africans, he becomes an exemplar to do our own self-critical reflection in order to take responsibility for our own part in transforming this wonderful world of which we are humble stewards."

—Vern Neufeld Redekop, *Professor Emeritus of Conflict Studies, Saint Paul University, Ottawa, Canada*

"Every now and then a new book opens one's mind and casts fresh light on a much-discussed issue. Belachew Gebrewold's book is of that kind. It argues forcefully that African migration to Europe is not primarily driven by poverty and violence but by a "mimetic desire" to be like the former colonizers and to live among them. Drawing on a wide range of scholarly and literary sources, Gebrewold tells a convincing story with stark implications. A must-read, not only for migration scholars and Africanists but for all who want to better understand the postcolonial condition."

—Rainer Bauböck, *European University Institute, Florence, Italy*

"Postcolonial African Migration to the West by Belachew Gebrewold is an important addition to the literature on African migration. The book meticulously and rigorously explicates a concatenation of economic and extra-economic factors that drive African migration to the geo-political West. The sui generis position taken by Gebrewold compellingly attributes Africans' migration motivations to the West to the non-materialist mimetic desire of Africans for recognition among their peers and social groups. Gebrewold's analysis of endogenous and exogenous causes of African migration to the West is controversial, yet provocative and laudatory. I hasten to recommend the book to those interested in understanding the human condition of Africa vis a vis the influence of slavery and colonialism on Africans' desire for the West."

—Charles T. Adeyanju, *Professor and Chair, Department of Sociology & Anthropology, University of Prince Edward Island*

# Contents

CHAPTER 1

# Introduction

It was in September 1988 that I left my region in the southwestern part of Ethiopia. I moved to Addis Ababa for studies. We were about 16 students coming from different parts of Ethiopia and living in a Catholic students' house, including Eritrea. Most of my colleagues were either from larger ethnic groups or cities. I was one of those few coming from the countryside and a minority ethnic group. In Ethiopia, to come from the countryside makes you a second-class citizen, "uncivilized", and "backward" as it is often called. Urban people are considered better people. If you are coming from a big city and an influential ethnic group (Amhara, Tigray, etc.), you are a first-class person.

My first two formative experiences in Addis Ababa were derogatory attitudes toward minority ethnic groups and the rural people and the rivalry between politically, economically, and culturally dominant ethnic groups in Ethiopia, especially the Amhara, Tigrayans, and Eritreans (before they became independent in 1993). The Oromo, though the biggest ethnic group in Ethiopia, were marginalized and did not play a critical role in the rivalry, at least at that time. The biggest rivalry was among the Amhara, Tigrayan, and Eritrean students. They are not only historically the most dominant groups in Ethiopia but also, they are the most similar to one another: the predominance of the Orthodox Church in all three ethnic groups, their predominantly Semitic and similar languages, and their role in the governance of Ethiopia for hundreds of years until the Ethiopian



B. Gebrewold, *Postcolonial African Migration to the West*, Politics of Citizenship and Migration,
https://doi.org/10.1007/978-3-031-58568-5_1

Monarchy ended in 1974. Often, discussions were going on about which ethnic group was more civilized, beautiful, educated, and influential politically and economically. What fascinated me even at that time was why these quite similar ethnic groups showed the highest potential for rivalry. I was not part of the discussion because my ethnic background was insignificant and I was from the countryside. When I came across the mimetic theory by René Girard during my studies in Innsbruck in the mid-1990s, I was able to better understand the rivalry between these ethnic groups. This theory will help me also in this book to better understand postcolonial African migration to the West.

Moreover, my colleagues in Addis Ababa were talking about their friends and family members who migrated to the West. I knew nobody who had migrated and could not be involved in the discussion. The discussion was characterized by a fascination for the West. My colleagues had more money and better clothes than I had. Until then, I did not have ambitions to go to the West. After having seen what they had and following their discussions, my interest in migration to the West started to increase steadily. Especially, after having finished our study period in Ethiopia and when the chance came to study abroad, all of us opted to study in the West (London, Rome, Paris, Innsbruck, Chicago) even if there were also possibilities to study in Nairobi, Bogota, Sao Paolo, and Lima. For all of us, our ideal option was the West. The West had something magical and fascinating. None of us were, however, ready to admit it. This fascination was completely unknown in my area. Thirty years later, as my discussions with many teachers in schools in my region showed me, the fascination for migration and the West is widespread at the expense of focusing on education. This is not only because the school quality has deteriorated or many graduates are without jobs, but also equally because the mimetic mechanism of migration has become widespread in the different regions of Ethiopia, as the book by Kefale and Gebresenbet (2021) shows, and in Africa in general. This is why I discuss in this book postcolonial migration from Africa to the West from the mimetic perspective. Through a mimetic approach to migration, I argue that postcolonial migration from Africa to the West is not always caused by poverty, conflicts, and climate change.

Since the mass migration to Europe in 2015, Europe has been intensifying its migration policy strategies toward Africa, even if Africa is not the main origin of Europe's migrants. The migration policy discussion is mostly about increasing development aid and economic and political cooperation to prevent or reduce migration from Africa to Europe. The Rabat Plan of Action (2006), the Global Approach to Migration and Mobility (2011), the EU Agenda on Migration (2015), the Valetta Action Plan (2015), the Marshal Plan with Africa (2017), and the Abidjan Declaration (2017) are strategies to curb African illegal migrants coming to Europe. As we shall in Chap. 2, various studies have shown that improving economic situations will not necessarily reduce emigration instead increases it. That is, migration is a sign of an improving economic situation.

Therefore, the main research question of this book project is "Why are postcolonial Africans even who are not acutely affected by poverty, conflicts, or climate change willing to migrate, especially to the West which is often criticized as neocolonial or imperialist?" The main argument of this book is that the African migration to the West is a part of the bigger picture of the postcolonial African desire for being through liberation, recognition, and equality.

Such a desire for being is not particularly African. In general, human relationship is characterized by at least three phenomena: ***dominance and recognition, scapegoating***, and ***mimetic desire***. These three phenomena play a key role in my analysis of the postcolonial African migration to the West. I will discuss ***dominance and recognition*** based on the works of such as Camus, Dostoevsky, Homer, Heraclitus, F. Douglas, Degruy, Achebe, Soyinka, Du Bois, etc. *Scapegoating* will be discussed against the background of Girard and Sophocles. Works by Goethe, Shakespeare, Dostoevsky, Fanon, Memmi, Cervantes, Girard, Lukasik, etc. will help us understand why the postcolonial African desire to migrate to the West is *mimetic*. However, the postcolonial African mimetic desire for being is at least to some extent peculiar because it strives for mental liberation, recognition, and equality against the background of slavery, colonialism, Western cultural hegemony, imperialism, and evangelization. The postcolonial African will to migrate to the West will persist irrespective of conflicts, poverty, and climate change. Postcolonial African migration to the West is not only a flight from poverty, the effects of climate change, and conflicts but also a flight from oneself, from perceived inadequacies; it is an expression of the adoration of the West and an entreaty for recognition.

The purpose of this book is to show how postcolonial African migration to the West is also a psychological liberation process through the imitation of the colonizer against the background of the dehumanizing historical experiences of slavery and colonialism. Based on mimetic theory, this book shows that socio-psychological causes are often disguised as material causes (poverty, conflict, and climate change). The legacies of slavery, evangelization, and colonialism play a key role as causes of migration, in addition to economic, environmental, and political causes or factors. Postcolonial migration from Africa to the West is primordially a mimetic mindset and only secondarily and epiphenomenally a spatial movement. Spatial movement is a historic physical migration from Africa to the West. However, the more important one is the postcolonial African West-oriented migratory mindset or inner disposition, which I call metaphorical migration. This is a postcolonial African inner disposition for a mimetic *desire for being* by "whitening" oneself, if possible, both physiologically and metaphorically. It is an expression of the general and pervasive African desire for liberation from the dehumanizing colonial legacies and experiences through imitation of the colonial white West and the rejection of hierarchy, difference, and inequality. Postcolonial Africans have been trying to gain their identity and subjectivity not through detachment from the dehumanizing colonizer but rather through imitation of the colonizer's lifestyle and by striving to live together with him. The colonization of Africa is not only a physical-historical but also a mental-psychological subjugation by the West with long-lasting effects even for non-colonized countries like Ethiopia. My analysis applies even to us Ethiopians who were not colonized, as we are also affected by the culturally, technologically, economically, and esthetically hegemonic West.

As a consequence of Western power and hegemony, we Africans have lost self-respect and dignity. We seek our liberation and dignity through the help of the West and not by ourselves or through detachment from the West. Africa's poor institutions that keep us in poverty, destitution, unending conflicts, and state failure are the outcome of the interaction of external (colonization, slavery, and Western hegemony) and internal cultural factors (patriarchy, patrimonialism, and autocracy). Colonization was a double tragedy for Africans. First, through colonization, Africans were dehumanized and exploited. Second, because of the colonized mind, Africans blame colonization and the West for almost all their current ills: a condition of perpetual victim-mentality and self-pitying. As a consequence of the dehumanization, the colonized personality is on one hand affected by an inferiority complex, and on the other hand it desires equality with the colonizer through imitation of the latter. Therefore, the relationship between the West and Africans in general and African migration to the West, in particular, is determined by three forms of mimetic and scapegoating mechanisms. First, the slaveholders and colonizers imitated one another in their brutality against Africans, and then the intra-Western rivalry was appeased during the Berlin Africa Conference (1884–85) in which the Western powers agreed to divide Africa among themselves. Mimetically speaking, the colonial violence against Africans was both a *conflictual mimesis* (conflict among the Europeans was transferred to the Africans) and an *acquisitive mimesis* (colonial powers competed with one another to exploit Africa predatorily) (Girard 2019: 25). Second, external powers (the West, China and Russia) compete with one another in the systematic exploitation of Africans. Third, Africans and the East (Russia, China, Arab Countries) collectively scapegoat the West for all the ills of Africa because of colonization. African leaders are often slavishly allied with Eastern powers (Russia, China, etc.) against the West economically and politically. They ignore the fact that now Africans are not mere victims of the international economic, security, and political system that pushes them to migrate to the West. Africans slavishly adoring the West on the one hand and scapegoating it for their poor economic and political performance on the other hand is paradoxical and mimetic.

The Western ideals of beauty, lifestyle, and consumption have become the models for postcolonial Africans. The model is delighted that he is being imitated but alarmed that he could be on the same level as his imitator. The imitator hates the model because he builds fences and borders to exclude him. At the same time, he loves and admires him because of his wealth, military power, technology, political stability, lifestyle, and beauty ideals. The colonizer hates the postcolonial African migrants because they endanger his superiority and security by attempting to be equal. At the same time, he loves them because, through their attempts, they prove the colonizer's superiority, admirability, and desirability. Even Dostoevsky questioned why all admirable things must come to Russia from Europe. Indirectly, through his criticism of Europe because of its individualism (which Russians eagerly imitate), Dostoevsky admits the temptation to imitate, adore, admire, and desire the West. In his *Winter Notes on Summer Impressions*, he admits Russian inferiority and inner contradiction:

> After all, everything, literally almost everything we can show, which may be called progress, science, art, citizenship, humanity, everything stems from there, from that land of holy miracles. The whole of our life, from our earliest childhood, is shaped by the European mold. Could any one of us have withstood this influence, appeal, and pressure? (Dostoevsky 2021 [1863]: xi)

Postcolonialism revisits, remembers, and interrogates the colonial past, which consists of colonial oppression and the compelling seduction of colonial power. It assesses the relationship between reciprocal antagonism and desire between colonizer and colonized, i.e., the ambivalent prehistory of the postcolonial condition. As a therapeutic or cathartic retrieval of the colonial past, postcolonialism is not only willing to make but also to gain a theoretical sense of the past (Gandhi 1998: 4–5). There is perverse longevity of the colonized in the status of "colonizedness" as the colonial aftermath does not end with the end of colonialism. The colonizer and the colonized are paradoxically dependent on one another, a perverse mutuality. The colonized has a strong desire for independence from the colonizer, but on the other hand, he still has a strong longing for the colonizer. The colonized hates the colonizer but at the same time admires him passionately (Memmi 1968: 45; Gandhi 1998: 11). Postcolonial theories study the simultaneous hate and desire of the colonized and their voyeuristic gaze upon Europe and schizophrenic behavior (Gandhi 1998: 11). Therefore, one has to study contemporary migration from Africa to the West in the postcolonial context. As Gandhi puts it, "… Europe they know and value so intimately is always elsewhere'. Its reality is infinitely deferred, always withheld from them. Worse still, their questing pursuit of European plenitude, their desire to own the colonizer's world, requires a simultaneous disowning of the world which has been colonized" (Gandhi 1998: 12).

Psychoanalystically, for the colonized mind, postcolonial migration from Africa to the West is a cathartic-liberating process or salvation that takes place through "baptism" in contradictory natural events such as the Sahara Desert and the Mediterranean Sea on the way to the Promised Land (the West). Postcolonial African migration leads specifically to the West. The money of corrupt African politicians is deposed in Western banks. The most desired material goods come from the West. Lifestyle is copied from the West. Ideally, one studies in the West. The West has something overpowering or fascinating, magical or supernatural. As Ashis Nandy would call it, this is an "intimate enmity" between the former colonizer and postcolonial Africa (Nandy 1983).

Two of the long-lasting impacts of European slavery and colonization in Africa are the European dehumanization of Africans on the one hand and the African elites' justifying all African institutional failures, poverty, and conflicts in Africa in the light of slavery and colonization, on the other. African political elites love to instrumentalize African historical suffering for their own benefit. The Chinese and Russians know how to take advantage of this and mobilize Africans against the West.

Against this background, the objective of this book is to discuss the postcolonial migration to the West both literally and metaphorically or psychoanalytically. In Chap. 2, I briefly discuss migration policies and theories. The motivation, causes of migration, and challenges at destination have been widely researched. What, however, I miss in the existing migration research is a deep-going interpretation of the hidden or social-psychological causes of postcolonial migration from Africa to the West. In this book, I am not interested in general African migration, but rather in postcolonial African migration to the West. Therefore, the sociopsychological legacies of slavery and colonialism are the center of this study.

In Chaps. 3 and 4, I will discuss the economic and political dimensions of postcolonial African migration to the West. The objective of these two chapters is to show that Africa's ills are based on our permanent scapegoating of the West for all our incapability, authoritarianism and patriarchy, intolerance for opposing views, indifference toward the well-being of our fellow Africans, our inability to overcome ethnicized politics and construct a common national interest and identity. We are incapable of managing our natural resources, our economies, and our environment. It is ridiculous to suggest all the time that our economic or political problems come from global economic and political systems. Slavery and colonization cannot explain everything. We beg around the world instead of thinking about how to improve our politics and economy in Africa. We are not able to reflect on how we are making ourselves ridiculous around the world as barbarians killing each other, letting our people starve, our women being treated as mere objects, and adoring the white West as a semi-god. We do not have the ambition and determination to be prosperous, to be globally respected, to prove the white racists wrong, to construct a collective identity, to leave behind this archaic ethnicized mentality, and to become global big players sustained by prosperous, peaceful, and democratic African societies. Our only strength is to beg, to blame the West and colonial time for everything, and to blindly follow China and Russia.

We kill each other in wars as no other region in the world does; capital flight is one of the highest in the world; corruption is the highest in the world; and most African countries are the wealthiest in the world in terms of natural resources. However, Africa remains the poorest, bloodiest, and most corrupt in the world. Our well-meaning European liberals try to explain that Africa is poor because of colonial exploitation and the global economic system. They try to console us that Africa is affected by many conflicts because of its colonial borders. This is simply not true. Africa is poor and conflictual only because Africans are not willing to do better. If Africans were able to think and work for Africans, with Africans, and not against Africans, the Mozambican civil war between 1981 and 1992 would have not killed about a million people, the Angolan civil war between 1975 and 2002 more than a million people, the Liberian civil war between 1989 and 2003 more than half a million people, the war between Ethiopia and Eritrea between 1998 and 2000 about 70,000 people, the civil war in Ethiopia between 2020 and 2022 more than half a million people, and in the Democratic Republic of Congo between 1996 and 2023 approximately 6 million people (Hanlon 2010; Council on Foreign Relations 2023; World Peace Foundation 2015). The list goes on. Some would counterargue, and say, Africa is not that bad; many countries have shown a lot of progress since their independence, etc. This is true, but Africa could do much better. Africa has the resources, enough intellectuals, economists, engineers, scientists, etc. Africa could do much better than other colonized regions are doing. However, we Africans do not have the mentality and self-respect to prove the white racists wrong and to make our people prosperous. The colonization of our minds, the patriarchy, ethnic thinking, autocracy, and scapegoating of the West all the time for our shortcomings keep us always in poverty and violent conflicts, and we look for our salvation always in the West through self-degradation.

The racists animalized us, and we have internalized this animality exactly as we were taught by them. Even today, we adore and admire the West to the point of self-subjugation and self-degradation. The enslavement and colonization of our minds have worked perfectly. Postcolonial Africans migrate mentally and physically to the West irrespective of how Europeans treat them at their destination or in transit. Even if they dehumanize us, treat us like animals, and despise us, we still migrate to the West. Not because our situation in Africa is hopeless. We do this because we are convinced that we are nothing if we are not in the West. We have made Africa hopeless, and we have made our people believe that Africa is hopeless. Beauty, power, technology, modernity, humanity, civilization, and civility are only found in the West. We love the West at any cost. We willingly dehumanize ourselves in our dream of being in the West. As Aimé Cesaire would say, we Africans "thingify" ourselves. Instead of solving our political and economic problems in Africa, we prefer to be refugees or beggars in the West as politicians, intellectuals, or ordinary citizens.

The origin of all this self-degradation is slavery and colonialism. Therefore, in Chap. 5, I discuss the legacies of slavery and colonialism and postcolonial African self-perception to show how it has contributed to postcolonial African migration to the West. I have chosen the sub-title of this book "A Desire for Being" against the background of the *nothingness*, *nullification*, *thingification* (Aimé Césaire), "Ver-*Nichtung*" (negation, destruction, making something to nothing in German) of African personality due to slavery and colonization by the whites. African migrants to the West attempt and desire for *being* by overcoming this negation and *becoming* a bit white. Therefore, postcolonial African migration to the West is a desire for liberation from these *thingifying* legacies of slavery and colonization by a paradoxical mixture of hatred and admiration of the violent white West simultaneously. Migration is the desire to become and to be. It is a rite of passage and a way of striving for recognition.

In Chap. 6, we will see how migration, migrants, and (non)successful returnees are perceived locally. The migrant's desire to be recognized at the origin as a successful imitator of the colonizer's lifestyle is a driving factor in migration motivation. A migrant is a mirror image of the "superior" colonizer and a model for those at the origin. Our African human dignity is dependent on our imitation of the West. All good things come from the West, and Africa is a symbol of failure. That was what the slaveholders and colonizers maintained. We have learned it from them obediently and internalized it. This mindset has become an intrinsic part of our nature. Living in the West gives us a higher social status and recognition, as I will discuss in Chap. 6.

In Chaps. 6 and 7, I will discuss postcolonial African migration to the West from the mimetic perspective to show that there is something beyond poverty, conflict, or climate change that drives postcolonial African migration to the West. The desire for recognition and to be equal with whites not through detachment but instead through the imitation of the Western way of life, if possible, by living in the West is a decisive motivation behind the postcolonial African migration to the West.

From the postcolonial point of view, the theoretical approach of the book focuses on the historical negation of African humanity by the colonizer and on African mimetic desire for being. As Frantz Fanon suggests, for postcolonial Africans "*to be*" means to be with and like the former colonizer. As Chap. 7 shows, colonized and migrating African subjectivity wants to *become* and *to be* through imitation of the white West.

From the existentialist perspective, migration is a liberation process, as a drive for freedom, to break the colonial and neocolonial bondage of the categorization of African subjectivity as uncivilized and backward (Mbembe 2001). Postcolonial African migration to the West provides hope for a new identity. Some of the theoretical literature I will be using for this part will be René Girard, Frantz Fanon, Albert Memmi, Aimé Césaire, Jean-Paul Sartre, W.E.B. Du Bois, Chinua Achebe, Wole Soyinka, Chimamanda N. Adichie, Fyodor Dostoevsky, Miguel de Cervantes, Albert Camus, Goethe, Shakespeare, and Ngugi wa Thiong'o, etc. I will interpret migration as a capability to choose, to reject an imposed identity, to desire the forbidden, to break down the historical confinement imposed by the colonizer, and at the same to imitate him. To migrate means to be able to choose, and to choose means to be, as Jean-Paul Sartre would say.

Mimetic theory is the cornerstone of my theoretical approach to migration analysis. The mimetic theory was developed by René Girard. The model, desired object, and imitator constitute the mimetic triangle. The imitator desires the model's desired object. The fact that the imitator desires the same object possessed by the model increases the value and desirability of the desired object, as Goethe, Cervantes, Dostoevsky, Shakespeare, etc. show us. The imitator and the model exchange imitations reflexively (Girard 1965, 2013, 2023).

Because of colonial and neocolonial experiences, the colonized (*the imitator*) has an ambivalent relationship with the colonizer (*the model*). The ambivalence is because the model is violent as a colonizer, but at the same time, he is idealized and mystified for his economic and technological achievements, his globalized culture, consumer goods, ideals of beauty, etc. (Degruy 2017: 128). He and his achievements are imitator's *objects of desire*. In the face of this ambivalent relationship, the colonized being simultaneously experiences a deep-going inner conflict between hatred toward the colonizer and fascination for him. The hatred leads to scapegoating, whereas fascination leads to adoration. Postcolonial African migration to the West is a continuation of this fascination. Migration is a process of becoming a bit Western by imitating the colonizer by consuming the objects of desire, and ideally living together with him.

The model (colonizer) builds the walls higher and higher and is omnipresent and attempts to keep the imitator (the migrant) in his place. The imitator attempts at any cost to surmount all forms of borders to be able to enjoy the objects of desire not necessarily at his origin but specifically together with the model in the West. Migration is the imitator's desire to be equal to the model. In the mimetic relationship, the higher the desire of the imitator to be with the model, the higher the separating walls the model erects. Moreover, the borders of the colonizer will be mobile. Therefore, the mimetic relationship between the model and the imitator consists of the mutual intensification of erecting and modernizing the walls by the former and the relentless effort to break through them by the latter.

The postcolonial person believes that its salvation is in the white West. We postcolonial beings adore the West instead of improving our political and economic institutions. Ethnic violence, Islamist terrorism, dictatorship, one military coup after the other, corruption, nepotism, autocracy, etc. devastate the postcolonial man. In confrontations with the West, Africans always boast "We Africans!", etc. but within our own countries, we do not show any respect for others' lives. The brutal genocide in Rwanda, the civil wars in Ethiopia, Sudan, Ivory Coast, Guinea, Liberia, Sierra Leone, the Democratic Republic of Congo, and the barbaric Jihadists in Somalia, Mali, Niger, Burkina Faso, and Nigeria show that African lives do not value much. We have internalized the colonizer's mindset that Africans are sub-humans. How can we accuse the colonizer of barbarity if we are not better? As Aimé Césaire and Frantz Fanon say, the colonizer saw blacks as animals and treated them as animals (Césaire 2000: 41; Fanon 2001: 111). The tragic thing is that we have internalized this animality. Our problem now is no longer the colonizer. We are our problem. How can we, as African politicians and citizens, imitate and adore the West in such a degrading way if we had self-respect? Why do we make life in Africa so hopeless, mismanage the economy so catastrophically, treat anyone who has a different view as an archenemy, and stay in power for decades or life as if the power of the state were private property? Why do we make ourselves so despicable on the global stage? Even liberal thinkers in the West think that we are incapable of governing ourselves. Even those who are not racists think that we are only a bit better than mere animals. Are we willing to disprove the white racists that we are not mere animals but instead equal human beings with the capability to achieve what the whites have achieved? Are we determined to stop scapegoating the West, self-pitying, and ready to shape our history by ourselves?

As Chap. 8 shows, our journey to the promised land—often to our disgrace—has been possible through the decreasing resource gap, liberty gap, and knowledge gap. These decreasing gaps between origin and destination are accelerating this postcolonial African mimetic process and migration to the West. The mimetic desire between the colonizer (model) and the postcolonial man (imitator) intensifies: the more the imitator is rejected by the model, the more the imitator adores the model; the more the model notices this adoration, the more he feels superior, strengthens the border, loves the imitator for the adoration. However, the imitator should stay in his place. Stendhal's "*The Red and the Black*" describes the interesting reciprocal relationship between rejection and adoration (Stendhal 2002). In this relationship, the model unconsciously becomes the imitator of his imitator (Girard 1965, 2023). This is exactly the Heraclitan *panta rei*; there is only fluidity and fluctuation between seemingly opposing roles and positions. In Chap. 9, I sum up my main arguments and the problem of this postcolonial African mimetic process.

Are we postcolonial Africans willing to admit our denied desire to be at least a bit white both literally and metaphorically? This is something shaming, therefore, we deny it. It was when I noticed this denial that I decided to change my research method for this book project. I started to conduct my field research in 2019 in Ethiopia and collect data on poverty, conflicts, and impacts of climate change as causes of migration. I had discussions on this with many young people eager to migrate to the West. However, during the interviews, in addition to poverty, conflicts, and impacts of climate change, I noticed a pervasive admiration of the West and a desperate desire to go there. All the young people admired the superiority of the West and the whites. Even their admiration and respect for me was high because I have lived in Europe. However, in formal research interviews, they always underlined the economic, political, economic, or environmental causes of their intention to migrate. It was only through informal discussions that I was able to observe the underlying and implicit adoration of the white West. Therefore, I changed my original research plan and shifted my interest to why Africans adore the West so much and what are the underlying causes for this. Similarly, in 2017, I had discussions with migrants from different African countries living in South Africa. Instead of classical interviews, I changed to informal discussions. I could observe how all of them would have preferred migrating to the West. In the discussions during my stay in Kenya in 2022, it was the same observation. Regarding the causes of migration, in formal interviews and informal discussions, I got different answers. In the former case, poverty, climate change, and conflicts are often mentioned as the causes. However, in informal discussions, what was conspicuous was the admiration and adoration of the West.

These implicit motivations of migrants in which I am interested could not be extracted from formal interviews. I could only extract the hidden reasons from different informal discussions sitting together during coffee; when I visited former school friends and relatives; and when I participated in conferences and workshops in different countries I visited for various reasons in Ethiopia, Kenya, South Africa, and Djibouti. I had some informal discussions with some employees at international organizations. In all those informal discussions with African colleagues, what I was able to notice was the hidden admiration of the West or the denied desire for "Whiteness". Therefore, I decided to focus on it and analyze it from a mimetic perspective.

In informal discussions as well as in the analysis of various migration-related surveys in Africa, I had opportunities to observe that we Africans suffer from a pervasive inferiority complex and entertain a slavish adoration, admiration, and fascination for the West. Some people I know who had jobs and had lived a respected life back home and went to the West as tourists threw away their passports, applied for asylum, and claimed that they were persecuted politically. However, in Europe, they had to live as beggars, homeless, and without money to buy food. Some went from Ethiopia to South Africa and became economically successful but their dream is still to move to the West.

Every time I go to Ethiopia, I meet people almost every time who call me "Ferenji" (White, European), even if I am completely black and was born and grew up in Ethiopia, even though I have been living in Europe since 1993. The hidden message is that even if you are black, as long as you are living in the West, you are almost "white", i.e. you are better, even if the reality of those blacks living in the West is often just the opposite. The objective of this book is to explore the underlying causes of this self-degrading and denied desire for being which manifests itself in the form of postcolonial African mental and physical migration to the West.

## Literature

Césaire, Aimé (2000): Discourse on Colonialism, Monthly Review Press: New York.

Council on Foreign Relations (2023): Conflict in the Democratic Republic of Congo, https://www.cfr.org/global-conflict-tracker/conflict/violence-democratic-republic-congo, 13.07.2023.

Degruy, Joy (2017): Post-Traumatic Slave Syndrome: America's Legacy of Enduring Injury and Healing, DeGruy.

Dostoevsky, Fyodor (2021 [1863]): Winter Notes on Summer Impressions: Alma Classics, London.
Fanon, Frantz (2001): The wretched of the earth, Penguin Books: London.
Gandhi, Leela (1998): Postcolonial Theory: A critical introduction, Columbia University Press: New York.
Girard, René (1965): Deceit, desire and the novel: self and other in literary structure, John Hopkins University: Baltimore.
Girard, René (2013): Violence and the Sacred, London: Bloomsbury.
Girard, René (2019): Things hidden since the foundation of the world, Bloomsbury Revelations: London.
Girard, René (2023): All desire is a desire for being, Penguin Classics: London.
Hanlon, Joseph (2010): Mozambique: 'The war ended 17 years ago, but we are still poor', *Conflict, Security & Development*, 10:1, 77–102, https://doi.org/10.1080/14678800903553902.
Kefale, Asnake and Gebresenbet, Fana (2021): Youth on the Move: Views from below on Ethiopian International Migration, Hurst and Company: London.
Mbembe, Achille (2001): On the Postcolony, University of California Press: Berkeley.
Memmi, Albert (1968): Dominated Man: notes toward a portrait, Orion Press: London.
Nandy, Ashis (1983): The Intimate Enemy: Loss and Recovery of Self Under Colonialism: Oxford University Press: New Delhi.
Stendhal (2002): The Red and the Black, Penguin Classics: London.
World Peace Foundation (2015): Mass Atrocity Endings: Angola: civil war, https://sites.tufts.edu/atrocityendings/2015/08/07/angola-civil-war/, 13.07.2023.

CHAPTER 2

# Challenges in Determining Migration Causes: Policies and Theories

## European Migration Policy Toward Africa

The migration surge to Europe has underscored the already migration policy-dominated approach of Europe toward Africa. The 2004 Hague Program for the period of 2005 to 2010 underlined the external dimension of migration policy, which means addressing the root causes as well as fighting irregular migration to the West. The EU Commission's Communique of 2006—"Comprehensive European migration policy"—outlined its objective to address the root causes of migration from sub-Saharan Africa within the context of the Cotonou Agreement (2000–2023) between the EU and the African, Caribbean, and Pacific States. In 2011, the European Commission issued a Communication on the Global Approach to Migration and Mobility. The third pillar of the Communication envisages promoting international protection and enhancing the above-mentioned external dimension. The EU Global Strategy of June 2016 places migration in the EU's external action priorities. The Euro-African Partnership for Migration and Development (Rabat Declaration 2006) underscored the importance of cooperation with countries of origin and transit to address and prevent the root causes of displacement and migration.

The European Union created three ways of cooperating with Africa on migration: first, on the continental level, Africa–EU Migration, Mobility,



B. Gebrewold, *Postcolonial African Migration to the West*, Politics of Citizenship and Migration,
https://doi.org/10.1007/978-3-031-58568-5_2

and Employment Partnership, and EU– Africa Summits as a framework. Second, on the regional level, there are the Rabat Process, the Khartoum Process (Migration Route Initiative with the Horn of Africa in 2014), and Regional Development and Protection Programs in North Africa and the Horn of Africa. Thirdly, bilateral mobility partnerships exist, such as with Cape Verde (2008), Morocco (2013), and Tunisia (2014); Common Agenda on Migration and Mobility, such as with Nigeria (2015) and Ethiopia (2015). The EU Horn of Africa Regional Action Plan adopted in October 2015 sets the EU Strategic Framework for the Horn of Africa. The Sahel Regional Action Plan 2015–2020 of April 2015 sets a framework for the EU strategy against the causes of migration. Similarly, the 2015 Agenda on Migration aims to reduce irregular migration by addressing the root causes, border management, protection of the vulnerable, and legal migration.

The Valletta Action Plan of 2015 established the EU Emergency Trust Fund for Africa with a general budget of €4.09 billion (in 2023 already more than €5 billion) to finance projects related to the root causes of migration in the Sahel and Lake Chad, the Horn of Africa, and the North of Africa regions in the following policy areas: economic development and improving employability; strengthening resilience for improved food and nutrition security; migration management by addressing the drivers of irregular migration; and enhancing synergies between migration and development (European Commission 2020).

The EU has been working with the countries of the Sahel and Lake Chad region in the improvement of governance and conflicts, migration management, strengthening resilience, and economic and employment opportunities in more than 100 projects with a budget of more than €1.9 billion. Similar projects have been established in the Horn of Africa. The Horn of Africa region faces challenges that go beyond country borders: climate change, forced displacement, demographic pressures, environmental stresses, various forms of conflict, trafficking of human beings, and smuggling of migrants, as well as organized crime and violent extremism. In this region, the EU has been supporting 99 projects funded by more than €1.9 billion. The projects include improved governance and conflict prevention, migration management, strengthening resilience, and economic and employment opportunities.

Tunisia and Morocco have played a significant role in preventing migration from Africa to Europe. In 2021, however, Morocco allowed migrants to enter Spanish territory, and within hours, Spain approved $37 million in aid to Morocco for border policing. Italy pushed the IMF in 2023 to release a stalled $1.9 billion loan to Tunisia to prevent African migrants from moving from Tunisia to Europe (Jackson 2023). On 11 June 2023, the European Commission tasked the Minister of Foreign Affairs, Migration and Tunisians Abroad and the Commissioner for Neighborhood and Enlargement to work out a Memorandum of Understanding on the comprehensive partnership package, to be endorsed by Tunisia and the European Union before the end of June 2023 (European Commission 2023).

The Union for the Mediterranean (all countries of the European Union and 15 countries of the Southern and Eastern Mediterranean) since its relaunch in 2008, the 5+5 Western Mediterranean Union (Algeria, Libya, Mauritania, Morocco, and Tunisia, the 5 EU countries are FR, IT, MT, PT, and ES), the Rabat Process that started in 2006, the Tripoli Process of 2006, the Khartoum Process of 2014, etc., have become the central instruments in European migration management. Consequently, facilitating legal migration, fighting irregular migration, and strengthening synergies between migration and development became the key issues of discussion in various multilateral meetings and cooperation agreements. Ministerial meetings in Paris in 2008, Dakar in 2011, Madrid in 2013, and Rome in 2014 prepared the way for the EU-Africa summits in Brussels in 2014 and Valetta in 2015, when African migration was the top agenda. Those African countries that cooperate with the EU's objectives will receive rewards with greater levels of financial support. (Collyer 2016: 614). The EU migration policy focuses on the so-called drivers and facilitators of migration, training, and capacity building to prevent African irregular migration to Europe (Collyer 2016: 618).

Between 2008 and 2015, approximately €300 million were spent by the European Union in West Africa on projects envisaged mainly to stop irregular migration (Bisong 2019: 1304). Additional sums had been offered to Nigeria, Senegal, etc. to curb the root causes of migration. Countries like Germany have been intensifying their cooperation, such as with Senegal, on migration, to create of better opportunities for returnees, skills development, vocational training, and promoting business contracts with German companies. Cooperation with Mali and Niger also focuses on reducing the number of transit migrants by providing support to transit communities and promoting sustainable agriculture and reconstruction, thus reducing the pressure to migrate. Before the situation started to get complicated after the military coup and increasing Russian influence, Mali was also supported by the German military to stabilize the uprising in the country. "All these measures are aimed at stemming the root causes of migration" (Bisong 2019: 1304).

The Frontex budget increased from €100 million in 2014 to €280 million by 2017 and to €400 million in 2020. For the period between 2022 and 2029, the EU has allocated to Frontex €5.6 billion. Frontex personnel is expected to grow to 10,000 by 2027 instead of 6500 by 2021. Due to migration, European borders are expanding as far as Senegal and Niger (The Economist 2021a: 19).

Muammar Qaddafi of Libya had blackmailed Europe and demanded €5 billion from the EU to stop African migrants from crossing the Mediterranean lest Europe become black (The Economist 2021c: 26). Since the North Africa region has been considered an area of origin, transit, and final destination for mixed migration flows from sub-Saharan Africa, West Africa, the Horn of Africa, and the Middle East, more than 30 projects are being financed with almost €750 million by the European Commission (European Commission 2020). The EU has been cooperating with different states in the origin as well as transit regions. The European Union has been training and equipping the Libyan coastguard to prevent irregular migrants from reaching Europe. Libyan detention centers are full of African irregular migrants who are mistreated and their human rights are horribly violated through torture, detention, slavery, forced labor, and sexual violence. The detention centers are often run by militias who make a lucrative business with the African migrants whom they force to ask their families to send money for their release, with an average amount of $500 per person. Some detainees tell their families back home to sell their farms to pay for their release. Some detainees are forced to work on construction sites, on farms, or forced into prostitution (The Economist 2022j: 32). Detainees attempt to kill themselves, and guards shoot and kill them. Some reports show that the EU tried to keep individuals and NGOs from aiding migrants. Italian ports closed ports to ships run by humanitarian groups such as the MSF. European navies halted their rescue operations (The Economist 2022j: 32).

Despite workforce demand in many European countries, new permits for sub-Saharan Africans to work in Europe decreased dramatically from 33,000 in 2008 to approximately half that number in 2018. In Europe, the average income is 11 times higher than the average African income. As most African migrants often do menial jobs in Europe, they earn on average only three times what they did back home (The Economist 2021c: 25). Nevertheless, remittances are key to Africa's economy. Whereas foreign direct investment was only $31 billion in 2019, almost $50 billion in remittances flowed into sub-Saharan Africa, including those from sub-Saharan Africa itself (40%). At the same time, European-funded anti-migration policies make campaigns and spread information in Africa about migrants' deaths and challenges they face, such as begging, and that everyone will hate them for it (The Economist 2021c: 27).

Migration started to become a business not only for migrants, human traffickers, or smugglers but also for example for the West African States themselves, who started to attract additional funding from Europe for security and economic development to negotiate better treatment through aid in return for combating irregular migration (smuggling and trafficking in persons) and readmission and return agreements (Bisong 2019: 1304).

In the remaining part of this chapter, I will discuss different migration theories. Before moving to the mimetic approach to postcolonial African migration to the West in the chapters that follow, I shall discuss the various theories that explain the causes of migration.

## Migration Theories

Migration is defined as an individual or collective change of abode through spatial movements after considering the advantages and disadvantages (Saunders 1956: 221; Eisenstadt 1954: 1). O'Rourke (2014: 528) defines migration as the relocation of people(s) from one place to another caused by economic, social, and ecological forces, which ultimately affect the genetic structure of the human population. The IOM defines a migrant as "a person who moves away from his or her place of usual residence, whether within a country or across an international border, temporarily or permanently, and for a variety of reasons" (IOM 2023).

The Cambridge Survey of World Migration by Robin Cohen (2010) shows migration movements beginning in the sixteenth century with European colonialism and covers the history of migration to the late twentieth century. Then, it shows the role of global communications and transport systems and how they accelerated the movements of labor migrants and skilled professionals, the flights of refugees, and illegal migration. This wide-ranging coverage of migration is an important contribution to understanding the causes of migration dynamics. The Encyclopedia of European Migration and Minorities: From the Seventeenth Century to the Present (2013) by Klaus J. Bade shows that migration movements, integration, and multiculturalism have always been part of Europe's history. This encyclopedia of migration contains survey studies of various regions and countries in Europe and presents information on different groups of migrants in different countries. An edited volume by Michael H. Crawford and Benjamin C. Campbell, *Causes and Consequences of Human Migration: An Evolutionary Perspective* (2014) provides an evolutionary perspective on human migration past and present. The volume edited by Johannes Knolle and James Poskett (2020), which provides new insights into the making of the modern world, raises important questions from multi-disciplinary perspectives and discusses what migration is, how it has changed the world, and how it will shape the future.

Even before the dramatic rise in immigration to Europe, African migration has been a key area of research. In the volume edited by Berriane and de Haas (2012), different migration researchers explore the diversity of African migration patterns, migration within and from Africa. Bets and Collier (2018) suggest that refugees need havens, whereas migrants hope for honeypots (2018: 30). In his book "*Exodus*", Paul Collier (2014: 37) argues, "The wide gap gave people in poor countries a powerful economic incentive to move to rich ones." In Chap. 8, I argue that not the wide gap but rather the decreasing gap between Africa and the West that spurns African migration.

In the volume edited by Ippolito and Trevisanut (2015), the authors show how international organizations and actors cooperate to prevent, control, and manage migration in the Mediterranean. Against the background of migration influx, especially since the Syrian conflict, migration governance has become an urgent policy area, as shown in the volume edited by Geddes et al. (2019). From the perspectives of transnationalism, decolonization, climate change, development, humanitarianism, bordering, technologies, and so on, the edited volume by Carmel et al. (2021) analyzes how governance systems play a key role in shaping migration. In the above-mentioned studies, the presumed drivers of migration, such as political insecurity, poverty, and climate change, are considered the main objectives of migration governance systems.

Stephen Castles' study on the interrelation between development and migration raises the question of which is the cause and which is the effect (Castles 2009). Similarly, the edited book by Bakewell (2021), which contains many very interesting case studies from around the world, discusses the linkages between migration and development, the impact of (under-) development on migration, the impacts of remittances, inequality, poverty, transnationalism, diaspora, etc. on migration.

Stephen et al. (2014) argue that migration is an integral part of social transformation. Similarly, Kane and Leedy (2013) make very interesting contributions regarding psychological, sociocultural, and political dimensions of African migration, trans-local and transnational connections, and show that migration is a factor of cultural change abroad and at home. In the volume edited by Gebrewold and Bloom, the authors show that migration causes and migration decision-making processes are multifaceted (Gebrewold and Bloom 2018; Castles 2010). Several driver complexes may interconnect to shape the eventual direction and nature of the movement (Van Hear et al. 2018; Clemens and Postel (2018). Various researchers argue that migration is not always caused by poverty, conflicts, or climate change but is also part of a social transformation process (Castles 2009; Castles et al. 2014).

The volume edited by Carrera et al. (2019) extensively examines EU migration and asylum policies in times of crisis; it assesses patterns of cooperation in the EU's migration management cooperation with third countries. One important aspect of this volume is the external dimension of EU migration policies. One of the most comprehensive recent books on migration studies is the Oxford Handbook of Migration Crises edited by Menjívar et al. (2019). This volume covers a wide range of historical, economic, social, political, and environmental conditions that contribute to migration. The Handbook of Migration and Global Justice by Leanne Weber and Claudia Tazreiter (2021) interrogates the intersections between migration and global justice from multidisciplinary and geopolitical perspectives. It explores the link between cross-border mobility and migration on one hand and rapid economic, cultural, and technological globalization, global injustice, failures in refugee protection, worker exploitation, and violence against migrants on the other.

For most researchers, migration causes are usually humanitarian crises, armed conflict, environmental catastrophes, poverty, social exclusion, and unemployment (Bets and Collier 2018). Grieveson et al. (2021: 10) suggested that the wealth gap, wars, and climate change will continue to "push" migration from Africa toward the EU. Melachrinos et al. argue that "conflict, economic hardship, poor governance, deteriorating political situations and social exclusion of marginalized groups have the potential to internally displace entire communities or force them to leave their homes to seek refuge in other countries" (Melachrinos et al. 2020). Conflicts, poor payment, poverty and lack of work are often suggested reasons for asylum seekers and migrants (Mixed Migration Centre 2019).

An African Union and IOM study from 2020 also underlines that climate change, environmental degradation, political, social, and economic tensions and conflicts, population growth, urbanization, improving economic development, and a higher level of education are push factors of African migration (African Union and IOM 2020). As Black (2011) suggest, migration can bring opportunities for coping with environmental change, but it also changes the social and ecological world (Goldin et al. 2011). Adepoju argues that wars, civil unrest, and superpower competition in postcolonial Africa generated economic migrants, refugees, and IDPs (Adepoju 2010: 166).

Some researchers have categorized migration causes into push (at origin) and pull (at destination) factors. The imbalance between the "desired standard of living" and the "actual scale of living" pushes people to migrate. As will be shown later in this book, the "desired standard of living" is relational, comparative, continuous, elastic, elusive, and therefore mimetic. Real or imagined opportunities at the destination have a great influence on the location choice of subsequent migrants who follow the paths of forerunners (Piore 1979; Massey et al. 2006: 36; Todaro 1976; de Haas 2010: 1589; Czaika and De Haas 2017). Therefore, this network is an essential social capital that provides newcomers with transportation, initial accommodation, and information on employment opportunities, and it contributes to the decrease of the economic, social, and psychological costs of migration (de Haas 2010: 1589; Massey et al. 1998). Economic discrepancies between origin and destination characterized by a widening gap in GDP per capita, rising unemployment, growing labor demand at the destination, higher levels of political freedom, education, mobility, and liberal family reunification laws, among others, play an important role (Natter 2014: 9–10). A favorable economic situation or absence of economic crisis at the destination also provides an opportunity for immigrants (Roos and Zaun 2016: 1585). It is essential to understand the structural forces behind African migration, motivations, goals, and aspirations, as well as the social and economic structures that connect origin and destination (Massey et al. 1998: 281). Therefore, decisions to migrate consist of complex processes involving economic, social, and political factors (conflicts and political insecurity) at the origin and destination, geography, migration and social policies at the destination, availability of information, networks, resources, communications technology, the prevalence of smugglers or agents in the irregular migration, etc. (De Haas, 2011c; Koser & McAuliffe, 2013; McAuliffe, 2013; Kuschminder et al. 2015: 13; 54;

Schapendonk and van Moppes, 2007: 2, 30). Similarly, as Cummings et al. suggest, a person's motivations and intentions may change frequently before the journey starts as well as throughout the journey and are therefore difficult to predict. There are various determining factors for this unpredictability, such as the quality, availability, and affordability of smuggling services, the availability, and reliability of the information that facilitates migration, a culture of migration, local and international networks, the amount of financial resources for the journey, the intensity of driving factors such as conflicts and social unrest, absolute and relative poverty, economic instability, lack of economic opportunities in the country of origin and the hope of greater opportunities at the destination, and environmental factors (Cummings et al. 2015: 6; 24). Migration culture means the self-perpetuation of migration facilitated by migrant networks that enable migrants to overcome obstacles to migration (de Haas 2011: 22). Foreign labor migration becomes integrated into the structure of values and expectations of families and communities and forces young people to consider migration (Cummings 2015: 27). Therefore, many migration researchers argue that it is difficult to categorize individuals as "economic migrants" or "asylum-seekers" as the factors behind migration are complex and fluid. Moreover, it is equally important to consider international and national migration policies, border controls at origin, transit, and destination, and the culture of migration at origin (Cummings et al. 2015: 17; Kuschminder et al., 2015: 13; Fargues and Bonfanti, 2014: 4). However, de Haas argues that tight border controls and strict migration policies are unlikely to substantially influence migration decisions (de Haas, 2011a: 26). Czaika and Hobolth (2014) suggest that increasing the restrictiveness of asylum policy could reduce the number of asylum applications; however, at the same time, it increases irregular migration (Czaika and Hobolth 2014: 19).

Fussell (2012) provides a good overview of migration theories from economic, political, and environmental perspectives. She analyzes four main migration theories: neoclassical economics, world system theory, the new economics of labor migration, and social capital theory. Capitalists move where labor is cheaper, and workers move where they can sell their labor force better. From the *world systems theory* perspective, migration is a movement process toward this meeting point because of capitalist development (Wallerstein 1974; Massey et al. 2006: 42; Massey et al. 1993: 445).

From the *neoclassical macroeconomic perspective,* migration is the result of a decision that depends on a cost-benefit analysis by calculating migration benefits minus migration costs (Massey et al. 2006: 37; Sjaastad 1962: 87). According to the *Decision Theory of Migration,* migration follows a variety of individual decisions weighing the costs, dangers, challenges, etc. against the material as well as immaterial benefits through risk minimization and profit maximization as (Thielemann 2003: 19). The *new economic theory of migration* suggests that decisions are made not only by individuals who migrate but also by households to diversify the economies of households, manage household risks and maximize benefits, and invest in productive activities (Stark 1991; Massey et al. 2006: 46).

How are migration decisions made? However, this simple question is not easy to answer. As HDR 2009 suggests, differences in living standards are decisive as three-quarters of international movers move to a country with a higher HDI than their country of origin. As a consequence, people from the poorest countries gain the most from moving across borders on average, such as a 15-fold increase in income (to US$15,000 per annum), a doubling in education enrolment rate (from 47% to 95%), and a 16-fold reduction in child mortality (from 112 to 7 deaths per 1000 live births) (UNDP HDR 2009b: 22–24). Many African regions have a long tradition of migration, particularly as short-term and seasonal workers, such as from Sudan, Mali, Niger, Chad, Nigeria, and Mauritania, partly due to recurring drought, moving from rural to urban areas or neighboring countries seeking jobs. However, as migrants increasingly perceive that job opportunities remain low in the region and the gap in revenue generated between the average local worker and a worker in Europe becomes increasingly pronounced, an increasing number of young people try to reach Europe against the background of lack of economic prospects and harsh conditions affecting the agricultural sector (Benattia et al. 2015: 22–25). The increasing scarcity of fish along the coast, such as in Senegal or Somalia, also encouraged emigration (Benattia et al. 2015: 33).

Some studies suggest that decisions are individual, whereas other studies argue that migration decisions are taken rather on household levels in which families, friends, and social and religious networks determine the decision to migrate as they support the migrants financially to pay for the services of smugglers or agents, who influence which destination is offered, promoted, or available, and the route taken (Loschmann and Siegel 2014: 13; Cummings et al. 2015: 24). Kinship, religious, and other social networks play a key role in the migration decision by providing prospective migrants with remittances and information on how to migrate. TV, mobile technology, internet-based technology, and social media facilitate ties with family and friends back home, with other migrants on their way, and at their destination. Moreover, these modern social media technologies have become indispensable for enquiring about routes, acquiring day-to-day subsistence, and sharing food and accommodation (Cummings et al. 2015: 29). This solidarity of friends, etc. deepens a "culture of migration" at the origin (Cummings et al. 2015: 24). Remittances are also sent to cover smugglers' fees and bribes as a key driver for further migration and family reunification in the destination country (Kibreab 2013: 27). However, the decision processes can vary from family to family and from individual to individual. Different individuals and families have different financial backgrounds and cultural settings. Is the migrant a woman or a man? What are the community feelings or solidarity of a family or a village? Can the individual or the family finance the migration? On what terms do family members, relatives, or friends lend money to a would-be migrant?

One important contributor to migration that is not adequately addressed by migration theories that primarily focus on poverty, conflicts, and climate change is relative deprivation (poverty). As the *cumulative causation theory* (Myrdal 1957; Massey et al. 2006: 46) suggests, individuals' or households' sense of relative deprivation increases the motivation to migrate. As the neighborhoods' income with migration background increases, the income and wealth difference as well as living costs increase, resulting in further migration from the community. Emigration is positively associated with inequality in the country of origin, especially as the early stages of development tend to cause rising inequality (Clemens 2014: 14; Thornton 2001; Frazer 2006). Historical, economic, cultural, and political linkages between origin and destination contribute to migration, as the unified migration systems theory suggests (Kritz et al. 1992; Mabogunje 1970; Zlotnik, 1998: 12–13). Migration "unifies" or links home and host countries linguistically, culturally, historically, politically, or economically and attracts migrants to the wealthy countries, especially to the former colonizers (Massey et al. 2006: 49; Gold 2005). Moreover, the number of persons traveling a given distance is directly proportional to the number of opportunities at that distance and inversely proportional to the number of intervening factors, including geographical distance and natural hindrances such as ocean, sea, rivers, and desert. (Stouffer 1962: 69–91).

Structurally speaking, warfare, colonialism, conquest, occupation and labor recruitment, shared culture, language and geographical proximity, settled migrants at the destination, etc., play a role as causes and facilitators of migration (De Haas 2010: 1589; Castles and Miller 2009; Natter

2014: 4). Governments of countries of origin are not always passive; instead, their policies and practices play a "constitutive role" in international migration (Natter 2014: 5). For example, Ethiopia's Prime Minister Abiy Ahmed announced that Ethiopia would send 50,000 people to work in the United Arab Emirates (UAE) in the 2019/2020 fiscal year to reduce unemployment in Ethiopia, to receive training in various sectors, including driving and nursing at the destination, earn higher wages, and "boost their capacity" (Africa Research Bulletin 2019a: 22606).

The UNDP Global Report 2016–2018 examines the causes of migration and displacement from the perspective of economic development. An increasing number of studies have suggested that development can cause more migration. According to a study by Clemens and Mendola, in low-income countries, people actively preparing to emigrate have 30% higher overall incomes than others (Clemens and Mendola 2020).

For example, a study by Clemens and Postel (2018: 9) shows that sustained economic development tends to encourage emigration, and emigration rates in middle-income countries are typically much higher than those in poor countries (Clemens and Postel 2018: 9; UNDP HDR 2009b: 24; UNECA 2017; Bakewell 2008: 1350). A study by Clemens, published in 2020, suggests that "emigration rises on average as GDP per capita initially rises in poor countries, slowing after roughly US$5,000 at purchasing power parity, and reversing after roughly $10,000" (Clemens 2020). This means that average emigration first rises and then falls with development, which Clemens calls the *emigration life cycle*. A study by Clemens and Postel (2018) argued that assuming that aid can increase economic growth and that economic growth would decrease emigration is not necessarily true. If today's poorest quintile of countries continued to grow at their historical rate of growth (over the last 24 years), would only reach PPP$8000 after the year 2100, and the deterrent effect of economic growth on emigration in poor countries is when they reach roughly PPP$8000–10,000 in GDP per capita. "If development aid could systematically raise their economic growth by one percentage point every year—more than doubling the historical rate—it would take until the year 2097. If aid could raise growth by two percentage points—a tripling—it would take until the year 2067" (Clemens and Postel 2018: 6). Similarly, Cummings et al. argued that trade and investment in a source country are likely to increase, not reduce, migration. It is not individuals from the poorest households who migrate to Europe, but rather those who have access to sufficient resources to pay for their journey (Cummings et al.

2015: 24). I will discuss this in Chap. 8 from the "decreasing resource gap" analytical perspective.

A study by the United Nations Economic Commission for Africa suggests that economic development in low-income countries is initially associated with increasing rather than decreasing levels of emigration, and emigration decreases only when societies become wealthy. "...in situations of poverty and constraints migration is generally part of deliberate, carefully planned, and largely rational strategies by families to improve their long-term social and economic wellbeing rather than a stereotypical 'desperate flight from poverty'" (UNECA 2017). "As development proceeds, human capital accumulates, connections to international networks increase, fertility shifts, aspirations rise, and credit constraints are eased. All of these changes tend to increase emigration. The most important of these factors appear to be rising education levels and international connections, which both inspire and facilitate emigration" (Clemens and Postel 2018: 9). I call this "decreasing knowledge gap" and discuss it in Chap. 8.

According to various research findings, aid raises net emigration from the average poor country to high-income OECD countries. Some studies corroborated that poverty is a constraint to emigration. A case study in Mexico showed that the probability of migration increased with higher income levels for household incomes lower than US$15,000 pa, and in Bangladesh, monetary incentives significantly increased the likelihood of migration. The UNDP studies show that giving emigrants an amount equivalent to a week's wages at their destination increased the propensity to migrate from 14% to 40%, which means that development in countries of origin will not necessarily reduce emigration (UNDP HDR 2009b: 25). Moreover, even if on the political level it is very often argued that conflicts, human smuggling, and trafficking play an important role in migration, not a large proportion of overall human movement is influenced by them (UNDP HDR 2009b: 25).

Castles (2010: 1565) points out the complexity, interconnectedness, variability, contextuality, and multi-level mediation of migratory processes in the context of rapid global change. He understands migration as part of the process of transformation of global political, economic, and social relationships. Flahaux and De Haas also suggest that increasing development levels and social transformation have increased Africans' capabilities and aspirations to migrate and shaped migration behavior (Flahaux and de Haas 2016a: 1; de Haas 2011a: 4). The notion that economic development and poverty reduction would increase migration in West Africa is well analyzed by Quartey et al. (2020: 270; de Haas 2006). Clemens and Postel argue that as economic growth increases in emigration, development aid should not aim to deter migration but to shape it for mutual benefit (Clemens and Postel 2018: 1). Improving economic growth and rising incomes generally increase demographic growth by decreasing child mortality, resulting in rising unemployment and thus emigration pressure (Clemens 2014: 10–11). Growth brings disposable income to finance the costs of migration (Gould 1980).

However, Lanati and Thiele (2018) argue that there is "evidence of a negative relationship between the total aid a country receives and emigration rates. According to them, this even holds for the poorer part of recipient countries, which suggests that the budgetary constraint channel does not play a significant role in shaping migration decisions as any positive welfare effects of foreign aid tend to manifest themselves in better services for the poor and encourage them to stay rather than migrate (Lanati and Thiele 2018: 60–66). However, Berthélemy et al. argue that due to "attraction effect", bilateral aid enhances migration as it provides would-be migrants with more information about the destination country and its labor market conditions. Moreover, they suggest that from the push effect perspective, total aid increases the expenditure capacity of migrants as it increases wages in the countries of origin (Berthélemy et al. 2009: 1589–1599).

In his aspiration/ability model, Carling distinguishes between a wish to migrate and the realization of this wish. Carling places the possibility of involuntary immobility at the center of his migration analysis because there are people who wish to migrate but are unable to do so (Carling 2001). Similarly, Schöfelberger et al. (2020: 87) analyze why and how migration aspirations translate into migration flows to Europe.

According to Castles, less inequality, less poverty, or less human insecurity would not necessarily lead to less migration (Castles 2010: 1568). Castles, therefore, argues that migration is a normal part of social relations. However, would those at the destination ever accept this social reality as a normal part of social relations? The answer is simply no. Therefore, societies and governments at the destination are desperately looking for a "Single Theory" of migration and a simplification of complex and diverse patterns of behavior (Castles 2008, 2010: 1568).

Hofmann suggests that "desirability" and accessibility of destinations play a role in migration decisions (Hofmann 2015: 814). By choosing their destination, migrants create a destination "hierarchy" between highly and less desirable destinations (Paul 2011: 1864). What makes certain destinations more desirable than others? Is it just the migration-related material gain? This is the central issue that I address in my mimetic approach to postcolonial African migration to the West.

Economic factors such as labor markets, wage differentials and low levels of competition, policy environments that enable permanent residence, perceived anti-immigrant sentiment, a well-developed migration industry, resources to pay for their journey, and the prospect of good schools, safe communities, and affordable housing, etc. determine the desirability of destination (Hofmann 2015: 814–815; de Haas 2007). Moreover, the more migration takes place, the more information will be available for would-be migrants regarding migration routes, facilitating factors, dangers, and opportunities at the destination. On the other hand, the more information is available, the more migration will take place, as prior migrants can provide information about job search, earning potential, cost of living, marriage partners, legal formalities, and extralegal channels of movement (Clemens 2014: 12; Greenwood 1969). Therefore, these expanding networks make migration easier, safer, and cheaper and help the probability of migration increase. As a result, migration perpetuates itself (de Haas 2010: 1590; Massey 1990: 8).

As Bakewell suggests, European migration policies are primarily interested in keeping African migrants in their place. Various development policies of the EU for Africa or the EU Trust Fund for Africa established in 2015 are primarily concerned with keeping African migrants in their place. Moreover, Bakewell suggests that there is a persistent sedentary bias in many migration theories as well as in the practice of development cooperation (Bakewell 2008: 1342).

Martin (2012) highlights the role of war and natural disasters in forced migration. In the migration discussion, the widespread assumption is that people migrate to optimize and maximize their individual economic and political gains. Paul Collier argues in his book *Exodus* that the economic growth gap between rich and poor countries is fundamental to understanding the origins of modern migration. Second, according to Collier, the financial capability of poor migrants enables migration, i.e., if aspiring migrants are too poor, they cannot afford migration. Third, the diaspora and networks at the destination are seen as causes of migration (Collier 2014: 37–38).

The social capital theory perspective mentioned above addresses the role of networks and diaspora discussed by Collier as important factors in migration decision-making and its realization. Facilitated by modern information and communication technologies, migration can be more easily organized than 30 years ago when the level and quality of migration-related communication and networking were less developed. However, it is not clear why social capital theory addresses networks and diaspora as causes of migration rather than as facilitating factors. In my view, facilitating factors are epiphenomenal and not causal. Even financial affordability or non-affordability of migration cannot be analyzed as causes of migration, but rather as facilitating factors.

## Conclusion

As discussed above, various theories explain the causes of migration. For some researchers and theories, poverty, climate change, and conflicts are the driving forces of migration. For others, an improving economic situation facilitates emigration; in other words, emigration is a sign of an improving, not deteriorating economic situation. Other researchers have explored the global economic system, interconnectedness, and its impact as a push factor.

In this book, I argue that in postcolonial migration from Africa to the West, the legacies of slavery and colonization are the fundamental causes of migration. We shall see in the following chapters that African emigration to the West has psychological, historical, and mimetic reasons. Therefore, I will discuss postcolonial African migration to the West against the background of slavery and colonial legacies. The postcolonial African migration to the West is caused primarily by a mimetic desire for being through recognition, and therefore, is only a part of the bigger picture of the postcolonial African person. In Chap. 8, I will discuss why the decreasing gap between the origin and destination is an important factor in understanding postcolonial African migration to the West. The following two chapters address global political and economic structures and their role in postcolonial African migration to the West.

## Literature

Adepoju, Aderati (2010): The Politics of International Migration in Post-Colonial Africa, in Robin Cohen, *The Cambridge Survey of World Migration,* Cambridge University Press: Cambridge: pp. 166–171.

Africa Research Bulletin, Economic, Financial and Technical Series, Volume 56 Number 6, June 16th 2019a–July 15th 2019.

African Union and IOM (2020): Africa Migration Report: Challenging The Narrative, International Organization for Migration (IOM).

Bakewell, Oliver (2008): 'Keeping Them in Their Place': the ambivalent relationship between development and migration in Africa, *Third World Quarterly*, 29:7, 1341–1358.

Bakewell, Oliver (2021) Unsettling the boundaries between forced and voluntary migration: in Emma Carmel et al. Handbook on the Governance and Politics of Migration, Edward Elgar: Cheltenham, pp. 124–136.

Benattia, Tahar et al. (2015): *Irregular Migration: Challenges and Solutions* Migration, Dialogue for West Africa 2015 Conference Research Paper, Altai Consulting: Abuja.

Berthélemy, Jean-Claude, Beuran, Monica and Maurel, Mathilde (2009): Aid and Migration: Substitutes or Complements? World Development Vol. 37, No. 10, pp. 1589–1599, https://doi.org/10.1016/j.worlddev.2009.02.002.

Bets, Alexander and Collier, Paul (2018): Refuge: Transforming a broken refugee system, Penguin Books: London.

Bisong, Amanda (2019): Trans-regional institutional cooperation as multilevel governance: ECOWAS migration policy and the EU, *Journal of Ethnic and Migration Studies*, 45:8, 1294–1309.

Black, Richard (2011): Migration as adaptation, *Nature*, 478, 447–9.

Carling, Joergen (2001): Aspiration and ability in international migration: Cape Verdean experiences of mobility and immobility, Dissertations & Theses No. 5/2001, University of Oslo.

Carmel, Emma, Lenner, Katharina, and Paul, Regine (2021): Handbook on the Governance and Politics of Migration, Elgar Handbooks in Migration) Cheltenham.

Carrera, Sergio et al. (2019): The external dimension of EU migration and asylum policies in times of crisis, in Sergio Carrera et al. (eds.), Constitutionalizing the external dimensions of EU migration policies in times of crisis, Edward Elgar: Cheltenham, pp. 1–20.

Castles, S. (2008) Development and Migration*Migration and Development: What Comes First? New York: Social Science Research Council, http://essays.ssrc.org/developmentpapers/wp-content/uploads/2009/08/2Castles.pdf.

Castles, S. and Miller, M.J. (2009): The Age of Migration. Basingstoke: Palgrave Macmillan (4th edition).

Castles, Stephen (2009): Development and Migration—Migration and Development: What Comes First? Global Perspective and African Experiences, *Theoria: A Journal of Social and Political Theory*, December 2009, Vol. 56, No. 121, pp. 1–31.

Castles, Stephen (2010) Understanding Global Migration: A Social Transformation Perspective, *Journal of Ethnic and Migration Studies*, 36:10, pp. 1565–1586.

Castles, et. al. (2014): The Age of Migration: International Population Movements in the Modern World. Fifth edition. Palgrave, Macmillan, London.

Clemens, Michael A. (2014): *Does Development Reduce Migration? Center for Global Development*, NYU Financial Access Initiative and IZA, Discussion Paper No. 8592, October 2014.

Clemens, Michael A. (2020): The Emigration Life Cycle: How Development Shapes Emigration from Poor Countries, Center for Global Development, Working Paper 540 August 2020.

Clemens, Michael A. and Mendola, Mariapia (2020): Migration from Developing Countries: Selection, Income Elasticity, and Simpson's Paradox, Center for Global Development, Working Paper 539, August 2020.

Clemens, Michael A. and Postel, Hannah M. (2018): "Deterring Emigration with Foreign Aid: An Overview of Evidence from Low-Income Countries." CGD Policy Paper. Washington, DC: Center for Global Development. https://www.cgdev.org/publication/deterring-emigration-foreign-aid-overviewevidence-low-income-countries, 1–30.

Cohen, Robin (ed.) (2010): *The Cambridge Survey of World Migration*, Cambridge University Press: Cambridge.

Collier, Paul (2014): Exodus: Immigration and multiculturalism in the 21st century, Penguin Books: London.

Collyer, Michael (2016): Geopolitics as a migration governance strategy: European Union bilateral relations with Southern Mediterranean countries, *Journal of Ethnic and Migration Studies*, 42:4, 606–624.

Cummings, Clare et al. (2015): *Why people move: understanding the drivers and trends of migration to Europe*, Overseas Development Institute, Working Paper 430, 2015s.

Czaika, Mathias and De Haas, Hein (2017): Determinants of Migration to the UK, The Migration Observatory, Briefing, Oxford University, 2017.

Czaika, Mathias, and Mogens Hobolth. (2014): Deflection into irregularity? The (un)intended effects of restrictive asylum and visa policies. Working Paper, Oxford: University of Oxford: International Migration Institute.

De Haas, H. (2006) Turning the Tide? Why 'Development Instead of Migration' Policies are Bound to Fail. Oxford: International Migration Institute, IMI Working Paper, 2.

De Haas, H. (2011a). The Determinants of International Migration, DEMIG Working Paper 2. International Migration Institute, University of Oxford.

De Haas, H. (2011c): Mediterranean migration futures: Patterns, drivers and scenarios, Global *Environmental Change* 21S S59–S69.

De Haas, H., (2007): Turning the Tide? Why Development Will Not Stop Migration. Dev. Change 38, 819–841. https://doi.org/10.1111/j.1467-7660.2007.00435.x.

De Haas, Hein (2010): The Internal Dynamics of Migration Processes: A Theoretical Inquiry, Journal of Ethnic and Migration Studies, 36:10, pp. 1587–1617.

De Haas, Hein (2011): "The determinants of international migration: Conceiving and measuring origin, destination and policy effects". IMI Working Paper Series. Oxford: International Migration Institute, University of Oxford.
Eisenstadt, Shmuel N. (1954): *The Absorption of Immigrants*, London: Routledge and Kegan Paul.
Englisch Ausgabe von Gebrewold, Belachew and Bloom, Tendayi (eds.) (2018): Understanding Migrant Decisions: From Sub-Saharan Africa to the Mediterranean Region, Routledge: Abingdon.
European Commission (2020): The European Union and Africa: Partners in Trade, https://trade.ec.europa.eu/doclib/docs/2022/february/tradoc_160053.pdf.
European Commission (2023): The European Union and Tunisia agreed to work together on a comprehensive partnership package, https://neighbourhood-enlargement.ec.europa.eu/news/european-union-and-tunisia-agreed-work-together-comprehensive-partnership-package-2023-06-11_en, 06.07.2023.
Fargues, P., Bonfanti, S., 2014. When the best option is a leaky boat: why migrants risk their lives crossing the Mediterranean and what Europe is doing about it.
Flahaux, Marie-Laurence and De Haas, Hein (2016a): African migration: trends, patterns, drivers; *Comparative Migration Studies* 4:1, pp. 1–25.
Frazer, Garth (2006), 'Inequality and development across and within countries,' *World Development*, 34 (9), 1459–1481.
Fussell, Elisabeth (2012): Space, Time and Volition: Dimensions of migration theory, in Marc Rosenblum and Daniel Tichenor, The Oxford Handbook of the politics of international migration, Oxford University Press: Oxford, pp. 25–52.
Gold, S.J. (2005) 'Migrant networks: a summary and critique of relational approaches to international migration', in Romero, M. and Magolis, E. (eds) *The Blackwell Companion to Social Inequalities.* Malden, MA: Blackwell, 257–85.
Goldin, I. et al. (2011): Exceptional People: How Migration shaped our world and will define our future, Princeton University Press: Princeton.
Gould, J. D. (1980): 'European Inter-Continental Emigration: The Role of "Diffusion" and "Feedback"', *Journal of European Economic History*, 9 (2), 267–315.
Greenwood, Michael J. (1969): 'An Analysis of the Determinants of Geographic Labor Mobility in the United States,' *Review of Economics and Statistics*, 51 (2), 189–194.
Grieveson, Richard et al. (2021): Future Migration Flows to the EU: Adapting Policy to the New Reality in a Managed and Sustainable Way, Policy Notes and Reports 49, The Vienna Institute for International Economic Studies.
Hofmann, Erin Trouth (2015): Choosing Your Country: Networks, Perceptions and Destination Selection among Georgian Labour Migrants, *Journal of Ethnic and Migration Studies*, 41:5, 813–834.

IOM (2023): IOM Definition of "Migrant", https://www.iom.int/about-migration, 27.05.2023.

Jackson, Lauren (2023): *Tunisia's Influence in Europe*, https://www.nytimes.com/2023/04/05/world/africa/tunisia-europe-migration.html, 06.07.2023.

Kane, Abdoulaye and Leedy, Todd (2013): African patterns of migration in a global era: new perspectives, in Abdoulaye Kane and Todd Leedy, African Migration: patterns and perspectives, Indiana University Press, Bloomington, pp. 1–18.

Kibreab, G., 2013. The national service/Warsai-Yikealo Development Campaign and forced migration in post-independence Eritrea. J. East. Afr. Stud. 7, 630–649. https://doi.org/10.1080/17531055.2013.843965.

Koser, K., & McAuliffe, M. (2013). Est*ablishing an Evidence-Base for Future Policy Development on Irregular Migration to Australia*, Irregular Migration Research Program. Irregular Migration Research Programme, Occasional Paper Series.

Kritz, M., Lim, L. L. and Zlotnik, H. (eds.) (1992): *International migration systems: A global approach* (Oxford: Clarendon Press).

Kuschminder, Katie et al. (2015): *Irregular Migration Routes to Europe and Factors Influencing Migrants' Destination Choices*, Maastricht Graduate School of Governance.

Lanati, Mauro and Thiele, Rainer (2018): The impact of foreign aid on migration revisited, *World Development 111 (2018)*, 59–74.

Leanne Weber and Claudia Tazreiter, eds. (2021): Handbook of Migration and Global Justice, Edward Elgar, Cheltenham.

Loschmann, C., and Siegel, M. (2014): The influence of vulnerability on migration intentions in Afghanistan. Migr. Dev. 3, 142–162. https://doi.org/10.1080/21632324.2014.885259.

Mabogunje, A. L. (1970): Systems Approach to a Theory of Rural-Urban Migration. Geographical Analysis, 2(1): 1–18.

Martin, Susan (2012): War, natural disasters, and forced migration, in Marc Rosenblum and Daniel Tichenor, The Oxford Handbook of the politics of international migration, Oxford University Press: Oxford, pp. 53–73.

Massey, D.S. (1990) 'Social structure, household strategies, and the cumulative causation of migration', Population Index, 56(1): 3–26.

Massey, D.S., Arango, J., Hugo, G., Kouaouci, A., Pellegrino, A. and Taylor, J.E. (1998) *Worlds in Motion, Understanding International Migration at the End of the Millenium*. Oxford: Clarendon Press.

Massey, D.S., Arango, J., Hugo, G., Kouaouci, A., Pellegrino, A. and Taylor, J.E. (1993) 'Theories of international migration: a review and appraisal', Population and Development Review, 19(3): 431–66.

Massey, Douglas S., Joaquín Arango, Graeme Hugo, Ali Kouaouci, Adela Pellegrino and J Edward Taylor (2006) Theories of International Migration: A Review and Appraisal. In: Messina Anthony and Lahav Gallya (Eds.) The Migration Reader. Exploring Politics and Policies (Lynne Rienner, Boulder), 34–62.

McAuliffe, M. (2013). *Seeking the views of irregular migrants: Decision-making, drivers and migration journeys.* Irregular Migration Research Programme Occasional Paper Series.

Melachrinos, Constantinos et al. (2020): Using big data to estimate migration "push factors" from Africa, in International Organization for Migration, Migration in West and North Africa and across the Mediterranean: Trends, risks, development and governance, Geneva, pp. 98–116.

Menjivar, Cecilia et al. (2019): Migration Crisis: Definitions, Critiques, and Global Contexts, in The Oxford Handbook of Migration Crisis, Oxford University Press: Oxford, pp. 1–20.

Mixed Migration Centre (2019): Navigating borderlands in the Sahel: border security governance and mixed migration in Liptako-Gourma. Available at www.mixedmigration.org.

Natter, Katharina (2014): Fifty years of Maghreb emigration : How states shaped Algerian, Moroccan and Tunisian emigration, Working Papers, Paper 95, July 2014, DEMIG project paper 21.

O'Rourke, Dennis (2014): Why do we migrate A retrospective, in Michael Crawford and Benjamin Campbell, *Causes and Consequences of Human Migration: An evolutionary perspective,* Cambridge University Press: Cambridge, pp. 527–536.

Paul, Anju Mary (2011): "Stepwise International Migration: A Multistage Migration Pattern for the Aspiring Migrant." American Journal of Sociology 116 (6): 1842–1886. https://doi.org/10.1086/659641.

Piore, M. J. (1979). Birds of passage: Migrant labour in industrial societies. Cambridge University Press, Cambridge.

Quartey, Peter et al. (2020): Migration across West Africa: development-related aspects, in International Organization for Migration, Migration in West and North Africa and across the Mediterranean: Trends, risks, development and governance, Geneva, pp. 270–278.

Roos, Christof and Zaun, Natascha (2016): The global economic crisis as a critical juncture? The crisis's impact on migration movements and policies in Europe and the U.S., *Journal of Ethnic and Migration Studies,* 2016, Vol. 42, No. 10, 1579–1589.

Saunders, H.W. (1956): 'Human Migration and Social Equilibrium', in: J.J. Spengler & O.D. Duncan (Eds), Population Theory and Policy, New York: Free Press.

Schöfelberger, Irene et al. (2020): Migration aspirations in West and North Africa: what do we know about how they translate into migration flows to Europe? Migration in West and North Africa and across the Mediterranean Trends,

risks, development and governance International Organization for Migration (IOM), Geneva, pp. 87–97.
Sjaastad, L. A. (1962): The costs and returns of human migration. Journal of Political Economy, 70(5): 80–93.
Stark, O. (1991): The migration of labor. Basil Blackwell, Cambridge, MA.
Stouffer, Samuel A. (1962): Social Research to Test Ideas Free Press: New York.
The Economist (2021a): EU border policy: New kings of the wild frontier, & March, pp. 19–21.
The Economist (2021c): African Odyssey: Many more Africans are migrating within Africa than to Europe, 30 October, pp. 25–27.
The Economist (2022j): China and Africa: Chasing the dragon, 19 February, pp. 29–31.
Thielemann, E. (2003): 'Does Policy Matter? On Governments' Attempts to Control Unwanted Migration', *European Institute Working Paper*, May 2003, available at: http://www.lse.ac.uk/collections/europeaninstitute/pdfs/Eiworkingpaper2003-02.pdf, accessed May 05, 2005.
Thornton, John (2001), 'The Kuznets inverted-U hypothesis: panel data evidence from 96 countries.' *Applied Economics Letters* 8 (1): 15–16.
Todaro, M. P. (1976), *Internal migration in developing countries* (International Labour Office: Geneva).
UNDP (2009b): HUMAN DEVELOPMENT REPORT 2009: Overcoming barriers: Human mobility and development.
Van Hear, Nicholas et al. (2018): Push-pull plus: reconsidering the drivers of migration, Journal of Ethnic and Migration Studies, 44:6, 927–944, https://doi.org/10.1080/1369183X.2017.1384135.
Zlotnik, H. (1998): The theories of international migration. Paper for the Conference on International Migration: Challenges for European Populations, Bari, Italy, 25–27 June 1998.

CHAPTER 3

# The Economic Dimension of the Postcolonial African Migration to the West

State failure, corruption, clientelism, poverty, external debt, unemployment, and weak economic institutions are seen in many migration discussions as root causes of postcolonial African migration to the West. Many researchers argue that Africa is exposed to the global economic system, which is not always benefitting Africa. Accordingly, in the face of global competition among global big powers, Africa is a junior partner, that is not able to shape the global economic structure. Those who critique the capitalist system argue that Africa is poor because of the neocolonial capitalist system and globalization. While it is fair to argue that colonialism and global economic structures have caused massive and long-lasting damage to Africa, Africa's responsibility, i.e. poor economic governance structures cannot be neglected (Mills 2010).

In this chapter, I argue that in postcolonial Africa, Africans are free economic players as they have options to choose the most beneficial economic partners in the globalized economic structure. These options did not exist during the slave trade or colonial times. Therefore, neocolonialism is a poor argument to explain current African socio-economic problems. Various studies argue that the rivalry between the West and China had damaged Africa politically and economically; that the trade imbalance between China and various African countries would disadvantage African economies; that Chinese-African economic relations have led to public sector debt like in Kenya, Zambia, etc. (Usman and Abayo 2022).



B. Gebrewold, *Postcolonial African Migration to the West*, Politics of Citizenship and Migration,
https://doi.org/10.1007/978-3-031-58568-5_3

What I would call the *postcolonial Africa syndrome* in many African political and academic discourse attempts to excuse African political (insecurity) and economic failures (poverty) in Africa by way of colonization and neocolonialism. It neglects African responsibility for Africa's failure. The main argument of Chaps. 3 and 4 in this book is that Africans are not mere victims of the international economic, security, and political causes that push Africans to migrate to the West. It is neither the West, the IMF, the World Bank, China, Russia, Turkey nor any external actor responsible for African economic ills. Africa's problems are primarily caused by the free choice of African poor economic decision-makers. Africans are the enablers of the global actors' exploitation of Africa. If poverty is a cause of African migration to the West, not global big players but Africans are primarily responsible for their poverty and poor economic institutions. As long as Africans scapegoat others for their bad policies, their poor economic and political performances will keep Africans in poverty and instability and keep up the postcolonial African mimetic adoration of, fascination for, and the consequential migration to the West. In this chapter, while discussing the role of global big players in Africa, my objective is not to justify their exploitative behavior of the global players in Africa, nor to downplay its impact, but rather to point out the counterproductivity of the African mentality of victimhood, self-pitying, blame-someone-else, and scapegoating. Such a mentality would keep us Africans weak, manipulable by global players, and push us to search for our fortune in the West often in a degrading way.

## Global Big Players in Africa

In international relations studies, the debate has been going on whether the structure dictates the behavior of the agent or units or whether the units determine the shape and quality of the international structure. For David Singer, the behavior of the state is not dictated by the international structure (Singer 1960), whereas for Waltz, the structure constrains the behavior of agents and operates independently of the units (Waltz 1979). However, Wendt from a constructivist approach suggests that systems and units are mutually constituted, and it is not possible to talk about one without the other (Wendt 1992). They are intersubjectively constituted and constrained (Buzan and Little 2000: 39–45).

This chapter argues that Africans as agents are not mere victims of the international structure. They are free actors with the power to determine the impact of the global economic structure on Africa. They cannot change the global economic structure but they have the power to shape or limit the impact of this structure in Africa. The poverty that allegedly pushes Africans to the West is not merely the outcome of an evil international economic structure. As Wendt would argue, there is a constitutive mutual conditioning and impact between Africans as agents on one and the international structure on the other (Wendt 1987). Global actors are competing in Africa often by keeping authoritarians in power. We Africans love scapegoating the West for almost everything and forget our responsibilities and accountabilities. Structure and agents are consolidating Africa's failure.

Various researchers and civil society activists have been complaining that foreign land investors and national politicians have been grabbing land from local people without due compensation and their consent and leasing it to foreigners for many decades. According to Action Aid, poor farmers in Kenya, Mozambique, Senegal, Sierra Leone and Tanzania were subject to forced evictions, human rights violations, lost livelihoods, divided communities, destruction of culturally significant sites, rising food insecurity and, ultimately, increased poverty (ActionAid 2014: 7). Similarly, a report by Martin-Prével et al. criticizes the World Bank, G8 members, Western donors including the Bill and Melinda Gates Foundation, the US, UK, Danish, Dutch governments, and agroindustry and agrochemical companies driven by the pro-corporate push favoring private sector-led and market-driven food systems (Martin-Prével et al. 2016: 4).

Oram (2014) discusses how the world is paving the way for corporate land grabs. OXFAM maintains that land investors have been targeting countries with weak governance, sustaining poor governance, and doing good business for themselves (OXFAM 2013). Martin-Prével et al. (2016) argue that Western donors have been shaping African agriculture for a pro-corporate agenda. Studies by such as Graham et al. (2010), Ndi and Batterbury (2017), Ndi (2017), Cotula (2014), Kleemann et al. (2013), Gerstter (2011), Hall (2011), Cotula et al. (2009), Fairhead (2012), Twomey (2014), etc. explore "land-grabbing" in Africa. However, none of these studies shows that African political elites were forced to make those concessions against their will.

If we would approach the issues from an environmental perspective, the global predatory economy has been contributing to climate change. Many poor people in African countries (up to 80% in some countries) are farmers who have contributed almost nothing to climate change. More than 60%

of sub-Saharan Africans are smallholder farmers, and about 23% of the GDP of these countries in this region comes from agriculture (Goedde 2019: 2). Irrigation systems are rarely available. Due to climate change, rainy and drought seasons have become completely unpredictable. By mid-summer 2022, in Ethiopia, Somalia, and Kenya due to drought and famine about 18 million people experienced extreme hunger; 7 million children were acutely malnourished; 1.5 million were displaced; and 7 million animals died (World Economic Forum 2022). Millions are affected by overflooding in these countries toward the end of November 2023. Farmers do not have insurance or any other means of support in such situations. Poor people, who have contributed so little to climate change, are the ones who are suffering the most (Olumuyiwa et al. 2019). Climate change, desertification, and invasion by locusts have made millions of people homeless in various African countries, especially in the Horn of Africa. Out of 821 million people globally affected by climate change and desertification-induced food insecurity in 2017, 31% were in Africa. In 2021, in sub-Saharan Africa, conflict and violence displaced more than 11,558,000 people and disasters more than 2,554,000 (Norwegian Refugee Council 2022). It is not surprising that so many young people have been leaving the countryside and moving to cities or looking for better opportunities abroad.

The root causes of displacement and migration are internal and external. According to the UN, the global economic structure is exacerbating global inequality and leading to migration (UNDP Human Development Report 2019). However, even in Africa, better-governed countries are less vulnerable to the effects of climate change in comparison with poorly governed, as the case of Botswana shows. The Comoros, Seychelles, and Botswana are equally affected by climate change but they are better governed. The fact that 63% of persons living within Nigeria (133 million people) are multidimensionally poor (Nigerian National Bureau of Statistics 2022) is not due to climate change or resource scarcity, but instead because Nigeria is poorly governed (Acemoglu and Robinson 2012: 344).

In October 2022, the Environmental Justice Foundation reported that Chinese crew treated local fishermen in Ghana like "slaves", that there was widespread abuse and neglect linked to Chinese fishing vessels operating off Ghanaian shores; that the Ghanaians were beaten by the Chinese working on China's Distant Water Fishing (DWF) fleet in Ghana; they spat on them, kicked them. According to the report, all 36 crew members interviewed by EJF had been forced to work more than 14 hours a day and received inadequate food. 94% received inadequate medicine, or witnessed verbal abuse; 86% reported inadequate living conditions; 81% had witnessed physical abuse; 75% had seen serious injury at sea (Environmental Justice Foundation 2022: 15). Instead of blaming the Chinese for this outrageous behavior, it is the responsibility of the Ghanaian government to protect its citizens from such deeds by external actors and make them accountable.

### *External Economic Powers and Their Trade with Africa*

Africa's engagement with external economic powers has improved significantly, especially before the pandemic. Ports built by foreigners, insurance sold, and new mobile phone technology, etc. have increased. The GDP per head in sub-Sahara Africa is two-fifths higher than it was in 2000 (The Economist 2019a). Since around 2000, Africa has been diversifying its external trade partnership. Moreover, as the expanding African trade volume with some countries shows Africa's active participation in the global political economy has been growing fast. Africa is no longer a passive spectator of global happenings.

The China-Africa trade value changed from $7.3 billion in 2000 to $93.3 billion in 2010 and $135.9 billion in 2015. It reached $282 billion in 2022. China's share in Africa's global trade in goods changed from 3.0 in 2000 to 11.4 in 2010 and 15.9 in 2015. China ranked eighth in 2000 among Africa's trading partners, 2nd in 2010, and 1st in 2015. Similarly, India's total value of trade in goods was $6.9 billion in 2000, $37.5 billion in 2010, and $51.1 billion in 2015. India's share in Africa's global trade in goods accounted for 2.9% in 2000, 4.6% in 2010, and 6.0% in 2015. Among Africa's trading partners, India ranked 9th in 2000, 4th in 2010, and 2nd in 2015. For the FY 2016–2017 India-Africa trade amounted $44.63 billion, 2017–2018 $53.79 billion, 2018–2019 $58.05 billion, 2019–2020 $55.70 billion, and 2020–2021 $46.82 billion (Indian Ministry of Commerce and Industry 2022). For FY22, India's export goods are expected to be worth $40 billion to Africa, and imports worth $49 billion (Pattanayak 2022). Moreover, between January 2015 and May 2021, Indian firms in Africa invested about $23.3 billion.

Whereas, France, which ranked 2nd in 2000, remained 3rd both in 2010 and 2015; and the USA fell back from number one in 2000 and 2010 to 4th place in 2015. The value of China-Africa trade in 2020 was

US$176 billion, down from US$192 billion in 2019 (China-Africa Research Initiative 2022). In 2021, it jumped to $254.3 billion, out of which Africa exported goods worth USD 105.9 billion to China (Global Compliance News 2022; Economist Intelligence Unit 2022). As this data shows, and in general, for example during the last 20 years, there has always been a trade surplus in favor of China (China-Africa Research Initiative 2022).

From 2006 to 2018, the trade value of the largest Africa's trading partner EU increased by 41%, Chinese trade by 226%, India's by 292%, Turkey's by 216%, Russia's by 335%, Indonesia by 224%, although the last three from low starting points, and the US trade with sub-Saharan Africa shrank by 45% (The Economist 2019a). However, for example, in 2020, Russian exports to Africa accounted for $12.4 billion worth of goods, and imports from Africa only $1.6 billion worth resulting in a trade deficit of $10.8 billion for Africa (Econfin Agency 2022).

There is an increasing battle of reputations between the West and China, and the West and Russia. Therefore, questions arise regarding the motives behind the re-engagement of the EU and the US in Africa. Is it a desire to counter Chinese and Russian influence rather than work with African business partners? China's ambition is to overtake the EU as Africa's biggest trade partner by 2030 (Ayeni 2022). At the February 2022 EU-African Union summit, the EU pledged an Africa-Europe investment package of €150 billion to support a common ambition for 2030. In 2019, trade between the European Union and Africa reached €280 billion, accounting for 28% of Africa's total trade (European Commission 2020). In 2021, this rose to €288 billion (Eurostat 2022). In "*Towards a Comprehensive Strategy with Africa*", the EU proposes working together with Africa in five key areas: the green transition and energy access; digital transformation; sustainable growth and jobs; peace, security, and governance, and migration and mobility.

US engagement in Africa has been declining in Africa except in the security sector. Trump administration showed almost "zero interest in Africa" (Usman 2021). Whereas Chinese trade volume has been significantly increasing almost from year to year, the US trade with Africa has been small and going back: $62 billion in 2018; $57 billion in 2019; $45 billion in 2020; $64 billion in 2021 (United States Census Bureau 2022). U.S. direct investment in Africa declined from $69 billion in 2014 to $45 billion in 2020. In 2010, bilateral trade between the U.S. and Africa was $113 billion, and in 2020 $44 billion (Schneidman 2022).

China's reliance on African oil has fallen from 32% in 2007 to 18% in 2020. In 2018, China contributed $25.7 billion of Africa's $100 billion investments in infrastructure projects such as railway projects, and telecommunications in various African countries. The Digital Silk Road, which was unveiled in 2015 as part of the BRI, gives China a lead role in promoting digital connectivity (RAMANI 2021a: 5). China strengthens its relationship with Africa based on South-South "solidarity" against globalization led by the hegemonic US. In its massive presence in Africa, China attempts also to demonstrate the "superiority" of China's economic policies, encourage African countries to imitate the Chinese model, and prevent Western domination in Africa (Taylor 2006: 72–73).

However, there have been also racist incidents involving Chinese nationals in some African countries like Kenya, Malawi, Zambia, Zimbabwe, Uganda, the Democratic Republic of Congo, etc. (Bartlett 2022). Protesters chanted slogans against the Chinese "Zambia for Zambians" (Alden 2007: 85). In 2008, the Chinese were beaten up, houses set alight and hostel windows of the Chinese workers smashed in northern Zambia demanding better working conditions.

China attempts to depict the West as a colonizer and neocolonizer while China is bringing benefits to Africa through its investment, trade, and loans (Alden 2007: 127–128). Chinese firms accounted for 31% and Western firms for just 12% of all infrastructure projects in Africa in 2020, which is up from 12% in 2013 for China and down from 37% for the West. China is also investing in African cobalt mines, which are highly important for electric cars. For the period 2021–2027 Europe has planned $150 billion global gateway project for investment in infrastructure in Africa to compete with China, and claims to be better than the BRI as it creates "links not dependencies" (The Economist 2022e: 29; The Economist 2022g: 3; China-Africa Research Initiative 2022). The G7 launched in 2021 the so-called Build Back Better World initiative to counter the Chinese dominance in Africa.

Still, China is Africa's largest trading partner, bilateral creditor, and a crucial source of infrastructure investment, accounts for an estimated one-eighth of Africa's industrial output (The Economist 2022g: 3). Though the stock of Chinese FDI in Africa with its $44 billion is lower than Britain's $66 billion, France's $65 billion, or slightly higher than US $43 billion, its trade with Africa has risen from $10 billion in 2000 to a record $254 billion in 2021 (The Economist 2022g: 6). Probably there are about

10,000 Chinese firms in Africa, and 12% of Africa's production is accounted by Chinese companies (The Economist 2022g: 6–7).

During the China-Africa summit in 2021, President Xi declared that China and Africa had "forged an unbreakable fraternity" founded on friendship and equality, win-win, common development, fairness justice, openness, and inclusiveness. By 2021 China has financed, developed, or operated 35 big African ports and thousands of miles of roads and railways, and the trade in goods between Africa and China amounted to almost $200 billion in 2019 (China-Africa Research Initiative 2022).

China as a large and fast-growing source of aid and the largest source of construction financing supported many of Africa's most ambitious infrastructure developments in recent years (Sun et al. 2017: 9). Despite all criticisms, some studies show that China's activities in sub-Saharan Africa have helped economic growth in Africa as China as emerged as a source of trade credits, direct investments, and official aid and help to transform "hopeless" Africa into "aspiring" Africa (Gill and Karakülah 2021: 2–3). "This has happened as official finance from advanced economies started to dry up, both because of growing dissatisfaction with its development impact in Africa and economic crises at home." (Gill and Karakülah 2021: 5). The study by Sun et al. shows that the three main economic benefits to Africa from Chinese investment and business activity are: job creation and skills development, transfer of new technology and knowledge, and financing and development of infrastructure such as the $1.2 billion Tanzania Gas Field Development Project in 2015; the $3.4 billion, 750-kilometer Ethiopia-Djibouti Railway in 2016; and the $3.8 billion, 750-kilometer Standard Gauge Railway in Kenya in 2017 (Sun et al. 2017: 11, 17). Africa-China trade was $13 billion in 2001. Then it grew to $188 billion in 2015 and $204 billion in 2020. FDI has grown even from $1 billion in 2004 to $35 billion in 2015, a growth rate of 40% (Sun et al. 2017: 20).

During the COVID-19 pandemic, whereas Western countries reacted in a very self-regarding or inward-looking way, China and Russia tried to show their solidarity with African countries by supporting them aggressively (Hainzl 2021: 228). For example, Driessen suggests that power relations between Chinese management and Ethiopian rank and file at the Chinese-run construction sites are less asymmetrical than often portrayed (Driessen 2022). China attempts to boost its image and spread its development and one-party model in Africa (Kapchanga 2020). However, according to an Afrobarometer Survey from 2021, a majority of Africans perceive the US and the West in general as better development and political model than China (Sanny and Selormey 2021a). Still, as the 2021 FOCAC Ministerial Meeting in Dakar shows, the Chinese-Africa relationship will be expanded. The documents from the meeting and the summit such as *The Dakar Action Plan (2022–2024); the 2035 Vision for China-Africa Cooperation; the Sino-African Declaration on Climate Change;* and *the Declaration of the Eighth Ministerial Conference of FOCAC* show that the relationship will be further strengthened. On the other hand, the potential Chinese takeover of Uganda's Entebbe airport, the potential seizure of Kenya's Mombasa port, and the recurring debt repayment issues with which Zambia struggles are still open.

The influence of Turkey has been steadily increasing in Africa. The Turkish construction industry, which had no projects in Africa except in Libya in 2012, has 99% of projects in Africa in 2022. Before 2008 only 0.3% of Turkish construction projects were in Africa and by 2021 there were already 17%. In Ethiopia, Tanzania, etc. Turkey has been expanding its investment in the railway sector (The Economist 2022a: 58). Turkey is involved in many diverse economic areas as building a hospital in Mogadishu (Somalia), called the Recep Tayyip Erdogan Hospital. They are renovating ports, running hotels and the international airport of Mogadishu, building the parliament building, and constructing roads (The Economist 2022b: 31). Turkish Airlines had flight services in 2004 only to four countries, in 2022 it flew to 44 destinations; trade with Africa grew eight-fold between 2003 and 2021 (The Economist 2022b: 31). Even if in comparison Turkey's two-way trade of $29 billion is dwarfed by China's $254 billion in 2021, its economic involvement is growing fast in Africa (The Economist 2022b: 32). Turkish exports to sub-Saharan Africa have grown impressively since 2007, and signed energy cooperation agreements with various African countries (Shinn 2015: 7–10). South Africa, Nigeria, Ghana, Ethiopia, and Sudan are the five top trading parts in sub-Saharan Africa in construction, agriculture, textile industry, and railway projects (Shinn 2015: 11). Turkey's bilateral trade volume grew with Africa three-fold (to $18.8 billion) between 2003 and 217. It stood at $6 billion in 2015. Turkish investment in Africa is estimated to have surpassed $6 billion by 2018 (Atlantic Council 2018a). However, there is an increasing trade imbalance between Turkey and Africa. In 2001, Turkish exports to and imports from Africa $0.371 billion and $0.704 billion respectively. However, by 2008, it was $3.212 billion and $2.503 billion, and in 2019, $5.794 billion and $1.811 billion (Madrid-Morales et al. 2021: 96).

Turkey has been expanding its influence through its media presence in Africa. Like China and Russia, Turkey's media in Africa seeks to undermine the image of Europe as a colonial power in Africa. It trains African journalists in Africa or Turkey; it provides technical support to media organizations in Africa; it attempts to actively influence public opinion in African countries in the interests of the Turkish government through influence multipliers, such as journalists or public figures who shape general opinion (Madrid-Morales 2021: 90). According to the Turkish government, unlike the Ottoman Empire that prevented the penetration of colonialism in Africa, the Europeans (France and Great Britain) hindered Africa from fully realizing its great potential in which they maintained the colonial structures and a win-lose situation for Africa (Madrid-Morales 2021: 94).

Though Russia's trade with Africa before the COVID pandemic stood at just by one-tenth of the more than $200 billion Chinese trade with Africa, Russian economic influence in Africa has been increasing. By 2019, Russia's Rosatom had signed cooperation agreements and memoranda of understanding to set up nuclear plants in Egypt, Nigeria, Sudan, Rwanda, Kenya, Uganda, Zambia, Ethiopia, and Tanzania (Mugisha 2019). Russia's metals giant Rusal is one of Guinea's biggest and most loyal investors. Guinea is Russia's main bauxite supplier. Gazprom, Lukoil, Alrosa, Rusal, Renova, Rosatom, Norilsk-Nickel and Sintez have been active such as in South Africa, Angola, Algeria, the DRC and Namibia, Nigeria, Egypt, Botswana, Côte d'Ivoire, Ghana and Togo. Moreover, Russia has been expanding its activities in the areas of energy, infrastructure, telecommunications, fishing, education, health, tourism, and military-technical assistance (Arkhangelskaya and Shubin 2013: 18). The government of Colonel Mamady Doumbouya of Guinea, which came to power in 2021 through a military coup, has got in Russia an important and unconditional ally. In the face of the confrontation with the West, Russian demand for minerals important for industry such as manganese, bauxite, and chromium will increase. Diamond mines in Angola, Lukoil projects in Cameroon, Ghana, and Nigeria, and the Republic of Congo, and bauxite in Guinea will enhance Russia's economic power in Africa (Ramani 2021a: 8).

Whereas Beijing's influence is predominantly based on its financial firepower, Russia has built influence in sub-Saharan Africa by supporting authoritarian strongmen, managing natural resource projects in war-torn countries, and building on longstanding defense relationships (Financial Times 2019). Russian mining concessions are concentrated in poorly governed and authoritarian states such as Sudan, Guinea, the Central African Republic (CAR), and Mozambique (EIU 2022a). Moreover, Russia skillfully combines its military and economic interests in Africa.

The Russian invasion of Ukraine and the subsequent Western sanctions have had negative economic impacts on Africa. Moreover, the increase in fuel prices, the fertilizer export ban by China and Russia, and the increasing costs of urea and phosphate—two major components of fertilizer—are substantially affecting the African agricultural economy and food production. When the head of the African Union, Macky Sall, told Vladimir Putin at a meeting in Russia in early June 2022, Mr. Putin said Russia was always on the side of Africa (BBC 2022). However, African countries are hit hard by wheat and fertilizer shortages as a consequence of Russian aggression against Ukraine. Russian blockade of Black Sea ports and Russian and Ukrainian mines along the coast have prevented the export of grains from Ukraine. For example, Libya and Eritrea get more than 40% of their wheat from Ukraine. Moreover, 38 of 45 sub-Saharan African countries are net oil importers. Even before the Russia-caused crisis started to bite petrol already accounted for about 20% of imports in Kenya and Ghana. Transport and food costs constitute up to 57% of the inflation index in Kenya, Rwanda, Ghana, and Nigeria (The Economist 2022d: 27).

By deploying Wagner mercenaries Russia has expanded its influence in Africa such as in Libya, CAR, Sudan, Mozambique, and Mali. Moreover, Russia is active in disinformation campaigns to undermine democracy; spread fake news linking protesters to foreign powers; use Africans in the UN voting; debt forgiveness and trade increase; gratis Sputnik V COVID-19 vaccines; the fight against Ebola; security cooperation (security and military cooperation agreements, arms export, joint naval exercises, secure port and base access to support naval operations); mineral, diamond, oil contracts, nuclear power deals; educational and cultural exchanges such as scholarships and increasing presence of the Russian Orthodox Church (Siegle 2021a: 81–85; Pildegovičs et al. 2021: 6–8; El-Badawy et al. 2022: 6–7); or secure economic ties like with South Africa. Russia is South Africa's second-largest market for apples and pears and its fourth-largest for citrus fruit... South Africa is also reportedly pursuing a $2 bn-a-year deal to buy gas from Gazprom..." (The Economist 2022c: 52). Russia strengthens its relations with African countries through military assistance, and access to the energy and mining sectors of African countries such as the Central African Republic, Libya, Sudan, Eritrea, Mozambique, Angola, and Madagascar. At the beginning of the 2019 Russo-African summit, President Putin proudly proclaimed that Russia is exporting to Africa US$25 billion worth of food—which is more than arms export, at $15 billion (Ljubas 2019).

Russia is also expanding its soft power diplomacy. In 2015, it created an alternative credit rating agency to counterbalance the Western agencies. In 2019, the Russian Namibian Cultural and Education Centre (RusNam) was launched in Windhoek to promote higher education; an agreement was signed between Zambia's Copperbelt University and the People's Friendship University of Russia (RUDN University), to set up a center offering Russian language courses to students in Botswana, Namibia, Tanzania, Zimbabwe, Mozambique, and Angola; language centers of education to promote Russian culture and language are built; scholarships are being facilitated by Russia to African students, and at least 17,000 young Africans were studying in Russia by 2019 (Signé 2019; El-Badawy et al. 2022: 19). Russia is also trying to spread culturally-conservative values in Russia's image in Africa as a defender of traditional mores against the imposition of decadent "European values" by the West (Stronski 2019a: 13). Russian Orthodox Church is also expanding its presence in Africa, partly in retaliation of Alexandria's recognition of the Orthodox Church of Ukraine, and partly as a continuation of the Kremlin's strategy to expand its influence in Africa. Sputnik and RT are trying to bring more news to Africa including fake news partly to combat Western narratives in media and public opinion, to promote support for African dictators, to discredit Western institutions and democracy, promote Russia's role in the international system, to influence African elections, news and media narratives to influence attitudes, perceptions and public opinion, and spread anti-French messaging on social media such as in CAR (European Parliament 2019:8; Clifford and Gruzd 2022: 6; Siegle 2021a: 83; Pildegovičs et al. 2021: 10), or by spreading an anti-colonial narrative and by training pro-Russian hackers in Africa working to de-legitimize the West, with a particular focus on France (El-Badawy et al. 2022: 20). Reportedly, anti-Western news in African social media are being increasingly supported by pro-Kremlin agencies (Clifford and Steven 2022: 10), for example some people in Ghana and Nigeria being secretly trained to fan the flames of racial division in the United States on Twitter and Facebook (El-Badawy et al. 2022: 20). Moscow seems to be committed to the "liberation of African countries from Western narratives in which journalists usually just report deaths, murder and mass protests in Africa, and neglect success and achievements" (Madrid-Morales et al. 2021: 114).

The pro-Russian and anti-West messaging and disinformation campaign makes it more difficult for ordinary African readers to identify inauthentic accounts (Siegle 2021a: 83). In its campaign in Africa, Russia has been trying to exploit the absence of a Russian colonial past in Africa as a keystone of Russia's information strategy (Pildegovičs et al. 2021: 6).

### *External Economic Powers and Africa's Debt*

Whereas Chinese involvement in infrastructure projects in Africa decreases, China eyes growing trade with Africa as its priority despite the significant trade imbalance of $113.2 billion in Chinese exports and $95.5 billion in Chinese imports in 2019, and $114.2 billion in Chinese exports and $72.7 billion in Chinese imports in 2020 (Sun 2021). Some studies suggest that about 50% of Chinese lending to poor countries is hidden from the World Bank and IMF so the state of public debt is not open (The Economist 2022g: 5).

Between 2000 and 2020, China lent $160 billion to African governments (The Economist 2022g: 4), and from 2007 to 2020 the Chinese infrastructure financing for sub-Saharan Africa was 2.5 times as big as all other bilateral institutions combined, and at the same time Chinese loans accounted for 17% of the stock public debt in sub-Saharan Africa. In seven of 22 countries suffering from debt distress, the figure is even more than 25% (The Economist 2022g: 4). In 2015 and 2018, China announced it would provide $60 billion in new financing. However, the question is when and under what conditions the Africans are going to repay their Chinese loans (The Economist 2022g: 6–7). By 2021, African countries with the largest debts to China were Angola ($25 billion), Congo Brazzaville ($7.3 billion), Ethiopia ($13.5 billion), Sudan ($6.4 billion) and Zambia ($6.6 billion) (Ramani 2021a: 7). In 2023, Ethiopia owed China more than US $14 billion of its $27.9 billion debt (Africa Research Bulletin 2023c: 24117). Whereas the average debt of emerging markets was 12.5% of revenues, African governments were spending 16.5% of revenues servicing foreign loans (The Economist 2022b: 29). Many experts have been warning that African countries that borrow from China may lose key assets if they are unable to pay back their loans (Brautigam 2019; Green 2019). As to the general government gross debt in relation to GDP in sub-Saharan Africa, as of 2021, debt accounts for 159.2% in Cabo

Verde, 151.9% in Eritrea, 102% in Mozambique, 98.9% in Mauritius, 84.6% in Ghana, etc. (Statista 2022a).

Hang Zhou (2022) argues that Chinese construction companies operating in Uganda were driven by profit and interested mainly in quick project turnaround rather than in systematically shaping sectorial governance in the host country. According to Hang Zhou, both China and Western actors in Uganda created favorable conditions for the Ugandan leadership's increasing authoritarianism (Zhou 2022: 28–29). A study by Cezne and Wethal (2022) about the Maputo Ring Road, implemented by the China Road and Bridge Corporation, and the Moatize Coal Project, led by the Brazilian mining company, Vale SA shows that the Chinese and Brazilian workplaces expose overlapping patterns of inequality, contention, and hostility. Mozambican workers were employed in low-skilled positions operated under the supervision of higher-skilled Chinese workers (Cezne and Wethal 2022: 353).

Regarding the 1860 km-long Tanzania–Zambia Railway Authority (TAZARA), which links Zambia's Central Province with the port of Dar es Salaam, Tim Zajontz analyzes whether the cooperation is dominated by all-powerful China from a structuralist perspective and African agency void of structure. He suggests that it is important to transcend both crudely structuralist accounts of an all-powerful China and ontologically shallow conceptions of African agency that fail to explain how agents are differentially constrained or enabled by their structural contexts (Zajontz 2022: 111–112, 131).

Some analysts argue that the Chinese debt trap is exaggerated. On average, 32% of African external public debt is owed to private lenders and 35% to multilateral institutions such as the World Bank. Even though China is the biggest bilateral lender its loans are just 20% of the total (The Economist 2019a). In the face of such a Chinese influence in Africa, the US doubled the lending capacity of its Overseas Private Investment Corporation to $60 billion in 2018 and in 2019 the EU announced it would give €40 billion in grants from 2021 to 2027, building on Germany's "Marshall Plan for Africa" launched in 2017. Such competition is welcome for many African political elites as Africans would not have gotten that much money from the West without China, and African leaders realize they have more choices than ever (The Economist 2019b).

Similarly, some studies suggest that China has restructured or refinanced approximately US$ 15 billion of debt in Africa between 2000 and 2019; that there is no evidence for "asset seizures" and despite contract clauses requiring arbitration, no evidence of the use of courts to enforce payments or application of penalty interest rates; and that between 2000 and 2019, China had canceled at least US$ 3.4 billion of African debt (Acker et al. 2020: 2). Some researchers argue that the debt Africans owe to China is generally a small portion of their total external public debt (Gill and Karakülah 2021).

However, The Economist suggests that African countries usually remain the weaker partners in military and economic agreements and lack better-trained negotiating teams, and many African diplomats lack adequate language skills until they negotiate with a single voice through the African Continental Free Trade Area agreement (The Economist 2019a). Such as global competition can end up corrupting the leaders instead of promoting democracy and transparency. I find this a surprising argument. African countries have enough economists and lawyers who can understand international financial and economic agreements. The problem is rather often those who come to power in Africa do not understand economics or do not listen to their economists who are often exiled or put in prison. Africans have to question, why China with its 1.4 billion population accounts for 20% of the world's economic output, and Africa with the same population size only 3% (The Economist 2022g: 4).

Even if poverty were the cause of African migration to the West the causes of African poverty are primarily homemade. Africans are not mere victims of any kind of global conspiracy. Russia and China do not care about African democracy in the name of non-interference in African internal affairs. China indeed supports despots like Denis Sassou Nguesso of Congo-Brazzaville, Russia supports Faustin-Archange Touadéra of the Central African Republic, or those who came to power unconstitutionally in Mali, Burkina Faso, or Guinea. Many observers argue that the West, too, has a long history of supporting its preferred "strongmen" on the continent, though hypocritically, it maintains that it stands for democracy and human rights (The Economist 2019a). However, the primary responsibility lies with Africans, not with China, Russia, or the West.

## Scapegoating the External Powers for Africa's Economic Ills

Most of Africa's economic ills are rooted in corruption, which has at least two main causes. Firstly, cultural or endogenous factors such as patriarchy, patrimonialism, autocracy, the perception of the infallibility of men, and intolerance toward opposing views are essential parts of African culture.

Opposing someone's view is often viewed as questioning one's authority and personality. Secondly, in many African cultures where social esteem has such high value, it is perceived that the easiest way to achieve it is through wealth accumulation. Migration is perceived as a fast track toward this goal. The will to create a national collective identity and interest and national wealth is poorly pursued or developed. We Africans are good at blaming the West and colonial times all the time but are not able to manage our natural resources or govern ourselves in an accountable way. We do not have the determination and the interest to prove wrong racist whites who always doubted African capability.

The best example of this African incapability is Nigeria. Oil-rich Nigeria failed to profit from an oil boom among others because of spluttering production, oil theft (5%–20% of Nigeria's oil), poorly managed and dysfunctional oil industry, poor economic management, more expenditure in fuel subsidies than in education, health care and welfare, etc. Whereas oil-producing Gulf States, Norway, etc. live in prosperity, Nigeria and Angola are marred by poverty and corruption. Around 40% of Nigerians live on less than the equivalent of $1.90 a day (The Economist 2022h).

Obiang, the son of Africa's longest-serving leader President Teodoro Obiang Nguema, between 2000 and 2011, acquired a collection of luxury assets and properties in France, such as the Square de l'Avenue Foch mansion (with a comfortable and prestigious interior, an indoor swimming pool, a home cinema, and a private night club) in Paris's wealthy 16th district (Davies 2022).

The 20 least electrified countries in the world are in Africa, South Sudan 93%, Chad 89%, Burundi 88%, etc. non-electrified (Auth 2023). Africans often scapegoat the West for their own poor economic and political performance. That was why during the US-Africa summit in December 2022 in Washington D.C., Ghana's President Nana Akufo-Addo urged African governments to stop begging from the West and invest and spend African money inside the continent. Africans govern themselves so poorly and beg around the world because they have internalized the colonial disrespect for Africans. Their poverty is not a destiny, an accident, or a bad fate, but made by Africans themselves.

Africans are not poor victims of external powers manipulating their economies or resources. As the article by Alden and Otele (2022) shows, both national and local elites in Kenya were active agents at the inception and policy formulation phase, as well as the implementation phase of the Chinese-funded Standard Gauge Railway, built between 2014 and 2017 (Alden and Otele 2022: 458).

Africa is not experiencing any new scramble. During the colonial scramble for Africa Europeans divided Africa among themselves. Europeans were convened by the German Chancellor, Otto von Bismarck, to the Berlin Africa conference of November 1884–February 1885 to systematize colonial competition for territorial establishment in Africa and manage the ongoing process of colonization to avoid the outbreak of armed conflict among rival colonial powers (Soulé 2020: 634).

There is a big difference between the scramble for Africa during colonial times and nowadays. During the former, Africans were not subjects of their history, and Africans and the African continent were divided among the Europeans at will, whereas nowadays Africans can choose their partners and reject external influences. African dependence on external powers is not due to neocolonialism, but instead, is caused by African politicians' incompetence, greed, and indifference toward the suffering of their population. Very often, foreigners in Africa are depicted as neocolonial exploiters, interested only in the continent's natural resources, not its people, and ready to bribe politicians (The Economist 2019a). According to the Transparency International Corruption Perceptions Index 2021, no African country is among the 38 best in the world. But rarely is it clearly expressed that Africans themselves are responsible for Africa's trade deficit, debt, and exploitation of natural resources. Neither the global economic system nor the climate change alone explains the fact that some farmers in the south of Ethiopia in 2022 doing an ox's job of pulling a plow after having lost their livestock because of the drought and were left without an option. Well-governed countries, peaceful, and accountable states are less vulnerable to even severe climate change.

The historical injustice and exploitation by external powers cannot justify the greed, incompetence, and indifference of African politicians. African political elites,—instead of working for their population degraded, exploited, and dehumanized by the colonizers and global political and economic exploiters in the past and present—are continuing the dehumanizing and degrading project of the colonizer. African political elites have been allying themselves blindly and slavishly with external actors like the US, Europeans, Chinese, or Russians to the detriment of their degraded population. For the African political elites, if their primary task is to enable peace and prosperity for their citizens—not wealth accumulation for themselves—they have to be eclectic in their choice of global partnership and cooperation. There is no reason to believe that Europeans are better or worse than Russians or Chinese for Africa. All of them are in Africa out of self-interest. In their slavish and blind partnership and alliance with external actors, African political elites reflect a kind of dependence syndrome driven by greed, incompetence, and indifference leading to physical and mental migration from Africa. This physical and mental migration is prevalent among postcolonial Africans as the desire to imitate the former colonizer and other global powers. Postcolonial African will to be dependent on and imitate the West is a major cause of physical and mental migration from Africa, which I understand as a pervasive imitation of the West. To afford themselves a Western lifestyle, in their mental migration to the West the African political elites are not willing to work for their people and with their people, instead, they work against their people. What is keeping Africans in poverty and insecurity is the betrayal of Africans by Africans and the forgetfulness of their African history.

Today Africans are not innocent victims of external powers competing to pursue their interests by ignoring Africans. Africans are part of the game; they make agreements freely with external actors like China, US, and Russia. The "new scramble" narrative is misleading from the perspective of both diplomatic presence and agenda-setting (Soulé 2020: 637). Africans are not coerced, displaced, or enslaved. Africans are now free agents. If Africans do not understand the contracts they are signing with these external powers, then it is their fault. If they understand them and make disadvantageous agreements, it is their fault. Contrary to colonial times and right after independence, Africans have now enough economists, engineers, and lawyers who can make good deals with external powers.

For example, for the political economy of state failure in Somalia, primarily it is the Somali political and religious elites responsible. The Somali state has failed because of intra-elite struggle, endemic political violence, and insecurity including the threat posed by Islamic extremism, and the neighboring countries. Its internal weakness led to interference from the outside. The failure of economic development and reform agendas in

Somalia is caused by endemic conflict and governance issues due to vested interests and elite fragmentation (Elder 2022). Aid practices from outside, and foreign interventions cannot change anything as long as the elite fragmentation continues.

For both 2022 and 2023, Ethiopia, Zimbabwe, and Sudan recorded an inflation rate of more than 30% (the latter two even more than 50%) (EIU 2022a: 7). African states are required to repay about US$75 billion of external borrowing both in 2023 and 2024 each (EIU 2022a: 8). As of 2023, 283 m Africans suffer from hunger. The African continent, on one hand, has enough arable land to feed 9 billion people, but on the other hand, it spends $75 billion each year importing more than 100 m tons of food (Africa Research Bulletin 2023b: 24149).

Acemoglu and Robinson (2012) argue that one of the key characteristics of states in danger of failure or which have failed is that they are affected by extractive political and economic institutions. According to them, "Extractive political institutions concentrate power in the hands of the narrow elite and place few constraints on the exercise of this power. The elites often structure economic institutions to extract resources from the rest of society. Extractive economic institutions thus naturally accompany extractive political institutions" (Acemoglu and Robinson 2012: 81).

Gains from cheap extraction of minerals indispensable for modern industries of mobile phones, laptops, tablets, electric cars, space technologies, and so on benefit the global economic system. Local people in Africa where these minerals come from stay poor; the powerful ones get richer with a cheap supply of precious minerals. The global economic system or external actors exploit Africa because Africans are not able to manage their politics and natural resources. It is primarily homemade weakness that makes them vulnerable to external actors going to Africa to pursue their national interests.

There are strong correlations between entrenched leadership and conflict or instability, stagnant or declining economies, and democratic backsliding. The probability of rights abuses, secret or arbitrary arrests and detentions, tight restrictions on freedom of expression, and police brutality, chronic underdevelopment with a long-struggling, kleptocratic, and extractive economy is the highest in the countries with longest-serving leaders, as the cases of Zimbabwe or Zaire / DRC show (Klobucista 2021). From the governance quality point of view, sub-Sahara belongs to the worst in the world. Of 188 countries ranked in 2020 for governance quality including indicators such as political stability and non-violence, control of corruption, regulatory quality, and rule of law, it was only Botswana ranked 45th as the best performing sub-Saharan country; 4 sub-Saharan African countries were among the ten worst performing (The World Bank 2022). No data was available for the Democratic Republic of Congo. Neocolonialist outsiders impose none of these maladies on Africans. As Doumbia (2020) demonstrates, there is a direct correlation between political, economic, and institutional features of good governance, especially the control of corruption and regulatory quality, on one hand, and income increase and poverty decrease on the other. African countries with natural resources are usually well known for their poor governance and kleptocracy, greed, incompetence, indifference, and forgetfulness of their history. No external power has enforced them on Africans.

Some argue that corruption is increasing because of China and in many countries people have started to show their discontent saying they are fed up with China and they would prefer America, Europeans, etc. For example, the Ghanaians perceive that their country is on the wrong track because of illegal gold mining and river pollution by the Chinese entrepreneurs, armed confrontations between the Chinese and the locals but the Chinese were rarely arrested and the Chinese were fishing illegally in the Gulf Guinea, etc. (The Economist 2022g: 8). But it is misleading to make the Chinese responsible for the mistake of the national government. It is the responsibility of the national government to guarantee that the Chinese do not exploit or mistreat Africans and to make them accountable.

In DRC, which possesses 70% of the world's cobalt, the Chinese are the uncontested dominant power. The population in these areas is exploited and displaced without profiting from the natural resources. There is also a huge trade deficit with China. Just Angola, Congo, and South Africa accounted for 62% of Africa's exports to China in 2021 (The Economist 2022g: 7). However, it is the responsibility of the Congolese authorities to change this imbalance.

Many African economies rely too much on raw materials. In 2022, 83% of African countries were dependent on the export of raw materials, up from 77% around 2011. The very limited raw materials processing capacity is one of the weaknesses of the African economies. African countries rich in minerals and exporting raw materials such as Angola, Nigeria, Sierra Leone, Guinea, and the Democratic Republic of Congo are prone to dictatorships, civil wars and conflicts, poor governance and human rights violations, and poor investment in infrastructure and education (The Economist 2022f: 24–25).

Europe's demand for natural gas and its determination to make itself independent from Russian energy would open opportunities for resource-rich African countries. For example, the European Union has a growing interest in Tanzania's gas reserves. Tanzanian President Hassan has intensified negotiations with energy companies to attract about $30 billion in foreign investment to revive the construction of offshore liquefied natural gas projects in 2023. In Senegal, 40 trillion cubic feet of natural gas were discovered between 2014 and 2017. On 16 February, Nigeria, Niger, and Algeria signed an agreement to develop the Trans-Saharan Gas Pipeline—estimated to cost $13 billion—and to export natural gas to European markets (Rosnick 2022). On the other hand, there are African countries such as Kenya, Ethiopia, and Sudan already affected by the shortage of fuel, fertilizer, and food because of the Russian invasion of Ukraine and augmented by the fertilizer export bans by China and Russia, affecting households, the agricultural sector, and food security severely (Rosnick 2022). Besides all these, what I find worrying is that African political elites have blindly started believing the Russian narrative that Russia empowers Africa; Russia brings Peace to Africa; the West destabilizes Africa; the West is a colonizer; Russia supports Pan-Africanism; Russia develops Africa; Africa should not trust Western media; and the West neglects Africa (Pildegovičs et al. 2021: 20–26).

African development cannot and will not come from outside. Despite their natural wealth, many African countries are incapable of managing their economies and abundant natural resources. For example, the DRC is one of the richest countries in natural resources. It is home to immense resource wealth (an estimated $24 trillion worth of untapped mineral resources key for clean technologies), significant hydropower potential, and great biodiversity. Despite this natural wealth, the DRC has historically been and continues to be, subject to political instability, violent conflict, humanitarian crises, poverty, and malnutrition. Rebel groups such as the March 23 Movement and 100 different active armed groups continued to terrorize the population because of which 4.5 million are internally displaced, more than 800,000 live as refugees in other nations, and about 60 million Congolese—about 73% of the population—lived on less than $1.90 a day in 2018. One of the main reasons has been weak governance, which has accelerated the armed conflict, as the leadership does not have control over the entire territory. For example, with natural resources such as 70% of the world's cobalt, 60% of the world's coltan, and the fourth-largest producer of copper, the population of the DRC could be one of the wealthiest in the world. The Congolese politicians lavish in luxury and are not capable of managing these resources and the politics of their countries for their people. Chinese companies backed by Beijing control these key minerals for clean technologies (Signé 2022).

Collier argues that Africans are poor because they are entrapped in bad governance, violent conflict, abundance of natural resources, and they are landlocked (Collier 2008). But what are the root causes of the bad governance and violent conflicts? They are man-made (African-made!) and not imposed by anyone from outside. Africans are poor or insecure not because of colonization, neo-liberalism, landlockedness, resource abundance, ethnic diversity, or unfavorable climate (Chang 2009). Africans are poor and insecure because they have poor institutions (Acemoglu and Robinson 2012). But where do these poor institutions come from? Africans have poor institutions because they do not have the will to make better institutions. The abundance of resources can exacerbate conflicts and bad institutions but it does not cause them. The resources are never the causes, they are only the means. Landlockedness only coincides with bad institutions, incompetence, and indifference. It does not cause poverty or conflict. Competent and willing societies can easily overcome the challenges of landlockedness.

Whereas about 43% of Africans are living in poverty, every year over US$60 billion worth of stolen assets flow out of Africa instead of investing in jobs and social services. In Corruption Perception Index, six out of the bottom ten countries are African where misappropriated funds account for a 25% loss of development resources. One in two African citizens report they pay a bribe for various issues. Several mining companies are involved in bribing high-ranking officials to win mining licenses such as in Sierra Leone and the Republic of Congo; Shell and Eni are accused of bribery in Nigeria. "Between 2010 and 2012, the DRC reportedly lost over US$1.36 billion from the underpricing of mining assets that were sold to offshore companies linked to Gertler." (Transparency International 2019). In 2016 and 2017, French authorities confiscated Vice President Teodorín Obiang's (Vice President to his own father—the president of Equatorial Guinea) assets worth US$35 million, while Switzerland seized 25 of his supercars (Transparency International 2019). The former president of The Gambia Yahya Jammeh had laundered his money by buying a luxurious mansion in the US for $3 million in the state of Maryland (Zane 2022). Dictator General Sani Abacha who ruled Nigeria between 1993 and 1998 took up to $5 billion of public money (Transparency International 2020).

Capital flight from Africa exceeds aid to the continent by ca. $40 billion each year (Stearns 2022: 155). UNCTAD estimates that almost $89 billion (3.7% of Africa's GDP) of capital flight from Africa is taking place through tax evasion by multinational companies, bribery, and embezzlement, which is almost equal to the combined total annual inflows of foreign aid, ca. $48 billion, and yearly foreign direct investment of $54 billion, received by African countries between 2013 and 2015. As a result, Africa lost around $836 billion in total illicit capital flight from 2000 to 2015, or an estimated $1 trillion between 1970 and 2020 (UNCTAD 2020: 25, 182).

Studies show that there is a huge amount of illicit financial flows from Africa: Angola $60 billion during 1986–2015; Cote d'Ivoire $32 billion during 1970–2015, and net cocoa export misinvoicing of $3.7 billion during 1995–2014; Democratic Republic of the Congo €9.95 billion during 2000–2010; Ghana Gold $6 billion and cocoa $4.3 billion during 2000–2017, and abnormally undervalued export of gold was $3.8 billion and of cocoa, $12.6 billion between 2011–2017; Nigeria Oil export misinvoicing at $44 billion and import misinvoicing at $45 billion during 1996–2014; Zambia $14.5 billion during 1995–2014, and net export misinvoicing of copper worth $14.5 billion during 1995–2014; and South Africa $198 billion during 1970–2015, net export misinvoicing during 2000–2014 for Silver and platinum, $24 billion; iron, $57 billion (UNCTAD 2020: 46).

Studies suggest that Africa loses about $148 billion to corruption every year (UNCTAD 2020: 22). Moreover, 20% to 30% of private wealth in many African countries is held in foreign tax havens. The beloved destinations of African stolen money by the African dictators and kleptocrats are Western bank accounts: Kenya's Moi and his family bought several multi-million-pound properties in London, New York, Australia; President of Gabon Ali Bongo, Omar Bongo's son, purchased a luxury housing estate and keeps luxury cars in Paris; the Panama Papers leaks revealed that Nigeria's political elite and public officials use money stolen from their people to buy expensive homes such as in London; certain Swiss and Italian firms have been accused of bribing Somalia Transition Government officials to secure contracts to deposit highly toxic industrial waste in the waters of Somalia; companies such as Shell, BP, Mobil, Total, Elf, Texaco, Mittal and Anglo-America Corporation caused catastrophic environmental damage in Nigeria, Ghana, Gabon, Equatorial Guinea, Angola, Congo-Brazzaville, Democratic Republic of Congo, South Africa, Guinea, Sierra Leone, Liberia, and Senegal. Some US corporations were accused of arming rebel groups and collaborating with them to traffic from the Democratic Republic of Congo gold, diamonds, timber, and coltan (a precious ore essential to videogame consoles, laptop computers, and cellphones) (TWN 2016). Many Africans complain that Western media reports about wars and the desperate faces of hungry refugees but are silent on the role played by Western financial institutions, corporations, and governments in facilitating the transfer and laundering of stolen African money (Adusei 2016).

I wonder why the West should care for Africa while Africa's political elites exploit it, degrade Africans, and enrich themselves. Why should Europeans be better for Africans than Africans themselves? I can better understand the European egoism, hypocrisy, and complicity with African kleptocrats than the Africans' greed, incompetence, indifference, slavish dependence on, fascination for the Western lifestyles, and forgetfulness of their history. The neocolonial argument does not help us as no African country or its leader is forced to divert the wealth of the country to foreign accounts. If Africans argue that Europeans are neocolonialists, why do they hide all those African billions of dollars and their wealth in Europe and in European bank accounts?

Africans love to imitate European lifestyles. By putting their money in European bank accounts, buying European real estate, making vacations in Europe, and consuming European goods they desire to become at least a bit of European. Africans try to become human beings by imitating the Europeans who denied their humanity. By imitating Europeans African political elites hope to get rid of the inferiority complex inculcated in them by Europeans.

In the global political economy, Africans are free to negotiate, change international economic partners, or cease unprofitable or disadvantageous partnerships. The wealth gap between African political elites and their citizens is much bigger than the one between the wealth of European politicians and their citizens. The income share of 10% Africans is 54%, whereas in Europe it is 10% to 37%. Only the Middle East is more unequal than Africa, 10% to 58% (Robilliard 2022). If Africans care so much about their citizens, why don't they narrow the gap, and live comparable life standards with their population? Where is the African culture of solidarity and communitarianism, which they adore so much? Neocolonialism is not responsible for African poverty and insecurity. Poverty, displacement, conflicts, hunger, forced migration since the end of colonization, etc. are caused by African political elites' greed, incompetence, indifference, and forgetfulness of their history. I think it is worth quoting in length what the Kenyan writer Ngugi says,

> One could quote other even more incredible episodes, of the callous massacre of children, of the equally callous genocide of part of a population and all by native leaders on behalf of imperialism. For the point is this: the Mobutus, the Mois, and the Eyademas of the ne-colonial world are not being forced to capitulate to imperialism at the point of an American maxim gun. They themselves are of the same mind: they are actually begging for a recolonization of their own countries with themselves as the neo-colonial governors living in modern fortresses. They are happier as the neo-slave drivers of their own people; happier as the neo-overseers of the U.S.-led economic hemorrhage of their own countries. How does a writer, a novelist, shock his readers by telling them that these are neo-slaves when they themselves, the neo-slaves, are openly announcing the fact on the rooftops? How do you shock your readers by pointing out that these are mass murderers, looters, robbers, thieves, when they, the perpetrators of these anti-people crimes, are not even attempting to hide the fact? When in some cases they are actually and proudly celebrating their massacre of children, and the theft and robbery of the nation? How do you satirize their utterance and claims when their own words beat all fictional exaggerations? (Ngugi 1986: 80)

Since I have been in Europe for the past three decades, one of the things I observed is that Europeans are very cautious in criticizing Africans lest they be denounced as racist. Europeans easily criticize other Europeans or US Americans for their global politics. Europeans whom I know personally as well as most of those who write about Africa try to show almost some understanding or excuse African failures because of slavery or colonialism. However, my impression is that even if they are convinced that the current African ills are at least partly caused by the Africans' incompetence, greed, and indifference, they don't dare to say it. For example, as a researcher originally coming from Ethiopia and living in Europe, working together with European universities and researchers, and trying to expand our cooperation with African universities, I observe many remarkable things. Very often it happens that when we try to establish inter-university cooperation with African universities, it is often difficult to get even the simplest information from the African side because of unnecessary bureaucracy there. Europeans in such cases try to pretend to show understanding for the Africans and they do not dare to criticize Africans.

There is a clear link between African culture, corruption, and clientelism. Patriarchy is a cultural mindset affecting African women in particular and African societies in general by reducing the schooling chances of girls and reducing them to childbearing machines. Firstly, the well-being of mothers and their children suffers; second, the productive and innovative capability of women is wasted; thirdly, it has contributed to an unsustainable population growth obliterating the economic growth; and finally, this archaic and anachronistic patriarchal culture denies the subjectivity of African women and keeps them as a second-class citizen.

In 2022, I participated in a workshop on food security that took place in Seville, Spain. Many professors and researchers from West African and European countries were present. One professor from a West African country suggested that overpopulation in Africa from food security perspective is not a problem because it is easy to produce enough food for everyone by using the right agricultural technology. Shockingly, most of the conference participants ignored the comment.

As various studies show us, overpopulation is a multifaceted problem. From an economic perspective, as Banerjee and Duflo suggest, countries with a very high fertility rate like Ethiopia (6.12 Children per woman) are fifty-one times poorer than the United States with a total fertility rate of 2.05 (Banerjee and Duflo 2011: 106). Sexual education and contraception are taboo in many African countries. Here, patriarchy and the cultural position of women are decisive. Paradoxically, men who do not pay most of the physical costs of bearing children make fertility decisions (Banerjee and Duflo 2011: 116). There are many who suggest that in poor households parents consider children as old-age insurance (Banerjee and Duflo 2011: 119). However, even uneducated parents are not too stupid to understand that the more children they have the less saving they are going to have, ending up in a zero-sum game. Here the decisive factor is rather patriarchy than economic rationality which leads to larger families.

Poor institutions keep people poor. Poverty can conversely produce and promote poor institutions (Banerjee and Duflo 2011: 236). It is possible to break this vicious circle only through good political and economic institutions (Acemoglu and Robinson 2012; Sharma 2016: 7). However, as Banerjee and Duflo suggest, "the rulers who have the power to shape economic institutions, do not necessarily find it in their interest to allow their citizens to thrive and prosper." (Banerjee and Duflo 2011: 238).

Powerful Africans (politicians) stand to lose from political and economic institutional reforms (Banerjee and Duflo 2011: 254). African leaders remain key agents of Africa's fall or rise. A competent, accountable, and transformational political leadership is crucial, a leadership that promotes collective interests permits new ideas and innovation, and builds conducive and accountable institutions and respects them (Signé 2017). Good institutions empower all citizens; they are participative; they give free space for innovation; new ideas and collective interests can be promoted; and individual properties, physical integrity, and the public good are safeguarded.

Aid to Africa is often criticized as one of the causes of African ills. The volume by Hagmann and Reyntjens (2016) deals with aid and authoritarianism in Africa and explores the motives, dynamics, and consequences of international aid given to authoritarian African governments; it addresses the collusions between donors and one-party regimes or dictatorships. The book explores the donor motives and the aid impact in authoritarian contexts. Donor motives and foreign aid may facilitate authoritarianism and kleptocracy but they do not cause them. Foreign aid can be good or bad depending on who uses it for what purpose. Aid is like any object that can be used or abused. It is important to understand not the giver or the gift, but the receiver and his motivation. We usually know the motivations of the giver. We can ignore aid if the giver has bad intentions. We know also that international aid is an object that cannot impose itself on anyone to be used this way or that way. So, Africa is not the victim of foreign aid or global conspiracy, but instead a victim of its leaders.

The lust for power, one-party democracies, fictionalization or romanticization of the African past, or the argumentation that democracy and human rights were foreign to African traditions, etc. are the outcome of Africans' free decision (Soyinka 2012: 14). Africanity, authenticity, etc. are fictions constructed by African politicians for personal benefits. This is the root of the African crisis, a crisis largely caused by the African leadership (Soyinka 2012: 48). As Soyinka says, as long as the criminality and brutality of the colonial past lead to the present impunity, and the African culture and its past are fictionalized, Africa is doomed to fail (Soyinka 2012: 66). Similarly, "resource curse" is one of the favored arguments as causes of African insecurity and poverty. Resources cannot be scapegoated. Resources are means which could be used in a good or bad way. It is the user responsible, not the used object. For those who can use them properly resources are a blessing, not a curse.

Both internally and externally, there is a widespread tendency to scapegoat development aid or natural resources. What is scapegoating? According to René Girard, scapegoats are beings considered extremely harmful to the whole society. A community is unified in its readiness to put the blame on that being made responsible for the ills. The community looks for an accessible cause of the trouble. The community members as potential persecutors unite to purge the community of impure elements made responsible for the internal ills. The crowds make a mob, *mobile vulgus*. The enemy will be identified through the mobilization of this mob (Girard 1986: 16). The victims can be chosen sometimes randomly, i.e. "the persecutors choose their victims because they belong to a class that is particularly susceptible to persecution rather than because of the crimes they have committed." (Girard 1986: 17). There are often stereotypes for the selection of scapegoats: ethnic or religious minorities, sick or mentally ill, people with genetic deformities, people with injuries, children, women, old people, disabled, poor or outsiders, rich and powerful, people with extreme characters such as beautiful or ugly, etc. In all this, the average defines the norm (Girard 1986: 18–19). According to Girard, different sacrificial practices in different cultures show the victimization of or scapegoating of these marginalized groups or the weakest segments of society. In many cultures, members of these groups were sacrificed to prevent or get rid of actual crises in society (Girard 2019: 123–126). What happens in such a mechanism is that not only the wrong people could be persecuted, but also—the most important thing in this case—that the persecutors overlook their own responsibilities for the ills of their society. Scapegoating can appease the community's feelings for a certain time but it cannot solve the underlying problem of society if the victims are wrongfully chosen whereas the real problem of the community comes from within. Scapegoating is a misjudgment or disguising the facts through translocation of own failures on an outsider. It is a psychological mechanism through which the persecutors transfer their failures to others and try to convince themselves of their innocence. Though the real causes of the problems are the persecutors themselves, repeatedly and almost ritually they have to look for scapegoats (Girard 2016: 308; Palaver 2003: 201–202). In this mechanism, they are wrongfully transferring their faults to others and ritually and subconsciously sweeping their own responsibility under the carpet.

This is to argue that externaal players, aid or natural resources are never the problem. Most human beings in their lifetime receive some kind of material or immaterial aid. However, this has not made us automatically corrupt. How we receive and spend that aid is our responsibility instead of pushing all our faults on the aid and making it our scapegoat. Even the giver of the aid is not responsible for the recipient's misallocation of aid or failure. Even if the giver has some mean intentions it is my responsibility to figure out why and under what conditions is the giver giving me the gift or aid. Mills argues that "Aid gives Africa more reason to externalize its problems rather than deal with them. Donors can make things worse..." (Mills 2010: 310). This is just a scapegoating of aid or donors. Aid can be used in a bad or good way depending on the user. If advisors or donors are bad, why do Africans allow them to do so? Are Africans so ignorant of the intentions of the donors? Why do the receivers expect donors to change their behavior when they themselves do not want to change theirs? Yes, "Aid can make a difference if it falls on fertile soil." (Mills 2010: 343).

There are very differing thoughts about the relevance and effectiveness of development aid or its euphemistic version, cooperation. For example, Jeffrey Sachs (2006), in his famous work, The End of Poverty pleaded for more development aid to end poverty in a short period. William Easterly (2007) critically analyzed the problem with development aid. Similarly, Dambisa Moyo (2010) criticizes very strongly the effectiveness of development aid. Acemoglu and Robinson (2012) emphasize the importance of institutions for the success of states. Similarly, Mills argues that Africa is poor because of its institutional weaknesses.

Africa suffered historical atrocities because of slavery and colonization with devastating and lasting impacts. However, African decision-makers are primarily responsible for the current institutional failures. Contrary to colonial times, Africans have today enough well-educated economists, scientists, and intellectuals in different fields.

Greg Mills in his book, *Why Africa is Poor*, argues that Africa has failed not because of external factors, though they have played a part, but because African leaders have made this choice (Mills 2010). Africa's high youth unemployment and poor or no education are not problems imposed by any external actor. Nigeria's enormous wealth coming from its oil is not taken away by any external actor. That 35 of 48 sub-Saharan economies import food despite fertile soil is not only because of climate change or neocolonialism, but instead, because African elites are indifferent, incompetent, and greedy. As Mills rightly argues, "...no amount of money was going to 'fix' African states if their leaders continued to make the wrong development choices" (Mills 2010: 9). Mills raises a very important question in his book: "But why do African elites not see the obvious advantage in growing their economies and uplifting their People? And why do African electorates allow them to get away with bad choices?" (Mills 2010: 11).

Most Africans stick to power because it gives them status and recognition, which is culturally so important. It guarantees them access to public goods, which they can divert into their pockets and therewith can imitate the lifestyle of the colonizer. This happens not only on the highest political level. As Bayart says, it is not the case that the masses are innocent and only the political elites are predators (Bayart 2009). There is a poor understanding of the public good in Africa. This starts already on the family level if we take patriarchy and autocracy seriously. It starts already on the local, village, or community level. Africanism, African socialism, and Negritude were some of the constructed romantic naiveties of the Africans.

The neo-colonial arguers suggest that Africa is poor because the external actors are exploiting Africa. But as Mills argues, "The world has not denied Africa the market and financial means to compete … Africa's poverty has not been because the necessary development and technical expertise is unavailable" (Mills 2010: 12–13). Africa has many more intellectuals, engineers, and highly qualified economists than it had after its independence (Mills 2010: 201). Neither is the West imposing any policy on Africa. Africans are freely agreeing to all financial or economic deals with the West.

The cultural argument of African poor institutional performance is avoided in development theories because it is perceived to contain some racist connotations, not because the idea itself is wrong. However, patriarchy is a disadvantageous cultural mindset, which undermines the well-being, creativity, and productivity of women and promotes their servility and fertility. For example, where I grew up, it is not only men who behave according to patriarchal expectations but even women themselves expect men to be dominant, macho, masculine, etc. Would any reasonable human being defend patriarchy as a propitious cultural value, which would promote economic and political institutions that support creativity, productivity, participation, and innovation? Countries that promote patriarchy and patrimonialism stay weak and poor politically, economically, and culturally.

Many Africans lament all the time that Africans are poor because the West is rich or because of colonialism (Mills 2010: 42). Other colonized people do not lament as much as the Africans (Mills 2010: 86). The

Chinese, Russians, Turks, Iranians exploit this African mindset that the West was colonial and neo-colonial whereas they are their real partners. The tragic thing is that Africans believe them. The West was and is as self-interested as the Chinese, Russians, Turks or Iranians are. No state is going to Africa out of charity, instead out of self-interest. It is utterly African responsibility to make the right choices with whom to cooperate to promote Africa's long-term interests.

Similar to the neo-colonial notion of Kwame Nkrumah, Langan understands neo-colonialism as the "continuation of external control over African territories by newer and more subtle methods than that exercised under formal Empire" (Langan 2018: 4). Langan argues that Africa's ills come from neo-colonialism but he does not explain why Africans allow this to happen. Unlike now, Africans did not have the political power to reject exploitation during the colonial time. Langan explains that Africans enjoy legal or juridical sovereignty, but not empirical sovereignty or self-determination (Langan 2018: 5). However, there is no proof whatsoever that any external power is forcing any agreement on any African country to enter into a political or economic agreement. Secondly, in the face of global power competition, the global actors are competing to gain Africans for themselves politically and economically. Langan himself, by citing Nkrumah, suggests that foreign companies support corrupt African governments (Langan 2018: 6). Then it is corrupt Africans themselves cooperating freely and willingly with the external powers who are depleting Africans. "... coopted African elites do more to serve their foreign benefactors than to effectively pursue the material and cultural wellbeing of their own citizens" (Langan 2018: 26).

Fanon describes the blacks who in the footsteps of the Europeans exploit their people whiter than the Whites pursuing their own interests and privileges, making use of the war to strengthen their material situation and their growing power at the expense of their own people. The iniquitous fact of exploitation wears now a blackface. The national black bourgeoisie learns lessons of negation and decadence, it identifies itself with the decadence of the bourgeoisie of the West. He spends large sums on cars, luxury houses, and private jets to be recognized as equal to the Whites whereas his people do not have enough to eat. He does not want to stop old rivalries, tribalism, and interracial hatred (Fanon 2001: 115–129). As Memmi suggests, "Such is the history of the pyramid of petty tyrants: each one, being socially oppressed by one more powerful than he, always finds a less powerful one on whom to lean, becomes a tyrant in his turn"

(Memmi 1967: 17). For example, North Africans despised by the Whites despise black Africans, the lowest of the lowest human beings. They are happy that they found someone whom they can compare themselves with and feel better, more beautiful, and civilized. They have a thousand-year-old tradition of culture, they are Mediterranean, nearer to Europe, and share Greco-Roman civilization unlike the uncivilized, savage, and brutal black Africans, the Negroes (Fanon 2001: 129–130).

Africans are poor and weak not because of external exploitation, instead their poverty and weakness make them prone to exploitation and rejection. Langan himself raises a good point regarding the freedom of the Africans to choose their external economic partners: "The dispute over the selling of Kosmos stakes should also be understood in this wider context of US geopolitical interests. Ghana's decision not to press ahead with the Chinese takeover of Kosmos' oil resources, and to allow the US company to retain its original stake (even after its proposal to sell to ExxonMobil) should be understood in the wider ambit of US-Ghana bilateral ties" (Langan 2018: 44). This example by Langan shows that Ghana was able to take the best bidder.

By taking an example from Uganda, Langan laments that oil companies are operating at the expense of environmental standards, that President Museveni is gaining largesse from the presence of foreign operators in terms of election spending, that he has militarized oil outlets on the pretext of defending resources, that there is increasingly less transparency in gains from the oil, that subsistence farmers are being removed from their land, etc. (Langan 2018: 46). But Langan does not show us convincingly why this should be the fault of foreign investors in Uganda. Do foreigners have any obligation toward Ugandans? The well-being of Ugandans is solely the responsibility of the Ugandan government.

Mills asks why have African leaders made "bad choices at the expense of hundreds of millions of their people?" Mills argues that this was because they have been able to do so thanks to their own people and external community and were able to externalize their internal problems, i.e. putting the blame on others (Mills 2010: 229). African autocrats put their financial interests before their people's well-being, and private interests before the public good (Mills 2010: 277).

As Mills suggests, for African success some of the most important factors are aligning government, the unions, and businesses, liberalizing access for trade and people, tax reform, increasing formal employment, food security and diversity, public sector reform and deregulation, empowering labor, good and accountable incentives for investors, encouraging export, protecting the vulnerable through comprehensive and universal social security, universal and quality education, and probably most importantly women empowerment and eradication of patriarchy and patrimonialism (Mills 2010: 429–435).

> All the ills that have hurt Latin America and the Middle East are exponentially compounded in Sub-Saharan Africa: bad government, unexpected sovereignty, backward technology, inadequate education, bad climate, incompetent if not dishonest advice, poverty, hunger, disease, overpopulation—plague of plagues. Of all the so-called developing regions, Africa has done worst: gross domestic product per head increasing, maybe, by less than 1 percent a year; statistical tables sprinkled with minus signs; many countries with lower income today than before independence. The failure is more poignant when one makes the comparison with other parts: in 1965, Nigeria (oil exporter) had higher GDP per capita than Indonesia (another oil exporter); twenty-five years later, Indonesia had three times the Nigerian level. (Landes 1999: 499)

Landes poses the question, "And no one paused to ask why the colonial powers were so quick to leave" (Landes 1999: 499). This is a clear suggestion that the Western colonial powers should have stayed longer in Africa, which is an implicit paternalism and racism. Landes suggests that "... Africans of one or another place realized that freedom was not an automatic gateway to happiness and prosperity" (Landes 1999: 500). This sentence implies that Africans are so stupid that they believe that all is well once they have achieved their independence. I think Africans knew very well that they had to build their countries economically, politically, and infrastructurally. African problem is not a problem of knowledge nor poor land fertility, material impediments, cheaper important goods than locally produced, or climatic variations (Landes 1999: 500). Failing African countries fail because of a failed policy choice and because they sustain counterproductive cultural values. African problems are complex but they are primarily homemade: the highest population growth in the world, large families as a result of patriarchy and the tabooing of sexual education and contraceptives, the decision to import than produce locally, the tastes that cannot be satisfied locally, the switch of tastes from "old", "boring" staples to new cereals (often imported), new urban elites habits leading to an increased demand for meat, the flight into slums, (Landes 1999: 500–501).

Landes puts almost all the blame on foreigners and suggests that foreign eager aid agencies, experts, and technicians who have wielded their power and have in their mind only the bigger-the-better approach have contributed to the African failure. This victimhood is eagerly internalized by many Africans:

> That is the fault of the West. The West told us to build power stations, bridges, factories, steel mills, and phosphate mines. We built them because you said so, and the way you told us. But now they do not work, you tell us we must pay for them with our money. That is not fair. You told us to build them, you should pay for them. We did not want them. (quoted according to Landes 1999: 504)

This is one of the problems in Africa, to see oneself as a victim, not to stand to one's failures, and blame always the West and the colonial time. There is no serious proof that the West forced postcolonial Africans to do something they did not want. If postcolonial Africa did not have an experience of self-government, if their rulers enjoyed a legitimacy bounded by kinship networks and clientelist loyalties, if they were not able to form a representative government, if the new political system was alien to their own tradition, if the passion for liberation and identity waned after the independence, if autocratic strongmen took their countries hostage, if one military coup followed another, if bureaucracies were inflated to provide jobs for henchmen, if foreign aid ended up in foreign accounts of these autocrats in Zaire / DRC, Equatorial Guinea, etc., if Africans blindly followed the capitalist or Marxist-Leninst propaganda, it was not the fault of the West or of USSR, the IMF, or the World Bank, it was Africa's choice (Landes 1999: 505–506). If Indonesia could do so well, why not Nigeria, Angola, or Zaire / DRC?

Landes makes an important conclusion in his argument: "The continent's problems go much deeper than bad policies, and bad policies are not an accident. Good government is not to be had for the asking" (Landes 1999: 507). Africans eagerly ask outsiders for everything. If they do not get it, they are furious. Aid will not be a solution for Africa, neither is it a problem by itself. What matters is how aid is being used.

Where does this deep-rooted reason for African failure lie? It lies in the psychological impact of colonization and in Africa's postcolonial perception of the self and the colonizer: the mindset that all the problems were caused by the West, that Africans are victims of neocolonialism, and the hidden belief that to become a full human being means to imitate the West and a simultaneous slavish fascination for and abomination of the West.

From the Post Traumatic Slave Syndrome perspective, one of the main causes of the current African political elites' violence, greed, and indifference toward their own people is rooted socio-genetically and psychoanalytically in the slavery and colonial experiences. "Slavery was an inherently angry and violent process. White people modeled anger and violence in every aspect of enslavement. ... Thus, Africans learned that anger and violence were key ingredients necessary to ensure that their needs were met. Anger and violence created and maintained the institution of slavery, and the anger and violence continued long after slavery was abolished." (Degruy 2017: 114).

This is a mimetic aspect of colonial anger. The political elites are angry at the former slaveholders and colonizers. But they take their own people as a surrogate victim of their anger, greed, and indifference. Since they cannot inflict anything on the Europeans, they can vent their anger and drive for violence and exploitation on their own people, the Africans. They imitate the slaveholder and the colonizer in two ways: his violence and his way of life. Two characters in the novel of the Tanzanian novelist Abdulrazak Gurnah, "*Memory of Departure*", discuss the African political problems and tribal violence and scapegoating: "... Which tribes will you start with? When will it be the turn of the Indians? When will you move on to the Arabs or the Somalis? And who will be your next scapegoat after that?' 'Scapegoat! That is the problem,' he returned. 'That is why we do not do anything. We all see ourselves as victims, waiting our turn. Waiting for somebody to come from out there and give a helping hand. We do not do anything for ourselves. Who will be next? Well, we will be next ... sooner or later. Unless we do something about it." (Gurnah 2021: 86).

## Conclusion

Africa is poor and failing because it is ruled by corrupt elites who organize society for their own benefit at the expense of the vast majority, and the political power has been narrowly concentrated and has been used to create great wealth for those who possess it (Acemoglu and Robinson 2012: 3). Deep-rooted African problems lie in at least three things: patriarchy and patrimonialism, colonization-induced inferiority complex, and postcolonial Africa syndrome (blame the West for all ills). The important issue is not how much those African countries owe to China or any external actor. Rather, why are they indebted? Does this indebtedness have anything to do with the quality of governance in those African countries? What are their records of corruption or accountability?

If the African political elites are competent enough and care for their population, they would choose only beneficial partnerships; they would make sound economic policies. There are enough educated Africans capable of differentiating between beneficial and detrimental economic partnerships and deals. External actors are not forcing the Africans to take counterproductive loans or to be indebted. It is the responsibility of the Africans to choose between risky and beneficial cooperation. It is a weak argument that Africans are poor because of neocolonialism. External actors do not take African natural resources away by force from Africa, as they did during colonial times. Be it China or Russia, they are invited by the African political elites themselves to invest in natural resources. Neither Russians, Chinese nor Europeans pretend to be charity organizations, nor are they neo-colonizers. They are indeed self-interested actors, hypocrites, and complicit with African kleptocrats, but they are pursuing their legitimate interests. Africans must stop their greed, indifference, and incompetence at least for the sake of tens of millions of Africans who died during the European slave trade and colonization. African political elites are free to choose among external actors and they are not under any form of coercion to make any economic agreement with any external actor.

African governments were allies of the West or the East during the Cold War. They let themselves be instrumentalized. African freedom fighters who promised freedom, equality, and prosperity to their citizens ended up as corrupt, inept, and dictators supported by the West or East depending on the ideologies they followed. With the end of the Cold War, the competition seemed to come to an end giving way to democracy, accountability, and economic growth. Since the increasing emergence of strongmen politics around the world in the mid-2006s (Rachman 2022a), African democratic governance has been declining and allegiance to external undemocratic powers has been increasing.

Africans have to stop indulging themselves in self-pity. Africans need mental freedom from the internalized inferiority complex and scapegoating of the West; they have to be politically, economically, and especially psychologically adults. They have to strive for dignity and self-respect, take responsibility in their own hands, and prove to the colonial racist that they are capable of proving to the colonial racist that he is wrong in depicting black Africans as inferior, and incapable of self-governance and self-development. Africans have the obligation—for the sake of those innocent black Africans who suffered under barbarous, hypocritical white Western Christians—to create better and accountable political and economic institutions that enable economic growth. Africans have the human capital, material resources, and favorable global conditions which they did not have during and at the end of colonialism.

The West has been also hypocritical when it declares commitment to democracy and human rights in its development cooperation with Africa but at the same time propping up dictators who stay in power by undemocratic means such as repression and changing constitutions. African politicians must know that human rights are universal human values, not merely European. If human rights are Western values, why did African countries ratify almost all international human rights law documents? Why did they argue during the anti-colonial war that colonialism was against human rights? Why is the majority of the African population in favor of civil and political rights? How can anyone from outside of Africa argue that human rights are Western values that must not be imposed on others?

Moreover, many in the West tried to mobilize Africans against the Chinese or the Russians on the grounds of neocolonialism. Even this is paternalistic. Africans themselves are responsible for freely choosing their international partners. African partnership with the Chinese or Russians is not the Chinese or Russian fault. Similarly, the Chinese and Russians mobilize Africans to scapegoat the West by suggesting the incessant neocolonial ambitions of the West.

At the same time, African political elites strive for the Western lifestyle and luxurious houses; they buy houses in the West; they open bank accounts in the West. They take vacations in the West. African politicians love the West for its lifestyles and modernity, but they pretend to hate it just to be able to easily mobilize the African masses against the West, an opportunistic pretentious hatred, but a surreptitious fascination, admiration, and imitation, as we shall see in Chaps. 6, 7, and 8. The following chapter discusses Africa's political scapegoating of the West.

## Literature

Acemoglu, Daron and Robinson, James (2012): Why Nations Fail: The origins of power, prosperity, and poverty, Currency: New York.

Acker et al. (2020): Debt Relief with Chinese Characteristics. Working Paper No. 2020/39. China Africa Research Initiative, School of Advanced International

Studies, Johns Hopkins University, Washington, DC. Retrieved from http://www.sais-cari.org/publications.

ActionAid (2014): The Great Land Heist: How donors and governments are paving the way for corporate land grabs, ActionAid International.

Adusei, Aikins (2016): Hiding Africa's looted funds: The silence of Western media, *Third World Resurgence* No. 309, May 2016, pp. 21–23.

Africa Research Bulletin (2023b): Dakar 2 Summit: Food Sovereignty For Africa, *Economic, Financial and Technical Series*, January 16th – February 15th 2023, p. 24149.

Africa Research Bulletin (2023c): Ethiopia – China, *Economic, Financial and Technical Series,* December 16th–January 15th 2023, p. 24117.

Alden, C. (2007), *China in Africa* (London: Zed Books).

Alden, Chris and Otele, Oscar M. (2022): Fitting China In: Local Elite Collusion And Contestation Along Kenya's Standard Gauge Railway, *African Affairs*, 121:484, pp. 443–466.

Arkhangelskaya, Alexandra and Shubin, Vladimir (2013): Russia's Africa Policy, Global Powers and Africa Programme, South African Institute of International Affairs, September 2013.

Atlantic Council (2018a): *Turkey's Growing Presence in Africa, and Opportunities and Challenges To Watch in 2018,* http://www.atlanticcouncil.org/events/past-events/turkey-s-growing-presence-in-africa-and-opportunities-and-challenges-to-watch-in-2018, accessed 05.09.2018.

Auth, Katie (2023): *How the U.S. Can Better Support Africa's Energy Transition*, Carnegie Endowment for International Peace.

Ayeni, Tofe (2022): reputational battle: China to overtake the EU as Africa's biggest trade partner by 2030, https://www.theafricareport.com/229297/china-to-overtake-the-eu-as-africas-biggest-trade-partner-by-2030/.

Banerjee, Abhijit and Duflo, Esther (2011): Poor Economics: A radical rethinking of the way to fight global poverty, Public Affairs: New York.

Bartlett, Kate (2022): Are Rights Abuses Tarnishing China's Image in Africa? https://www.voanews.com/a/are-rights-abuses-tarnishing-china-s-image-in-africa-/6560353.html.

Bayart, Jean-Francois (2009) The State in Africa: The Politics of the Belly, Polity.

BBC (2022): Ukraine war: Hungry Africans are victims of the conflict, Macky Sall tells Vladimir Putin, https://www.bbc.com/news/world-africa-61685383.

Brautigam, D. (2019). A critical look at Chinese 'debt-trap diplomacy': The rise of a meme. Area Development and Policy, 5(1), 1–1430 October.

Buzan, B. & Little, R. (2000), *International Systems in World History: Remaking the Study of International Relations, (*Oxford: Oxford University Press).

Cezne, Eric and Wethal, Ulrikke (2022): Reading Mozambique's mega-project developmentalism through the workplace: evidence from Chinese and Brazilian investments, *African Affairs*, 121: 484, pp. 343–370.

Chang, Ha-Joon (2009): Economic History of the Developed World: Lessons for Africa, The lecture delivered in the Eminent Speakers Program of the African Development Bank, 26 February 2009.

China-Africa Research Initiative (2022): China-africa trade, http://www.sais-cari.org/data-china-africa-trade.

Clifford, Cayley and Gruzd, Steven (2022): Russian and African Media: Exercising Soft Power, South African Institute of International Affairs, African perspectives, Global insights.

Collier, Paul (2008): The Bottom Billion: Why the Poorest Countries are Failing and What Can Be Done About It, Oxford University Press: Oxford.

Cotula, Lorenzo (2014): Addressing the human rights impacts of 'land grabbing', Directorate-General for External Policies of the Union Directorate B, Policy Department, European Parliament.

Cotula, Lorenzo et al. (2009): Land grab or development opportunity? Agricultural investment and international land deals in Africa, IIED/FAO/IFAD, London/Rome.

Davies, Alys (2022): Equatorial Guinea seeks to block sale of confiscated Paris mansion, https://www.bbc.com/news/world-europe-63105426, 24.05.2023.

Degruy, Joy (2017): Post-Traumatic Slave Syndrome: America's Legacy of Enduring Injury and Healing, DeGruy.

Doumbia, Djeneba(2020): The role of good governance in fostering pro-poor and inclusive growth, https://www.brookings.edu/blog/africa-in-focus/2020/07/01/the-role-of-good-governance-in-fostering-pro-poor-and-inclusive-growth/.

Driessen, Miriam (2022): Pidgin Play: Linguistic Subversion on Chinese-Run Construction Sites in Ethiopia, *African Affairs*,119: 476, pp. 432–451.

Easterly, William (2007): The White Man's Burden: Why the West's Efforts to Aid the Rest Have Done So Much Ill and So Little Good, Penguin Books: London.

Econfin Agency (2022): Africa imports seven times more Russian products than it exports to Moscow, https://www.ecofinagency.com/public-management/0903-43450-africa-imports-seven-times-more-russian-products-than-it-exports-to-moscow.

Economist Intelligence Unit (2022): A new Horizon for Africa-China Relations: Why Co-Operation will be Essential, The Economist Intelligence Unit Limited 2022.

EIU (2022a): Africa outlook 2023: the challenges ahead: Resilience amid disruption, The Economist Intelligence Unit Limited.

El-Badawy, Emman et al. (2022): Security, Soft Power and Regime Support: Spheres of Russian Influence in Africa, Tony Blair Institute for Global Change.

Elder, Claire (2022): Logistics Contracts and the Political Economy of State Failure: Evidence From Somalia, *African Affairs*, 121: 484, pp. 395–417.

Environmental Justice Foundation (2022): On the Precipice: Crime and corruption in Ghana's Chinese-owned trawler fleet, A report by the Environmental Justice Foundation.
European Commission (2020): The European Union and Africa: Partners in Trade, https://trade.ec.europa.eu/doclib/docs/2022/february/tradoc_160053.pdf.
European Parliament (2019): Russia in Africa: A new arena for geopolitical competition, Briefing.
Eurostat (2022): Africa-EU trade in goods: €4 billion surplus, https://ec.europa.eu/eurostat/web/products-eurostat-news/-/edn-20220217-1.
Fairhead, James et al. (2012): Green Grabbing: a new appropriation of nature?, Journal of Peasant Studies, 39:2, 237–261, https://doi.org/10.1080/03066150.2012.671770.
Fanon, Frantz (2001): The wretched of the earth, Penguin Books: London.
Financial Times (2019): Putin seeks friends and influence at first Russia-Africa summit, https://www.ft.com/content/f20dbcc2-f17b-11e9-ad1e-4367d8281195, 20.06.2022.
Gerstter, Christiane (2011): An Assessment of the Effects on Land Ownership and Land Grab on Development, Policy Department DG External Policies.
Gill, Indermit S. and Karakülah, Kenan (2021): Is China Helping Africa? Growth and Public Debt Effects of the Subcontinent's Biggest Investor, Duke Global Working Paper, Paper 3, March 2019, Duke University Center for International and Global tudies.
Girard, René (1986): The Scapegoat, The Athlone Press: London.
Girard, René (2016): Violence and the Sacred, Bloomsbury Revelations: London.
Girard, René (2019): Things hidden since the foundation of the world, Bloomsbury Revelations: London.
Global Compliance News (2022): While challenges remain, continental free trade will further boost Africa's trade partners, https://www.globalcompliancenews.com/2022/06/14/africa-chinas-trade-ties-with-the-continent-continue-to-strengthen-31052022/.
Goedde, Lutz et al. (2019): Winning in Africa's agricultural market, February 2019, Agriculture Practice, McKinsey & Company.
Graham, Alison et al. (2010): Land Grab study, CSO Monitoring 2009–2010 "Advancing African Agriculture" (AAA): The Impact of Europe's Policies and Practices on African Agriculture and Food Security, www.europafrica.info, 02.06.2014.
Green, M. (2019). China's debt diplomacy. Foreign Policy. 25 April.
Gurnah, Abdulrazak (2021): Memory of Departure, Bloomsbury, London.
Hagmann, Tobias and Reyntjens, Filip (2016): Aid and Authoriatarianism in Africa, Zed Books: London.

Hainzl, Gerald (2021): The People's Republic of China's Presence in Africa, In Frank/Vogl (eds.): China's Footprint in Strategic Spaces of the European Union. New Challenges for a Multi-dimensional EU-China Strategy. Schriftenreihe der Landesverteidigungsakademie No. 11/2021.

Indian Ministry of Commerce and Industry (2022): Foreign Trade (Africa), https://commerce.gov.in/about-us/divisions/foreign-trade-territorial-division/foreign-trade-africa/.

Jeffrey Sachs (2006): The End of Poverty: Economic Possibilities for Our Time: Penguin Books: London.

Kapchanga, Mark (2020): Africa may learn from china's political path to revitalize its own development plans, https://www.globaltimes.cn/content/1189460.shtml, 2020/5/25.

Kleemann, Linda et al. (2013): Economic and ethical challenges of "land grabs" in sub-Saharan Africa, Kiel Policy Brief, No. 67, Kiel Institute for the World Economy (IfW), Kiel.

Klobucista, Calire (2021): Africa's 'Leaders for Life', https://www.cfr.org/backgrounder/africas-leaders-life.

Landes, David (1999): The Wealth and poverty of nations: why some are so rich and some so poor, Norton and Company, New York.

Langan, Mark (2018): Neo-Colonialism and the Poverty of 'Development' in Africa (Contemporary African Political Economy), Palgrave Macmillan: New Castle.

Ljubas, Zdravko (2019): Russia Enters Africa with Soft Power, https://www.occrp.org/en/daily/10959-russia-enters-africa-with-soft-power, 22.06.2020.

Madrid-Morales, Dani et al. (2021): It is about their story: How China, Turkey and Russia influence the media in Africa, Konrad-Adenauer-Stiftung Regional Media Program, Sub-Sahara Africa.

Martin-Prével, Alice et al. (2016): The Unholy Alliance: Five Western Donors Shape a Pro-Corporate Agenda for African Agriculture, The Oakland Institute, Oakland.

Memmi, Albert (1967): The colonizer and the colonized, Boston: Beacon Press.

Mills, Greg (2010): Why Africa is Poor: And What Africans Can Do About It?, Penguin.

Moyo, Dambisa (2010): Dead Aid: Why aid is not working and how there is another way for Africa, Penguin: London.

Mugisha, Ivan (2019): Rwanda approves nuclear power deal with Russia, https://www.theeastafrican.co.ke/business/Rwanda-approves-nuclear-power-deal-with-Russia/2560-5318000-view-asAMP-815wfl/index.html?__twitter_impression=true, 20.06.2022.

Ndi, Frankline A. (2017): Land Grabbing, Local Contestation, and the Struggle for Economic Gain: Insights From Nguti Village, South West Cameroon, *SAGE Open*, 1–14.

Ndi, Frankline A. and Batterbury, Simon (2017): Land Grabbing and the Axis of Political Conflicts: Insights from Southwest Cameroon, in *Africa Spectrum*, 52, 1, 33–63.

Ngugi (1986): Decolonising the Mind: The Politics of Language, in African Literature: James Currey: Nairobi.

Nigerian National Bureau of Statistics (2022): *Nigeria Launches its Most Extensive National Measure of Multidimensional Poverty*, https://nigerianstat.gov.ng/news/78, 06.07.2023.

Norwegian Refugee Council (2022): Global Report on Internal Displacement 2022, Children and youth in internal displacement, Internal Displacement Monitoring Center.

Oram, Julian (2014): The Great Land Heist: How donors and governments are paving the way for corporate land grabs, ActionAid International.

OXFAM (2013): Poor Governance, Good Business : How land investors target countries with weak governance, OXFAM MEDIA BRIEFING, 7 February 2013 Ref: 03/2013.

Palaver, Wolfgang (2003): René Girards mimetische Theorie, LIT: Münster.

Pattanayak, Banikinkar (2022): India-Africa trade ties offer huge potential but challenges remain, https://www.financialexpress.com/economy/india-africa-trade-ties-offer-huge-potential-but-challenges-remain/2603269/.

Pildegovičs, Tomass et al. (2021): Russia's Activities in Africa's Information Environment: Case Studies: Mali and Central African Republic, NATO Strategic Communications Centre of Excellence.

Rachman, Gideon (2022a): The Age of the Strong-Man, The Bodley Head, London.

Ramani, Samuel (2021a): Russia and China in Africa: Prospective Partners or Asymmetric Rivals? South African Institute of International Affairs, Policy Insights 120, December 2021.

Robilliard, Anne-Sophie (2022): What's New About Income Inequality in Africa? World Inequality Lab Issue Brief 2022-09, November 2022.

Rosnick, Danielle (2022): What does the war in Ukraine mean for Africa? https://www.brookings.edu/blog/africa-in-focus/2022/02/25/what-does-the-war-in-ukraine-mean-for-africa/?utm_campaign=Global%20Economy%20and%20Development&utm_medium=email&utm_content=205330541&utm_source=hs_email, 20.06.2022.

Ruth Hall (2011) Land grabbing in Southern Africa: the many faces of the investor rush, *Review of African Political Economy*, 38:128, 193–214, https://doi.org/10.1080/03056244.2011.582753.

Sanny, Josephine Appiah-Nyamekye and Selormey, Edem (2021a): Africans welcome China's influence but maintain democratic aspirations, Afrobarometer Dispatch No. 489, 15 November 2021.

Schneidman, Witney (2022): Will Biden deliver on his commitment to Africa in 2022?, https://www.brookings.edu/blog/africa-in-focus/2022/01/10/will-biden-deliver-on-his-commitment-to-africa-in-2022/.

Sharma, Ruchir (2016): The rise and fall of nations: forces of change in the post-crisis world, Norton and Company: New York.

Shinn, David (2015): Turkey's Engagement in Sub-Saharan Africa: Shifting Alliances and Strategic Diversification, Chatham House, the Royal Institute of International Affairs, Research Paper.

Siegle, Joseph (2021a): Russia and Africa: Expanding, Influence and Instability, in Graeme P. Herd, ed., Russia's Global Reach: A Security and Statecraft Assessment (Garmisch-Partenkirchen: George C. Marshall European Center for Security Studies, 2021).

Signé, Landry (2017): Innovating Development in Africa: The role of international, regional and national actors, Cambridge University Press: Cambridge.

Signé, Landry (2019): Vladimir Putin is resetting Russia's Africa agenda to counter the US and China, https://www.brookings.edu/opinions/vladimir-putin-is-resetting-russias-africa-agenda-to-counter-the-us-and-china/, 20.06.2022.

Signé, Landry (2022): US Secretary of State Blinken to visit Africa as tension with China and Russia intensifies, https://www.brookings.edu/blog/africa-in-focus/2022/08/05/us-secretary-of-state-blinken-to-visit-africa-as-tension-with-china-and-russia-intensifies/?utm_campaign=Global%20Economy%20and%20Development&utm_medium=email&utm_content=222851075&utm_source=hs_email.

Singer, J. D. (1960), International conflict: Three levels of analysis, *World Politics*, 12:3, 453–461.

Soulé, Folashadé (2020): 'Africa+1'summit Diplomacy And The 'New Scramble'narrative: Recentering African Agency, *African Affairs*,119:477, pp. 633–646.

Soyinka, Wole (2012): Of Africa, Yale University Press: New Haven.

Statista (2022a): General government gross debt in relation to Gross Domestic Product (GDP) in Sub-Saharan Africa as of 2021, by country, https://www.statista.com/statistics/1223393/national-debt-in-relation-to-gdp-in-sub-saharan-africa-by-country/.

Stearns, Jason K. (2022): Rebels without a Cause: The New Face of African War, *Foreign Affairs*, May/June 2022, pp. 143–156.

Stronski, Paul (2019a): Late to the Party: Russia's Return to Africa, Carnegie Endowment for International Peace, October 2019.

Sun, Irene Yuan et al. (2017): Dance of the lions and dragons: how are Africa and China engaging, and how will the partnership evolve? JUNE 2017, McKinsey & Company.

Sun, Yun (2021): FOCAC 2021: China's retrenchment from Africa? https://www.brookings.edu/blog/africa-in-focus/2021/12/06/focac-2021-chinas-retrenchment-from-africa/, 29.06.2022.
Taylor, Ian (2006): China and Africa: engagement and compromise, London: Routledge.
The Economist (2019a): The new scramble for Africa: This time, the winners could be Africans themselves, 7 March 2019.
The Economist (2019b): A sub-Saharan seduction. Africa is attracting ever more interest from powers elsewhere: They are following where China led, 7 March 2019.
The Economist (2022a): Business in Africa: Ottomanpower, May 7th 2022, p. 58.
The Economist (2022b): Turkey and Africa: The call of the South, April 23rd, p. 31.
The Economist (2022c): Africa's debt crisis: Debt and denial, April 30th April, p. 29.
The Economist (2022d): Geopolitics: Friends like this, April 16th April, p. 50.
The Economist (2022e): Mercenaries: Vladimir's Army, April 9 April, p. 48.
The Economist (2022f): The ripples of Putin's war: Bread and Oil, 12 March, pp. 26–28.
The Economist (2022g): Jihadists in the Sahel: French Leave, 19 April, p. 31.
The Economist (2022h): African Economies: When you are in a hole…, 8 January 2022, pp. 24–25.
The World Bank (2022): Net official development assistance received (constant 2020 US$) – Sub-Saharan Africa, https://data.worldbank.org/indicator/DT.ODA.ODAT.KD?locations=ZG.
Transparency International (2019): Where are Africa's billions? https://www.transparency.org/en/news/where-are-africas-billions.
Transparency International (2020): A $100 million question for Nigeria's asset recovery efforts, https://www.transparency.org/en/blog/a-100-million-question-for-nigerias-asset-recovery-efforts.
TWN (2016): Hiding Africa's looted funds: The silence of Western media, https://www.twn.my/title2/resurgence/2016/309/cover04.htm, 21.07.2023.
Twomey, Hannah (2014): Displacement and dispossession through land grabbing in Mozambique: The limits of international and national legal instruments, The Refugee Studies Centre (RSC) Working Paper Series.
UNCTAD (2020): *Tackling Illicit Financial Flows for Sustainable Development in Africa*, Economic Development in Report 2020, United Nations: Geneva T.
UNDP (2019): Human Development Report: Scaling Fences: Voices of Irregular African Migrants to Europe, UNDP.
United States Census Bureau (2022): Trade in Goods with Africa, https://www.census.gov/foreign-trade/balance/c0013.html.

Usman, Zainab (2021): How Biden Can Build U.S.-Africa Relations Back Better, https://carnegieendowment.org/2021/04/27/how-biden-can-build-u.s.-africa-relations-back-better-pub-84399.

Usman, Zainab and Abayo, Aline (2022): Will U.S.-China Competition Shape Kenya's Trade Trajectory? https://carnegieendowment.org/2022/09/15/will-u.s.-china-competition-shape-kenya-s-trade-trajectory-pub-87919?mkt_tok=ODEzLVhZVS00MjIAAAGG4YA0W9biEIfzLVVwEhsPL1bXJrDH0qqXzxqwgyeb4_DrZSaVTjZDq3PoOpIPxJwEbMKSIj0K-4fjIR-ayvrw6W7Ziri86MB9ez7AGsyC, 24.05.2023.

Waltz, K. N. (1979), *Theory of international politics* (New York: McGraw-Hill).

Wendt, A. (1987): The agent-structure problem in international relations theory, *International Organization*, 41, 3, pp. 335–370, https://doi.org/10.1017/S002081830002751X.

Wendt, A. (1992), Anarchy is what the states make of it: The social construction of power politics, *International Organization*, 46:2, 391–425.

World Economic Forum (2022): The Horn of Africa is facing an unprecedented drought. What is the world doing to help solve it? https://www.weforum.org/agenda/2022/07/africa-drought-food-starvation/.

Zajontz, Tim (2022): 'Win-win' contested: negotiatingthe privatisation of Africa's Freedom Railway with the 'Chinese of today', *Journal of Modern African Studies*, 60: 1, pp. 111–134.

Zane, Damian (2022): How ex-Gambia President Yahya Jammeh's US mansion was seized, https://www.bbc.com/news/world-africa-58924630.

Zhou, Hang (2022): Western and Chinese Development Engagements In Uganda's Roads Sector: An Implicit Division of Labor, *African Affairs*, 121:482, pp. 29–59.

CHAPTER 4

# The Political Dimension of the Postcolonial African Migration to the West

As we have discussed in Chap. 2, political instability is perceived as one of the decisive push factors in migration theories. Various researches show that there is a link between political insecurity in Africa and emigration. According to a study by Giménez-Gómez et al., there is a link between rising political persecution, ethnic cleansing, human rights violations, political instability, and civil conflicts in African countries and increased migration into European countries (Giménez-Gómez et al. 2017). African political and security problems displace people and send them abroad as migrants. The postcolonial Africans are continuing the brutal violence against their people through dictatorship, autocracy, ethnicized politics, or Islamist fanaticism. Africans kill each other as if they were mere animals. In 2023, out of the ten most fragile states in the world, six were in Africa (Somalia, South Sudan, Democratic Republic of Congo, Sudan, Central African Republic, and Chad) characterized by mutual killings. This is one of the reasons why many Africans leave their countries and flee to the West. The causes are not the West, China, Russia, or any external power. The causes of such devastating African conflicts are Africans themselves.

In line with Acemoglu and Robinson (2012), Moyo (2010), Mills (2010), etc. I argue that Africa's problems are primarily internal and institutional. Where do the institutional problems emanate and why particularly Africa has poor institutions? I argue that Africa has poor institutions because of various internal reasons: scapegoating the West for all ills;



B. Gebrewold, *Postcolonial African Migration to the West*, Politics of Citizenship and Migration,
https://doi.org/10.1007/978-3-031-58568-5_4

imitating the West at all costs; inability to overcome ethnicity as a political category and to construct a collective national identity and interest; and African patriarchal and autocratic culture that dehumanizes the opponent, mythologizes authority's power, and discriminates and degrades women. I will discuss in this chapter that even though there are external factors and actors that contribute to it, Africa's poor political institutions are determined primarily by internal factors ultimately leading to postcolonial migration from Africa to the West. Like in Chap. 3, the main argument of this chapter is that scapegoating global players for Africa's ills is not only wrong but also counterproductive. If political instability causes African migration to the West, the main causes of the political instability are we Africans, not external powers. We do not only kill each other, displace our citizens, and send them as refugees and migrants, but also enable external influence and destabilization by making ourselves politically and economically weak and dependent on external powers.

## External Powers and Insecurity in Africa

African insecurity has intensified the presence of foreign powers in Africa. Burkina Faso, Mali, and Niger have been destabilized by the Jihadist groups such as Ansaroul Islam, Boko Haram, Islamic State in the Greater Sahara, and Jamaat Nusrat al-Islam wal-Muslimin. Russia (which happily replaces the West in Africa) was involved in the military coup in Burkina Faso in September 2022 by Captain Ibrahim Traoré who believes that he can fight the Islamist militant groups more efficiently not with the West but with the Russians who have been intensifying their presence in Mali, Niger, and Burkina Faso. Like in the Central African Republic, also in Burkina Faso, the Russian Wagner group has been involved in the exploitation of natural resources such as gold, diamonds, and timber.

Russian influence has grown along with U.S. disengagement across the continent especially during the Presidency of Donald Trump (Ramani 2022). Russia is teaching Africans a "Syrian model" of counterinsurgency, which prioritizes authoritarian stability by sending "private contractors to fragile states such as the Central African Republic, Libya, Mali, and Mozambique and to authoritarian partners such as Guinea and the Democratic Republic of the Congo" (Ramani 2022). Since the invasion of Ukraine, Russia succeeded in strengthening its diplomatic muscle to challenge the West for the hearts and minds of African governments, as he visited Egypt, Ethiopia, Uganda, and Congo-Brazzaville in 2022. Lavrov was able to gain Uganda's President Yoweri Museveni to the Russian side when he toured Africa in 2022. Museveni assured Lavrov that he does not believe in being enemies of somebody's enemy.

Russia and China emphasize the neocolonial discourse to mobilize Africans against the West and denounce its interference in the name of democracy and human rights. However, as an Afrobarometer survey from 2019 to 2021 shows, seven in 10 Africans (69%) prefer democracy to any other kind of government. Large and steady majorities in Africa consistently reject authoritarian governments, including military rule (75%), one-party rule (77%), and one-person rule (82%). Moreover, Support for core democratic institutions is strong and consistent in Africa: 63% for multiparty competition, 67% for parliamentary oversight of the president, and 77% for presidential term limits is also (Gyimah-Boadi and Asunka 2021).

African autocrats imitate Russia eagerly. If international law does not apply to Russia, why should it apply to Africans? In disputed territories, autocrats with expansionist ambitions could do the same in Africa. "At stake are established notions of sovereignty, territorial integrity, and the independence of member states, which suddenly become more temporal, arbitrary, and open to violent contestation" (Siegle and Smith 2022). While, in 2014, the United States refused to sell fighter jets and advanced helicopters to Nigeria on human rights grounds, Russia filled the gap and sold its jets to the Nigerian military. Russian military cooperation with African countries has been rapidly increasing. Between 2016 and 2019, Russia signed a military cooperation agreement with Nigeria, Chad, Gambia, Ghana, Guinea, Mali, Niger, and Sierra Leone (Ramani 2022). Russia deepened its ties with the authoritarian President João Lourenço of Angola, with Colonel Mamadi Doumbouya of Guinea, with Assimi Goïta who came to power after the military coup in Mali, with Sudanese military government which overthrew the civilian government, and with non-democratic governments in the Republic of the Congo, the Democratic Republic of the Congo, and Gabon (Siegle 2022). Russia tries to be the principal powerbroker in this geostrategically important Libya, to gain naval port access in the Red Sea, especially Port Sudan, to become a pivotal actor in the Sahel region such as Mali, to refocus its diplomatic efforts on propping up the military junta of Colonel Mamady Doumbouya of Guinea, and to cement its political sway and unencumbered access in the mining sector in Guinea, to expand its influence in the Republic of the Congo, the Democratic Republic of the Congo, and Gabon and to become a more significant player in the lucrative oil and mineral networks of Central Africa, and to deepen its relationship with Angola and have access to its diamond, oil, gold, and mineral resources (Siegle 2022: 117–118).

When the African Union president Macky Sall discussed with President Putin that the Russian invasion of Ukraine has led to food and economic insecurity in Africa, Putin responded that Russia was always on the side of the Africans. When Putin said "Russia is always on the side of Africa", he meant the Africa of dictators who can be easily mobilized against the liberal West (The Economist 2022n). In its global isolation because of the invasion of Ukraine, Russia needs Africans as an important diplomatic bloc. How can Russia be on the side of the Africans while sending mercenaries to Africa, propping up dictators, and flooding Africa with its arms?

Russia accounted for 87% of Sudan's major conventional weapons purchases in the period 2003–2007 when the Darfur conflict and violence caused displacement and emigration of hundreds and thousands of people (Holtom 2008: 315). Between 2004 and 2008 Russia was the leading supplier that provided Central, North, and West Africa with 74% of all major arms (Wezeman 2009). In 2008, Russia not only wrote off $4.6 billion in Libya's debt but also concluded at least $2 billion in military-technical cooperation (Africa Research Bulletin 2009: 17910). In 2015–2019 Russia accounted for 36% of arms imports by states in sub-Saharan Africa (Wezeman 2019: 7). Russia as the largest arms supplier to Mali in 2017–21 delivered four armed transport helicopters and four combat helicopters (Wezeman et al. 2022: 7–8). Between 2013 and 2017, Russia accounted for 39% of arms exports to Africa. Algeria's 78% of arms imports came from Russia. Among those who abstained or did not show up during the first UN vote against Russia countries such as the Central African Republic, Madagascar, Mali, Mozambique, Algeria, Angola, Uganda, and Sudan are clients of Russian arms trade with Africa (The Economist 2022d: 52). The USA accounted for 11% of arms exports to Africa in 2013–2017 (SIPRI 2017). Besides these major global arms exporters, there are new emerging suppliers of arms to different war-torn African countries: UAE to South Sudan and Nigeria; Turkey to different African countries; Brazil to Mali and Angola (Béraud-Sudreau et al. 2020: 5ff).

According to a pro-Kremlin think tank, Russia enjoys "a well-established image of a successful negotiator in bringing together irreconcilable parties and of a "guarantor" of peace and security", and it brings to Africa various advantages related to the military-industrial complex; defense, and security; high technologies; geological exploration; nuclear energy; processing industry equipment, etc. (Kortunov et al. 2020: 29; El-Badawy et al. 2022: 10). Russia's commitment to Africa is not necessarily for the promotion of African interests, but instead to isolate the West in Africa. A paper by Balytnikov et al. suggests that Russia promotes Afro-optimism and does not view Africa as an insurmountable challenge, as a source of a migration threat, or as a failed continent; and that Russia does not see Africa or Africans as "inept students" who do not know anything and are unable to learn (Balytnikov et al. 2019: 4). Interestingly, the paper reads that Russia does not see Africans as "inept students" who do not know anything and are unable to learn (Balytnikov et al. 2019: 4). As a psychological premise says—a denied desire is the strongest desire—for the Russians maybe Africans are indeed "inept students".

On the 2019 Russia-Africa summit, $12.5 billion in business deals were declared, in arms and grains from Russia. Russia has sent its so-called Wagner mercenaries to Libya, the Central African Republic, Mali, Mozambique, and Sudan. The mercenaries were paid in gold and diamond or as much as $4000 per month by the local African governments (Stronski 2019: 16). It is estimated that Mali paid $10 m a month to the Wagner group (The Economist 2022e: 48). About 2000 Wagners were protecting President Touadéra of the Central African Republic by 2022. When France attempted to arm the Central African government forces Russia blocked the move through the UN security council. But after France left Russia offered the government to arm them. "Mercenaries from Wagner soon showed up and a Wagner-linked company won concessions to mine gold and diamonds" (The Economist 2022e: 48). The CAR government made the Russian language even a compulsory subject for students at the University of Bangui.

The Wagner Group has been active in at least 21 countries since 2014 mainly in African countries where it tripled between 2014 and 2018 and actively involved in training Rapid Support Forces in Darfur in Sudan, in Libya in support of Khalifa Haftar's Libyan National Army (LNA), in the Central African Republic to guard diamond mines, train the army and provide bodyguards for embattled President Faustin Archange Touadéra, and in Mali to depose France (El-Badawy et al. 2022: 8). By April 2022, there were around 1000 Wagners in Mali, where Western countries are increasingly unwelcome and the national government is unable to contain the Islamist terrorist.

Some of the juntas of Colonel Goita, President of Mali, were trained in Russia. Russia backed the junta and blocked a UN Security Council resolution against Mali (The Economist 2022j: 34). When President Conde of Guinea declared unconstitutionally for a third term, the Russian position was that constitutions adapt to reality, not vice-versa. Therefore, Russia is propping up dictators in Africa as long as its interests are at stake, for example by appointing its citizens to national security advisors to the Central African Republic (Rachman 2022: 184). Popular support for Russia has been increasing in Burkina Faso, the Central African Republic, Guinea, and Mali. Russia sees itself as an alternative force that enables the shaky African governments to subdue the insecurity such as in the Central African Republic or Comoros (Alba and Frenkel 2019). By fostering instability, disrupting elections, exporting arms, and potentially fueling migration from Africa, Putin's "grand strategy" works to threaten the West away from Africa (El-Badawy et al. 2022: 4). Russia is using its media outlets to mobilize Africans against the West by depicting the latter as modern-day colonialists (Atanesian 2023).

Since Russia's invasion and annexation of Crimea in 2014 and against the background of the Western sanctions, sub-Saharan Africa and Russia have signed at least 19 military cooperation consisting of the supply of weapons and cooperation on producing military products, maintenance, countering terrorism and piracy, joint training of troops, training and cooperation on peacekeeping, search, and rescue at sea, deliveries of military equipment, spare parts, and components, facilitating meetings of military experts and cooperation on military education, joint training, and research and development. The military agreements include: in 2015 with Cameroon and Zimbabwe, in 2016 with Rwanda, Ghana, and Gambia; in 2017 with Central African Republic, Nigeria, Niger, Mozambique, Eswatini, Chad, Zambia, in 2018 with Tanzania, Sierra Leone, Ethiopia, Burundi, Burkina Faso, and again in 2022 with Cameroon (Reuters 2018; El-Badawy et al. 2022: 10). By 2019, Russia had 20 defense agreements with African countries, and it accounted for 39% of the continent's defense imports. Russia is trying to establish its military base at Port Sudan.

Between 2020 and 2021, conflict intensified in the Sahel because the Jihadists killed more than 6000 and displaced more than 3.5 million people in Burkina Faso, Mali, and Niger, and within two years. The 5000 French and European and 12,000 UN forces could not change the tide against the Jihadists (The Economist 2022j: 34). About 2700 Jihadists were killed but they were still able to recruit new ones and raise their number (The Economist 2022h: 31). In Burkina Faso, Russian flags decorated the streets, whereas the chants "Down with France!" denounced France. Russia jumped in to protect the unconstitutional government of Mali that came to power by coup in 2021.

Russia declared 2022 a "Year of Africa. During the 2019 Russia-Africa summit, 43 African leaders showed up, a higher turnout than British or French African summits. During the African liberation struggle, countries like Angola, Mozambique (whose flag features a Kalashnikov), Namibia, South Africa, and Zimbabwe sought and got Soviet support in their fight to end white rule (The Economist 2022e: 28). Russia sees Africa as a way to balance Western influence and as a geostrategic playing field. Putin's Russia strives to challenge Western dominance of global governance (El-Badawy et al. 2022: 5). It attempts to spread in Africa the idea that all political systems hold moral and governance equivalence. This way, the perception that democracy offers a more effective, equitable, transparent, or inclusive form of governance will be discredited (Siegle 2021: 80–81). In the Horn of Africa, Russia sees an opportunity to enter Africa, to project its power into the Red Sea, the Gulf of Aden, and the Persian Gulf. Russia wants to exploit new commercial opportunities and secure diplomatic support for its positions in multilateral institutions such as the United Nations, especially against the background that the US interest in the continent dropped (Stronski 2019: 3).

Of seven countries that voted against the UN General Assembly resolution vote against Russia on 23.02.2023 demanding *Russia's* immediate withdrawal from Ukraine, two were African countries (Eritrea and Mali). Of 32 abstentions, almost half were from Africa: South Africa, Ethiopia, Algeria, Angola, Burundi, Namibia, Central Africa Republic, Congo-Brazzaville, Gabon, Guinea, Mozambique, Sudan, Togo, Uganda, and Zimbabwe. 28 African countries abstained on a General Assembly motion condemning the annexation of Crimea. While the West condemned the Russian invasion of Ukraine for its utter violation of international law, the majority of African countries refused to criticize Russian aggression. Only 28 out of the 54 African countries voted in favor of the 2 March 2022 UN resolution condemning Russia (others stayed away, abstained, or voted against), whereas 81.29% of non-African countries voted in favor of the resolution. Furthermore, of the 35 countries that voted to abstain, 17 (48.6%) were African. On the UN General Assembly resolution in early April 2022 calling for Russia to be suspended from the Human Rights Council, nine African countries voted against the resolution, 23 abstained, and 11 did not vote, despite human rights being a key objective of the African Union Constitutive Act.

The stance of Africa regarding the UN Resolution on Russia is caused not only by sympathy and friendship with Russia but also it is because their alliance with the West has been diminishing increasingly. Many African countries are indeed refraining from criticizing Russia owing to commercial incentives, ideological commitments, strategic ambitions, or out of fear (The Economist 2022d: 50). The visit by the French and German foreign ministers to Ethiopia in January 2023 was primarily a reaction to the increasing Russian influence in Ethiopia and the latter's abstention and absence in various UN Resolutions against Russia (Hoffmann and Lanfranchi 2023).

According to the Declaration of the Beijing Summit of the African Countries and China in 2018, China pledges to support African countries in their efforts to independently resolve "African problems in the African way" (FOCAC 2018). Chinese officials visited 79 times between 2008 and 2018 (The Economist 2019b). China is leading in arms sales (27% in 2013–2017, compared with 16% in 2008–2012) and defense-technology ties with sub-Saharan Africa. China contributed to the UN Mission (about 500) in Mali in late June 2013. The contribution consists of police officers, medical forces, engineering troops, and combat troops; China is already the largest contributor to UN peacekeeping missions among the P5.

China has already a military base in Djibouti and is planning others in Tanzania, Kenya, Angola, Seychelles, and Equatorial Guinea; its participation in UN-led peacekeeping operations has grown; its contribution to the UN global peacekeeping budget has been significantly increasing. "Half of the missions and two-thirds of the budget are located in/allocated to Africa" (Hainzl 2021: 221). Since 2018, China has been trying to deepen its military and security cooperation with Africa. In 2018, a China-Africa security and defense forum took place in Beijing in which almost 50 African countries participated, and China conducted military exercises in Cameroon, Gabon, Ghana, and Nigeria. The following year they conducted a joint exercise with Tanzania, and trilateral naval exercises with South Africa and Russia.

China has become in between one of the biggest arms suppliers to sub-Saharan Africa by constituting 20% of sub-Saharan African imports value, just behind Russia (30%), and it dwarfed the US (5%). From 2016 and 2020, China delivered 41.7% of the total Sudanese weapons, 32.8% of the Nigerian, and 9.7% of the Angolan. China has built military training facilities and barracks in Tanzania and Equatorial Guinea (The Economist 2022k: 11). China provided 90% of all Sudan's small arms acquisitions between 2004 and 2006, totaling more than $50 million (Gunpolicy 2008), while the Darfur war was devastating the country. Between 2015 and 2019 China accounted for 19% of arms transfers to sub-Saharan Africa mainly to Angola, Nigeria, Sudan, Senegal, and Zambia, some of the leading migrant-sending countries (Wezeman et al. 2019: 7). China's arms exports to Africa rose by 55% between 2008–2012 and 2013–2017, and its share of total African arms imports increased from 8.4% to 17%. A total of 22 sub-Saharan African countries procured major arms from China in 2013–2017, and China accounted for 27% of sub-Saharan African arms imports in that period (compared with 16% in 2008–2012). Russia and China have been the main suppliers of major arms to Nigeria between 2017 and 2021: 272 armored vehicles from China, 7 combat helicopters from Russia, 3 combat aircraft from Pakistan, 12 light combat aircraft from Brazil (via the USA), and 9 patrol craft from France.

China is involved in financing, building, or operating 61 ports in 30 African countries as of 2022 to promote its economic interests and global influence and win the competition against the West. The West and Western financial institutions are often facing Chinese policies in Africa, which are often opposed.

Are the infrastructure programs of the West such as Build Back Better World of the US or the Global Gateway infrastructure-for-Africa of the EU going to make the West more attractive than China in Africa (The Economist 2021d: 33)? Alternatively, shall the West imitate China and put aside the demands for democracy and accountability? In April 2022, The Economist survey by *Premise* shows that in the surveyed seven African countries, the Chinese influence was favorable (The Economist 2022k: 12). Moreover, China's global economic and political influence has challenged the West militarily, economically, and technologically, especially in artificial intelligence (Jones 2020: 4).

In February 2022, China appointed a Special Envoy for the Horn of Africa Affairs. On June 20, 2022, the First China-Horn of Africa Peace, Good Governance and Development Conference took place in Addis Ababa, Ethiopia. China's security policy in Africa emphasizes non-interference, underlines the economic drivers of conflict, and opposes sanctions on human rights grounds (Ramani 2021: 5). The Chinese "non-interference-into-internal-affairs" model is appealing to repressive African regimes (Alden 2007: 60). China argued that violence such as in Darfur was entirely an internal affair. China's ambassador to Zambia warned that Beijing might cut off diplomatic relations with that country if its voters elected a Taipei-leaning opposition candidate in the presidential elections of 2006 (Alden 2007: 21).

African citizens increasingly perceive Chinese influence as ruthless politics, whereas for African authoritarian regimes Chinese patronage and its ties to the political elite are sources of strength that help politicians stay in power. The Chinese investment may help win elections, but it weakens democracy and corrupts politicians (The Economist 2022k: 8). China promises to support African politicians if they support China by rejecting the UN resolutions on Chinese violation of the human rights of Uyghurs in Xinjiang. Accordingly, thirty-four African countries supported the Chinese counter-statement in defense of China against the June 2021 UN Human Rights Council criticism. Some researchers calculated that a 10% increase in voting alignment with China yielded a 276% increase in aid and credit (The Economist 2022k: 9). "If China can help them win elections in Africa by building roads, it can also cost them elections by not building them" (The Economist 2022k: 9). Some African influencers and even academics are being paid to promote China's line, to suppress negative stories on China, etc. Chinese media outlets started a Swahili-language service, and African journalists are being trained in a more government-friendly approach to news and "cyberspace management" or "public opinion management" (The Economist 2022k: 9–10). Chinese media in Africa seeks to portray China and Africa in a positive light, and African countries have been playing a central role (Madrid-Morales et al. 2021: 10).

China actively protects its African autocratic friends, as it did when it put pressure on the International Criminal Court to refrain from prosecuting former President Omal Al-Bashir of Sudan. China systematically instructs them on how to stay in power, how to produce effective propaganda, manage opposition, and monitor dissent (Rachman 2022: 182). African despots like Mugabe, lauded shamelessly China for its "remarkable achievements in the field of human rights". The increasing number of Chinese scholarships for Africans is expected to contribute to drawing Africans to the Chinese orbit utilizing soft power.

China feels at ease in Africa not only because of economic opportunities but also because it is a politically welcoming environment where controversies with civil society do not take place, and human rights and democracy are not on any agenda (Madrid-Morales et al. 2021: 10). From the soft power perspective, Xinhua, among the largest news agencies in the world, with 28 offices in Africa, more than any international news agency, has been trying to broadcast "good news" from Africa, "instead of disaster reports from the West" (Thussu 2016: 228). Going hand-in-hand with this, Huawei has built about 70% of Africa's 4G infrastructure. Still, there is a discrepancy between the African population and the political elites: as a survey shows regarding the media, just 9% of Tanzanians watched the Chinese state media CGTN, whereas 73% watched BBC, in Congo CGTN 28% and 73% BBC, in Ghana only 4% watched CGTN. The worrying development, however, is that increasingly African countries are making use of Chinese know-how to go after dissidents and journalists, to help crack WhatsApp communications and arrest opposition figures and demonstrators (Rachman 2022: 183). For these anti-democratic measures, African governments are using the media infrastructure built by China. China knows how to exploit the contradictions in Western human rights policies. China criticized the hypocrisy of the West for its better treatment of the Ukrainian refugees after the Russian aggression while ignoring refugees from the Middle East, Africa, and South Asia (The Economist 2022d: 49).

Saudi Arabia, and the United Arab Emirates on one side, and Iran, Qatar, and Turkey on the other are competing against each other, especially in the Horn of Africa. Between 2003 and 2019, President Erdogan of Turkey visited a total of 23 African countries 39 times (Atlantic Council 2018a). Turkey has opened in Somalia its biggest foreign military base and trained more than 5000 Somali soldiers and police commandos by 2022. Its interest especially in sub-Saharan Africa is new. It seems that Turkey decided to expand its relations with Africa after its ambitions to become a member of the EU started to dwindle. The number of Turkish Embassies in Africa between 2009 and 2021 has more than doubled (Shinn 2015: 6; The Economist 2022b: 31). By May 2023, Turkey had 44 Embassies across Africa. Its public broadcaster TRT has launched a digital news platform in French, English, Swahili, and Hausa. Through its soap operas, Turkey has expanded its cultural impact in several African countries—from Ethiopia to Senegal, and the Turkish Cooperation and Coordination Agency funds schools, religious institutions, and medical facilities; built hospitals in Somalia, Libya, and South Sudan; increased humanitarian aid to Nigeria, Mauritania, and Niger; its annual trade with Africa increased from $5.4 billion to $34.5 billion between 2003 and 2021; and agreed with 30 African states security-related cooperation (Daras 2023).

Though Turkey cannot compete with the Russian 30% and Chinese 20% of all arms sold to sub-Saharan Africa, the speed of the Turkish arms export especially drones to countries like Ethiopia, Libya, Morocco, Tunisia, etc. is considerable (The Economist 2022b: 32). Turkey's military pacts are increasing such as with countries like Nigeria, Senegal, Togo, etc. Its arms sales to Africa surged seven-fold to $328 m in 2021, and in January and February 2022, they were already almost $140 million.

By repeatedly designating the West as colonial oppressors of Africans, Turkey knows how to exploit the historical painful experiences of the Africans. The number of African Embassies in Turkey has tripled to 37 between 2003 and 2022. Even some of the operating costs of the embassies of poorer African countries are covered by Turkey, and the costs of Somalia probably entirely (Shinn 2015: 7). Between 2010 and 2021, more than 14,000 African students received scholarships to study in Turkey; Africans are buying properties in Turkey; African diaspora such as from Somalia is increasing steadily in Turkey (The Economist 2022b: 32). Turkish participation of UN Missions in Africa and its naval vessels to African ports have been increasing steadily (Shinn 2015: 8). Turkey has built its largest overseas military base in Somalia, and its first in Africa (The Economist 2019b).

Saudi Arabia and Iran have extended their sphere of competition as far as sub-Saharan Africa and trying to cement their influence through financial incentives, exportation of conservative forms of Islam Islamic extremism, religious sectarianism, diplomatic ties, and strategic alliances. Iran has been struggling to break out of the isolation imposed in response to its nuclear weapons programs. Around 2015, Eritrea and Somalia became important partners for Iran in its confrontation with Saudi Arabia about the Yemeni civil war in which Saudi Arabia and Iran supported the Yemeni government and Houthi rebels, respectively (Feierstein and Greathead 2017: 2). However, since 2016 using their economic power the Saudis have been able to force Sudan, Djibouti, Somalia, and Eritrea to break with Iran. As a result, in 2017, Saudi Arabia and Djibouti formally agreed to the construction of a military base, the U.A.E.'s new military installation in Somaliland, and the jointly operated military base in Eritrea. The competition between Iran and Saudi Arabia has been continuing also in Western Africa. Both powers have been using soft power strategies like investments, particularly in the oil and gas fields, and increased cooperation between the two states in various economic, cultural, scientific, and political fields in sub-Saharan Africa. Iran has been trying to deepen its relationship with Uranium-producing countries such as Malawi and Namibia (Feierstein and Greathead 2017: 4–8).

Alarmed by the Russia-Africa summit held in Ethiopia in October 2022, US Special Envoy for the Horn of Africa Michael Hammer visited Egypt and Ethiopia, while the US Ambassador to the UN, Linda Thomas-Greenfield, visited Uganda and Ghana, and France's President Emmanuel Macron visited Cameroon, Benin, and Guinea-Bissau. *Europe, which created the African Peace Facility in late 2003 is strongly involved in peace and security policies in Africa. Between 2003 and 2019 it provided almost €3.5 billion to strengthen* conflict prevention, management, and resolution structures and mechanisms of the African Peace and Security Architecture; to establish a continental Human Rights and international humanitarian law compliance framework; for conflict Early Response Mechanism; for African-led peace support operations, such as the Multinational Joint Task Force (MNJTF) against Boko Haram, the African Union Mission to Somalia (AMISOM) or the G5 Sahel Joint Force; and to promote gender and human rights principles and practices in peace support operations (European Commission 2019).

However, at the same time, the Western powers pursue policies that contradict these strategies. The US ignored the fraud and endorsed the rigged election of the Democratic Republic of Congo in 2019 (Wilson et al. 2019). France has been an unconditional supporter of Chadian President Idriss Déby whose regime ranked as one of the most corrupt in the world in 2007 (ranked 172 out of 179 countries) and in 2005. Since 2012, it has scored only 20 points out of 100 and it is placed at the rank of 162 among 198 countries in the Corruption Perception Index (Transparency International 2019). Thanks to French military support, the Chadian government survived the rebel attack in 2006 as well as in late 2007 and early 2008. In Burkina Faso, Cameroon, or former Zaire, France played an important role in keeping dictators in power. Now, those juntas who came to power by armed forces in Africa are turning to Russia and China.

When the US accuses China and Russia of taking advantage of African states to increase their own power and influence (Bolton 2019: 5), it is afraid that it would be outcompeted by Russia and China in Africa. In 2019, the US provided $118 million in military aid to Uganda's Museveni who rejects the values of human rights and the rule of law. France intervened militarily in Chad in 2008 and 2019 to keep the authoritarian President Deby in power (Stearns 2022: 154). Authoritarian regimes in

Chad, Djibouti, Equatorial Guinea, Ethiopia, Rwanda, and Uganda have been backed by the US and European powers at different times.

African insecurity is not caused by external actors. As the EU Commission says in its "Comprehensive Strategy with Africa", 2020, "African states … bear the main responsibility to act, as they are the foremost guarantors of their own security" (EU Commission 2020: 10). However, instead of non-interference and African ownership, external powers are competing against each other in a mimetic rivalry with negative outcomes for Africa. The mimetic rivalry and mimetic scapegoating driven by the underlying African disinterest and incompetence to construct a national interest and identity are creating insecurity in Africa ultimately pushing people to migrate. Jihadist and ethnic violence in Africa show that Africans kill one another as if they were not human beings because they have internalized from colonizers that African lives are less worthy. They accuse the West of racism and propagate pan-African solidarity but their practical policies are just the opposite (Ochonu 2020). The intensity of the detrimental external rivalry in Africa depends on African weakness. Globally, Africa is not an equal competitor but still a free actor in the international system. As will be discussed in the following section, it is neither neocolonialism nor the international system that is the main problem of Africa, but rather the African inability and disinterest to create good political institutions.

## Don't Blame Outsiders!

One of the African problems is the scapegoating of the West for their own bad policies. Neither the West, the Russians nor the Chinese are responsible for African insecurity. External actors have been exacerbating African problems but they seldom have caused them. The weaker the African political and economic institutions are the more vulnerable the Africans become to external intrusions, exploitation, and coercion. As a result, Africans become even weaker.

Undeniably, the African problems are rooted in slavery, colonialism, patriarchy, patrimonialism, and autocracy. The first two will be discussed in the next chapter. The African patriarchal cultural structures have not only discriminated against and sidelined African women, but they have also given men absolute power, which disadvantages the African women in particular and the society in general. Freedom and equality of women are essential at least for four reasons: firstly, for their own wellbeing and the wellbeing of their children; secondly, for economic growth and innovative potential of women; thirdly, for the reduction of population growth that is going to destabilize Africa socio-economically and environmentally; and finally, for eradication of patriarchal structures.

Many Africans idealize the African spirit of community, mutual aid, and good neighborliness. However, as the daily situations betray, we Africans are not as communitarian as we claim to be. Ujamaa, Ubuntu, and African socialism as African cultural values are just a construction of African political elites to stick to power by idealizing African culture and identity. The longest-serving presidents of the world are in Africa. They believe that they are the only ones who can guarantee the stability and security of their countries. Patriarchy, hierarchy, and admiration for the rich and powerful are part of the African culture.

Africans argued in the independence struggle that the Europeans denied Africans human rights. Africans supported the freedom struggle of South Africans against the Apartheid regime. Moreover, Africans signed the African Charter of Human Rights, which is almost identical to the Universal Declaration of Human Rights or the European Convention on Human Rights. But how is it that Africans argue now that human rights are a European concept?

Uganda's President Yoweri Museveni argues that gay rights are Western concepts and weird cultures that they try to impose on other people. People who identify as gay in Uganda risk life in prison, including the death penalty in certain cases. The law would force friends, family, and members of the community to report individuals in same-sex relationships to the authorities. There are reports that this anti-gay law has led to increased blackmail and people are receiving calls that "if you don't give me money, I will report that you are gay" (Atuhaire 2023).

It is difficult to find African leaders who criticize Putin, Xi Jinping, Erdogan, etc. Even Lukashenko has friends in Africa, as the agriculture and defense agreements between leaders of Zimbabwe and Belarus toward the end of January 2023 show. South Africa, which has the potential to be a leader of the African continent toward peace, security, human rights, democracy, and prosperity was one of 17 African nations which abstained against condemning Russia's invasion of Ukraine and blatant violation of the international law and state sovereignty.

It is important to underline that Africans are not just unfortunate victims of competing global actors. Military coups like in Burkina Faso, Mali, and Guinea; brutal armed groups like in the Democratic Republic of

Congo, Al-Shabab in Somalia, armed factions in South Sudan, and Islamic State in the Sahel region, ethnic violence in many countries, etc. show devastating violence being committed by Africans themselves against their own people. Moreover, as Jason Stearns argues, politicians, military commanders, rebels, and entrepreneurs prospered on war and expanded it, and they did not want the end of violence, which would have meant the end of their livelihoods (Stearns 2022: 148). Many governments in African countries are unwilling to contain conflicts and address the causes of violence. Transnational mutual destabilization, internationalization of civil wars, interstate rivalry using military means, etc. are widespread common African problems (Twagiramungu 2019). Armed factions in South Sudan, ethnicized violence in Ethiopia, Islamist terrorists in Somalia or the Sahel region, armed beneficiaries of wars in the Democratic Republic of Congo or the Central African Republic, unconstitutional governments in Chad, Guinea, Burkina Faso, Mali or Sudan, authoritarian governments like in Uganda or Rwanda have caused widespread suffering to their citizens, cross-border conflicts, displacement, and migration. Between 2003 and 2009, the Sudanese civil war in the Darfur region caused the death of 300,000 Darfurians and the displacement of 2 million (Sikainga 2009). The armed conflict between Sudanese Armed Forces and Rapid Support Forces killed thousands of Sudanese since 15 April 2023 (ACLED 2023). External global actors such as Russia, and Gulf nations have also been contributing to the prolongation of the Sudanese conflict. According to some reports, the Wagner Group and the UAE are allegedly involved in RSF-controlled Sudanese sizeable gold deposits (Africa Research Bulletin 2023: 23977). According to some estimates, in the Democratic Republic of Congo between 1998 and 2008 as many as 5.4 million people were killed (Pilling and Schipani 2023). Similarly, about 600,000 people were killed during the brutal civil war in northern Ethiopia between 2020 and 2022.

Civilians have suffered for years of violence by government forces and jihadists such as the Islamic State in the Greater Sahara (ISGS), the Al Qaeda-affiliated Jama'at Nusrat Al Islam Wal Muslimin (JNIM), Boko Haram, armed Arabs, and Tuareg rebels, civilians armed for self-defense, and militias among ethnic Arab, Djerma, and Tuareg communities in Niger, Burkina Faso, and Mali. Since 2012, these jihadists, government forces, and various armed groups have become a security threat to the countries of the Sahel. In several instances, jihadists filled a vacuum and replaced the State, which was often either absent or whose forces were perceived as predatory (Cherbib 2018). This jihadism is spreading as far as Ghana, which is usually one of the stable countries in Africa. In Burkina Faso, armed group activity, military operations, political instability, two coup d'états in 2022, and human rights abuses led to the displacement of about two million people by mid-2023. UNICEF estimated that 970,000 children under five faced severe wasting in 2023 in Burkina Faso, Mali, and Niger.

The war between the Central African Republic armed forces and their Russian allies from the Wagner group on the one hand and the Coalition of Patriots for Change (CPC) rebels on the other continues to kill and persecute the civilians. Since late 2020, Wagner Group has been directly supporting the Central African Republic armed forces and the Wagner Group has become one of the dominant agents of political violence in the country in counter-offensives to reclaim rebel-held territory (ACLED 2022). According to a report by ACLED in August 20222, "Wagner Group mercenaries operated independently of state forces in at least 50% of political violence events each month since May 2021" (ACLED 2022).

The main cause of these brutal and incomprehensible wars and destructions has been taking place not because of the Western powers, Russia, China, or any other external power instead because of Africans themselves. It would not be difficult to prevent all this violence and killings, and destruction of the economy, infrastructure, and mutual trust of the population, if Africans care about African lives. As postcolonial African history shows, most Africans are not capable of valuing the lives of their co-citizens. The devaluation of African lives that started during slavery and colonization has been continued by Africans themselves even after the end of slavery and colonization. Even colonial barbarity is more understandable than the African one. Be it religious or political violence, for Africans African lives do not count. Africans kill each other instead of making compromises. Anyone who opposes them has to be eliminated. If Africans themselves do not value the lives of their co-citizens, why should the outside world care about it?

Contrary to colonial times, now African political elites have the freedom and the power to make decisions that address the needs of Africans. The neocolonial discourse, which represents Africans as victims of the global system dominated by the West (Europe and North America), or the East (China and Russia) is counterproductive. Africans are free to choose their partners; they are free to pursue their national interests. They did not have this freedom during colonial times. The neocolonial argument is dangerous because it downplays the violence, dehumanization, and exploitation of Africans during slavery and colonialism.

African political elites imitate blindly the West or East instead of looking for solutions that relieve the sufferings of the Africans who have been suffering for centuries from the consequences of slavery, colonialism, and exploitation. Secondly, African political elites instrumentalize black Africans' historical sufferings for their own personal benefits such as to stay in power for example by scapegoating the West. Thirdly, the Chinese and Russians are exploiting African history for their own interests by mobilizing Africans against the West, and Africans are letting themselves be instrumentalized by them. Africans' approach has to be independent and pragmatic, i.e. to focus on the question of which international cooperation is promoting African interests, and which is not. African elites' interests are not equal to African socioeconomic and political security, economic development, and environmental security. Moreover, in the name of national security, many African governments have been undermining democratic values and curtailing human rights.

The neocolonial argument plays down paradoxically the historical dehumanizing brutality of European slavery, forced labor, and colonialization of Africans by equalizing the past and the present. In today's international political and security system there is no forced labor, no enslavement, no systematic and institutionalized expropriation of Africans of their natural resources, and there is no forced displacement of Africans by the Europeans. Africans are free to join any international agreement with any global actor. Moreover, the rule of law, democracy, and human rights are not European or Western values; they are human values. They are values, which can never be rejected by any reasonable human being be it in Asia, Africa, or anywhere. To cling to power and delegitimize these values the political elites in Africa opportunistically claim that these are Western values.

If African political elites are interested in the economic development of Africans, they must work for Africans and with Africans, not against Africans. Africans should care for their hungry, poor, and unemployed people and ultimately contribute to the reduction of displacement and emigration. A new form of slavery and colonization is continuing now in Africa by Africans themselves who reject the civil and political rights as well as the social and economic rights of their citizens.

Many African countries have not managed to stop the blind imitation of the East or the West, to find their own way, to think about their poor, unemployed, and downtrodden citizens, uneducated children, marginalized girls and women, destroyed environment, displaced population, desolate infrastructure, etc. African political elites romanticize African culture and denounce the Western ways of life, culture, individualism, and decadence. But at the same time, they migrate to the West, imitate Western lifestyles, hide the wealth of their nations in Western bank accounts, and live in luxury whereas their citizens go hungry.

External arms transfer, military presence, and intervention are not the causes of Africa's problems but the consequences of African weaknesses. They help suppress the demands for human rights, freedom, and social and economic development. Western actors in Africa have been supporting their African undemocratic partners like Chad, Zaire (DRC), Cameroon, Ethiopia, etc. Russia, China, Turkey, Iran, Saudi Arabia, and UAE do not care about democracy and human rights and support any country as long as it is on their side.

The African political elites' behavior is cynical in the sense that they claim to promote the national interests of their citizens but they repress human rights and democratic values and threaten the physical security of their population. Their lifestyles, accumulated wealth, etc. are more similar to their external counterparts than to the African reality of impoverished, uneducated, displaced, unemployed, hungry masses. For example, in the Democratic Republic of Congo, there has been a collusion between rebels and administrators of state power in which official bureaucrats actively sustained state institutions and recruited rebel support during the war (Sweet 2020).

In 2021, sub-Saharan Africa scored the highest conflict and violence-related internal displacement (80.4% of the global total) (Internal Displacement Monitoring Center 2022: 26). Military coup is again increasing in Africa. Unconstitutional governments are becoming more common than a decade ago. In the 1990s and 2000s, the main focus of Islamist terrorists was the Horn of Africa or the Sahara region. In the meantime, the Sahel Region, Central Africa, Western Africa, and South-Eastern Africa are all affected. Nigeria, Niger, Somalia, Kenya, and Mozambique are affected by active Jihadist insurgency, and Chad, Cameroon, Uganda, and the Democratic Republic of Congo by low-level Jihadist insurgency. Jihadist cells are expanding in Sudan, Tanzania, Rwanda, Ethiopia, and South Africa (The Economist 2022d: 32). The Sahel has become the world's Islamic terrorism hotspot, accounting for more than a third of all terrorism deaths in 2021. In Mali nearly 2700 people were killed in conflict in the first six months of 2022, almost 40%

more than in all of 2021. In Burkina Faso, in the first half of 2022 about 2100 people were killed by the Jihadists. Al-Qaeda and Al-Shabab have been spreading fast and devastating the already unstable countries, especially Somalia. (The Economist 2022p: 27–28). Even here, it is primarily the responsibility of the Africans to put the security, human rights, and prosperity of their people first, not religion. These Islamic terrorists are Africans themselves, not any external power.

The Constitutive Act of the African Union Article 3(h) states that the objective of the Union is to promote and protect human and peoples' rights in accordance with the African Charter on Human and Peoples' Rights and other relevant human rights instruments. Moreover, the Preamble of the African Charter of Human Rights states that "Convinced that it is henceforth essential to pay particular attention to the right to development and that civil and political rights cannot be dissociated from economic, social and cultural rights in their conception as well as universality and that the satisfaction of economic, social and cultural rights is a guarantee for the enjoyment of civil and political rights". Africans argued in the name of human rights when they fought against the barbaric Western slavery and colonialism in Africa. Now, the very Africans argue that human rights are Western values that cannot be imposed on Africans. The argument is self-contradictory.

Violation of press freedom in Africa has also been increasing in the past years. According to Reporters Without Borders 2022, many African countries belong to the worst-performing group: Eritrea (179th), Djibouti (164th), Angola (99th), Zimbabwe (137th), Ethiopia (114th), Mali (111th), Benin (121st) (RSF 2022). Many African countries such as South Sudan, Tanzania, Central African Republic, Benin, Burkina Faso, etc. belonged in 2023 to the worst performing countries in terms of Freedom and Democracy (Freedom House 2023: 12). In 2020, sub-Saharan had only one "full democracy"—Mauritius -, and other "six best performing" African countries belong to "flawed democracies" on the global level (EIU 2021: 47). According to Freedom House 2023 report, only 7% of 1.4 billion Africans and 17% out of 54 countries are free. Out of 16 countries with the worst aggregate scores for political rights and civil liberties around the world, five are from the African continent (Freedom House 2023).

Africa is unfortunately famous for letting presidential term limits slip. Yoweri Museveni of Uganda, Gnassingbé in Togo, Obiang in Equatorial Guinea, Bongo in Gabon, Paul Biya in Cameroon, Afewerki in Eritrea,

Mugabe in Zimbabwe, Mobutu in Zaire / DRC, etc. are the best example, though the list is much longer. Jailing academics and opposition groups, arresting, shooting, and torturing opposition activists and journalists, etc. Around the mid-1990s Africa's wave of democratization, there were almost 50 new constitutions passed and more than 30 presidential term limits. Theoretically, there are still more than three-quarters of sub-Saharan African countries have term limits. But the number of leaders backtracking has been increasing. Since 2015, the leaders of no fewer than 13 African countries have sidestepped or weakened term limits (The Economist 2021e).

Afrobarometer survey results from 2019 covering 45,823 respondents in 34 African countries between 2016 and 2018 found that 75% of the interviewed in 2016/2018 said they preferred to use regular, open, and honest elections to choose their country's leaders. However, African presidents are all too ready to abandon presidential term limits although almost all countries have introduced a regular cycle of elections (usually every five years), and many have placed constitutional two-term limits that African presidents can serve (Bratton and Bhoojedhur 2019). However, as Reyntjens argues, effective term limits are less prevalent in practice than in constitutional texts (Reyntjens 2020).

A quite surprising counterargument by many African political elites is that Western demand for democracy is racist. Are those African populations demanding democracy, human rights, and the rule of law also racists? Maybe the African political elites have forgotten the brutal history of European slavery, colonialism, and racism. One may reject Western influence but it is irresponsible to exploit for political purposes the term racism, which was brutal, forced, dehumanizing, and degrading.

As of 2021, six out of 10 of the current longest-ruling non-royal national leaders in the world were in Africa: Teodoro Obiang Nguema Mbasogo in Equatorial Guinea (42 years), Paul Biya in Cameroon (46 years), Denis Sassou Nguesso in the Republic of Congo (37 years), Yoweri Museveni in Uganda (35 years), Isaias Afewerki in Eritrea (30 years), Idris Deby in Chad (31 years) (Klobucista 2021). Eritrea since 1991 has been governed by a government, which rounds up "thousands of people, including minors, for mandatory military service and punishing the families of suspected draft evaders" (Africa Research Bulletin 2023b: 23920). The kleptocratic José Eduardo dos Santos ruled Angola from 1979 to 2017 and died in a hospital in Barcelona on 8 July 2022. In 2021 and 2022, there were military seizures of power in Chad, Guinea, twice in

Mali, Sudan, and Burkina Faso, two unsuccessful coup attempts in Guinea-Bissau and Niger, defiance of constitutional term limits to win third terms in office in Togo by Faure Gnassingbé, in Côte d'Ivoire by Alassane Ouattara, and in Guinea by Condé, before his ouster in 2021 (Gyimah-Boadi 2022; Fragile States Index 2022). According to the Afrobarometer surveys conducted in 18 African countries (26,777 interviews) and published in early 2021, a majority of citizens say corruption has increased in their country, with police being the worst offenders and even many healthcare providers demanding bribes against which the government is doing too little. Activists working to uncover this issue feel a high risk of retaliation (Keulder 2021).

African poor institutions are also caused by a problematic mindset. It is not only the colonizers who suggested that the "superior" whites take the Africans by the hand and help them to rise out of fraternal love, help them progress toward an improved physical, social, and moral well-being, and elevate them from their inferiority (Soyinka 2012: 50). Even we Africans themselves believe in the inferiority and immaturity of our people. We believe that Africans are politically underendowed and that they cannot be permitted to participate in the process of governance or contribute their voice to the choice of their leaders (Soyinka 2012: 50). We still find it hard to overcome our ethnicized mindset according to which a master ethnic group rules the rest of society, the divinely appointed shepherds among his mindless folk (Soyinka 2012: 52). As Soyinka say, the postcolonial African governance is about actualizing (assuming) power and fictionalizing (mythologizing or romanticizing) African culture and people as if democracy and human rights were alien to them (Soyinka 2012: 52).

The colonial Western white brutality against Africans has left extremely unfortunate legacies in Africans. For the postcolonial African political elites, only the atrocities committed by the Western whites were real atrocities. Their own injustices and their neglect of African humanity are always legitimate. When the West criticizes this, their reaction is "mind your own business. This is a sovereign state, one that must be left alone to exercise its prerogative of ownership over its human possessions." (Soyinka 2012: 86).

Whereas the slavery and colonial chains were put around the Africans, the postcolonial chains have been put by the African constrictive, tenacious, and implacable religious and political absolutists like Idi Amin, Museveni, Afewerki, Mobutu, Bokassa, Siad Barre, Mengistu Hailemariam, Meles Zenawi, Paul Kagame, Idris Deby, Paul Biya, Kampaore, Obiang

Nguema Mbasogo, etc. who entrenched themselves in power and mythologized their "strong leadership" by being brutal to their own people but often subservient to outside interests (Soyinka 2012: 94).

African autocrats will manipulate or ignore the constitution to stay in power indefinitely while eliminating their political opposition despite the Lomé Declaration of July 2000 and Charter of the African Union, Art. 4(p), which condemn and reject the unconstitutional change of government. The number of military coups has increased and the African Union seems to be powerless (Siegle and Smith 2022). Foreign dictators may be friends of some African political elites, but not necessarily of all Africans. President Museveni's son Lieutenant-General Muhoozi Kainerugaba, whose security forces allegedly abducted hundreds of opposition activists and tortured them, boasts "Uganda belongs to Team MK!" (short form for his name—Muhoozi Kainerugaba). Political dynasties are not new in Africa. The sons of dead dictators rule Togo, Chad, and Gabon. In Cameroon, the son of Paul Biya is coveting the top job (The Economist 2022m).

Neocolonialism is not the best theoretical concept to explain African problems or the causes that force Africans into displacement and emigration. In the West, China and Russia are seen as neocolonial powers in Africa; China and Russia accuse the West of neo-colonialism. Some Africans perceive either side as neocolonial power depending on sympathy or antipathy. But Africans themselves have been contributing to the causes that push their citizens into displacement or emigration. One example is the case of DR Congo. More than 120 armed groups, with vague political aims, mostly fighting for precious minerals and armed with AK-47 s, have made the lives of the Congolese people miserable. One of these brutal armed groups is M23, led by ethnic Tutsis and backed by Rwanda. After Congolese and UN troops defeated M23 in 2013, the group splintered and fled into Rwanda and Uganda. However, neither Rwanda nor Uganda was willing to extradite the group's commanders, enabling them to escape justice for war crimes, and in 2017, M23 members slipped back into Congo actors. Child soldiers slaughter their own people with bullets and machetes, others die of disease or hunger. For example, according to the UN Joint Human Rights Office in the DRC, between April 2019 and April 2022 at least 3126 cases of torture, sexual violence, cruel, inhuman, arbitrary killings, or degrading treatment or punishment perpetrated by armed groups and the defense and security forces, and at least 5.97 million Congolese displaced, and in 2022 armed groups, which have been exploiting the absence or weakness of state authority killed at least 2400 people in the eastern provinces of the DRC (GlobalR2P 2023). Besides the Congolese army, Rwanda and Uganda, rebel groups supported by them and other armed groups fought by these states but operating in the Congolese territory have inflicted for decades unimaginable suffering on the civilians in the East Congo (Africa Research Bulletin 2023d: 23845). Rwanda and Uganda, have been fighting in Congo and looting the minerals to enrich themselves and to finance their wars. Hunger, diseases, and war have killed up to 5 million Congolese between 1998 and 2003. Neither neocolonialism nor the competition of the global powers alone explains this tragedy. It is primarily, the Africans themselves, the Congolese, and the neighboring countries, responsible for this tragedy. Greed and incompetence of Africans themselves denied Africans a decent life. There is no neocolonialist nowadays in Congo who forces the Congolese to work for him or to leave their land, nor anyone from outside who expropriates the Congolese of their natural resources as King Leopold II of Belgium did. The Congolese government is always free to cooperate or not with external actors (The Economist 2022l).

Another example: Somalia collapsed in 1991. It is estimated that between 1991 and 2011 between 450,000 and 1.5 million died, 800,000 were refugees, and more than 1.5 million were internally displaced people (Norris and Bruton 2011: 1). Floods, droughts, and famine killed more than 250,000 people in 2011. In 2022, again more than 18 million people were affected in Somalia, Kenya, and Ethiopia, but Somalia is affected the most because of its political fragility. Somalia, which imports nearly 80% of its food, has become highly vulnerable to Russia's invasion of Ukraine, which caused higher fuel prices and fanned food inflation. As a result, more than 40% of Somalia's 7 million population is struggling to find food, and around 1.4 million children are severely malnourished. Both drought and anarchy since 1991 are the causes of the Somali famine (The Economist 2022o). Due to internal greed and incompetence, the terrorist al-Qaida-linked al-Shabab brutality, and external cynicism and hypocrisy the sufferings of innocent Somalis are unimaginable. It is primarily, the Somali clan elders, corrupt politicians, and the pseudo-Muslim Al-Shabab, etc. persecuting their own population. There is no external power forcing its will, interest, or ideology on the Somali population or taking away their resources by force. Nor the climate change is the main problem.

## Conclusion

Conflict leads to displacement and migration. In 2021, out of 14.4 million internal displacements, 11. 6 million were in Africa (Internal Displacement Monitoring Center 2022). The causes of this displacement are primarily internal. Africans are not mere victims of the international system. Africans are free actors. They are free to choose their partners; they can be eclectic; they can be allies at the same time with two antagonistic parties. However, their guiding principle should be always, respect for international law and the interests of their citizens. These interests can never be materialized without civil and political rights and economic, social, and cultural rights.

Africans themselves contribute to the global mindset that African lives do not count as much as the lives of non-blacks. Africans themselves contribute to the devaluation of African lives. They slaughter each other; they exploit each other. Why should others respect African lives while Africans themselves do not do it?

The priority of African countries must be to clearly define and pursue the security interests of their citizens in the environment of international insecurity caused by competition among global superpowers and emerging powers. Any African argument of neo-colonialism against the West is an abuse of the historical sufferings of Africans under slavery and colonialism, which denied all freedoms to Africans. Russia, China, and Turkey underline the European colonial legacy in Africa only to mobilize Africans against the West and to keep them in their orbits. In the face of the massive presence of China and Russia in Africa, U.S. attention to African partners is primarily about boxing out Russia or China. Such global competition will keep African autocrats in power. If colonization is measured by the damage it has done to the African population, African political elites are not better than the colonizers. Scapegoating the West and abusing the experiences of slavery and colonization for their own interest is unfair and immoral.

African autocrats are unfortunately prone to imitate the likes of Putin, Xi, or Erdogan and together with them scapegoat the West for all Africa's ills. The external actors are not causing African conflicts or the migration of young Africans to the West. They are contributing to the causes, which are a combination of poverty, climate change, and bad government. The question should not only be on whether poverty, conflicts, and climate change are the causes of African migration but also on who are the actors behind them.

African success does not depend on a romanticized African culture, but instead on working for Africans, with Africans, and not against Africans. For our success, we Africans need our imprisoned, exiled, intimidated intellectuals and African girls in schools unconditionally with a quality education. African women are not the sex objects of their men. There are enough top-talented Eritreans, Rwandans, Congolese, Cameroonians, etc. to lead their countries. There is no need for a patriarchal autocrat to stay in power forever to the detriment of the nation.

Africans love to hate Europeans because they are fascinated by them. The African mind is a captive of a simultaneous abomination and fascination for the West. That is why Africans through the help of Russians and Chinese scapegoat the West for their own failures. Postcolonial African migration to the West primarily as a mindset and only secondarily and epiphenomenally a spatial movement to the West is the outcome of this mental slavery and colonization. The origin of this mindset and inferiority complex that pushes the Africans to love and hate Europe, to scapegoat and admire it at the same time is the legacy of slavery and colonization, as I will discuss in the following chapter. As we shall see in the following chapters, without denying the role of poverty, conflicts and climate change for migration, we focus in the following chapters on a psychoanalytical approach to the postcolonial African migration to the West.

## Literature

Acemoglu, Daron and Robinson, James (2012): Why Nations Fail: The origins of power, prosperity, and poverty, Currency: New York.

ACLED (2022): Wagner Group Operations in Africa: Civilian Targeting Trends in the Central African Republic and Mali, https://acleddata.com/2022/08/30/wagner-group-operations-in-africa-civilian-targeting-trends-in-the-central-african-republic-and-mali/#s4, 31.05.2023.

ACLED (2023): Fact Sheet: Conflict Surges in Sudan, https://acleddata.com/2023/04/28/fact-sheet-conflict-surges-in-sudan/, 30.05.2023.

Africa Research Bulletin (2009): LIBYA–US: Improved Ties Bring Better Cooperation 46 (3): p. 17910.

Africa Research Bulletin (2023): Sudan: Civil War Looms, *Political, Social and Cultural Series*, April 1st–30th 2023, p. 23977.

Africa Research Bulletin (2023b): Dakar 2 Summit: Food Sovereignty For Africa, *Economic, Financial and Technical Series*, January 16th – February 15th 2023, p. 24149.

Africa Research Bulletin (2023d): DR Congo: December Security Incidents, *Political, Social and Cultural Series*, December 1st – 31st 2022 p. 23845.

Alba, Davey, and Sheera Frenkel. 2019. "Russia Tests New Disinformation Tactics in Africa to Expand Influence." The New York Times, 30 October.

Alden, C. (2007), *China in Africa* (London: Zed Books).

Anette Hoffmann & Guido Lanfranchi (2023): Europe's re-engagement with Ethiopia Prioritising the prerequisites for lasting peace over geopolitical concerns, Clingendael – the Netherlands Institute of International Relations.

Atanesian, Grigor (2023): Russia in Africa: How disinformation operations target the continent, https://www.bbc.com/news/world-africa-64451376.

Atlantic Council (2018a): *Turkey's Growing Presence in Africa, and Opportunities and Challenges To Watch in 2018*, http://www.atlanticcouncil.org/events/past-events/turkey-s-growing-presence-in-africa-and-opportunities-and-challenges-to-watch-in-2018, accessed 05.09.2018.

Atuhaire, Patience (2023): Uganda Anti-Homosexuality bill: Life in prison for saying you're gay, https://www.bbc.com/news/world-africa-65034343.

Balytnikov, Vadim, et al. 209: Russia's Return to Africa: Strategy and Prospects, The Foundation for Development and Support of the Valdai Discussion Club, 2019, Moscow.

Béraud-Sudreau, L. et al. (2020): Emerging Suppliers in the Global Arms Trade, SIPRI Insights on Peace and Security, No. 2020/13 December 2020.

Bolton, John (2019): A New Africa Strategy: Expanding Economic and Security Ties on the Basis of Mutual Respect, The Heritage Foundation: Washington.

Bratton, Michael and Bhoojedhur, Sadhiska (2019): Africans want open elections especially if they bring change, Afrobarometer Policy Paper No. 58 | June 2019.

Cherbib, Hamza (2018): Jihadism in the Sahel: Exploiting Local Disorders, *European Institute of the Mediterranean*, https://www.iemed.org/publication/jihadism-in-the-sahel-exploiting-local-disorders/, 31.05.2023.

Daras, Marina (2023): Why Turkey's election is being closely followed in Africa, https://www.bbc.com/news/world-africa-65671723, 26.05.2023.

EIU (2021): Democracy Index 2020: In sickness and in health? A report by The Economist Intelligence Unit.

El-Badawy, Emman et al. (2022): Security, Soft Power and Regime Support: Spheres of Russian Influence in Africa, Tony Blair Institute for Global Change.

European Commission (2019): African Peace Facility: African Union Peace & Security Operations boosted by an additional €800 million from the European Union, https://ec.europa.eu/commission/presscorner/detail/it/ip_19_3432.

European Commission (2020): The European Union and Africa: Partners in Trade, https://trade.ec.europa.eu/doclib/docs/2022/february/tradoc_160053.pdf.

FOCAC (2018): Beijing Declaration – Towards an Even Stronger China – Africa Community with a Shared Future, retrieved 02.03.2021, from http://www.focac.org/eng/zywx_1/zywj/t1594324.htm

Feierstein, Gerald and Greathead, Craig (2017): The Fight for Africa: The New Focus of the Saudi-Iranian Rivalry, Middle East Institute, Policy Focus 2017-2, https://www.brookings.edu/blog/africa-in-focus/2022/03/09/figure-of-theweek-african-countries-votes-on-the-un-resolution-condemning-russias-invasion-of-ukraine/.

Fragile States Index (2022): Breaking the Cycle: Military Coups in West Africa, https://fragilestatesindex.org/2022/07/08/breaking-the-cycle-military-coups-in-west-africa/.

Freedom House (2023): Freedom in the World 2023, *Marking 50 Years in the Struggle for Democracy*, Freedom House.

Giménez-Gómez, José-Manuel et al. (2017): Trends in African Migration to Europe: Drivers Beyond Economic Motivations, Discussion Papers, Center for European, Governance and Economic Development Research, Georg-August-Universität Göttingen, ISSN: 1439–2305.

GlobalR2P (2023): Democratic Republic of the Congo, *Global Centre for the Responsibility to Protect*, https://www.globalr2p.org/countries/democratic-republic-of-the-congo/, 01.06.2023.

Gunpolicy 2008: Guns in North Africa, retrieved 12 March 2009, from http://www.gunpolicy.org/Topics/Guns_In_North_Africa.html.

Gyimah-Boadi, E. (2022): West Africa's Authoritarian Turn. Democratic Backsliding, Youth Resistance, and the Case for American Help, https://www.foreignaffairs.com/articles/west-africa/2022-07-11/west-africas-authoritarian-turn.

Gyimah-Boadi, E. and Asunka, Joseph (2021): Do Africans want democracy – and do they think they're getting it?, https://summit4democracy.org/do-africans-want-democracy-and-do-they-think-theyre-getting-it/.

Hainzl, Gerald (2021): The People's Republic of China's Presence in Africa, In Frank/Vogl (eds.): China's Footprint in Strategic Spaces of the European Union. New Challenges for a Multi-dimensional EU-China Strategy. Schriftenreihe der Landesverteidigungsakademie No. 11/2021.

Holtom, Paul 2008: International arms transfers. In: SIPRI Yearbook 2008: Armaments, Disarmament and International Security, 313–317.

Internal Displacement Monitoring Center (2022): *Children and youth in internal displacement*, Internal Displacement Monitoring Center; Norwegian Refugee Council.

Jones, Bruce (2020): China and the Return of Great Power Strategic Competition, Global China: Assessing China's growing role in the world, The Brookings Institution, February 2020.

Keulder, Christiaan (2021): Africans see growing corruption, poor government response, but fear retaliation if they speak out, Afrobarometer Dispatch No. 421.

Klobucista, Calire (2021): Africa's 'Leaders for Life', https://www.cfr.org/backgrounder/africas-leaders-life.

Kortunov, Andrey et al. (2020): Africa-Russia+: Achievements, Problems, Prospects: Report № 53/2020, Russian International Affairs Council (RIAC), Africa Business Initiative Union.

Madrid-Morales, Dani et al. (2021): It is about their story: How China, Turkey and Russia influence the media in Africa, Konrad-Adenauer-Stiftung Regional Media Program, Sub-Sahara Africa.

Mills, Greg (2010): Why Africa is Poor: And What Africans Can Do About It?, Penguin.

Moyo, Dambisa (2010): Dead Aid: Why aid is not working and how there is another way for Africa, Penguin: London.

Norris, John and Bruton, Bronwyn (2011): Twenty Years of Collapse and Counting the Cost of Failure in Somalia, A Joint Report from the Center for American Progress and One Earth Future Foundation.

Ochonu, Moses E. (2020): South African Afrophobia in local and continental contexts, *Journal of Modern African Studies*, 58: 4, pp. 499–519.

Pilling, David and Schipani, Andres (2023) War in Tigray may have killed 600,000 people,peacemediatorsays,https://www.ft.com/content/2f385e95-0899-403a-9e3b-ed8c24adf4e7, 30.05.2023.

Rachman, Gideon (2022): The Age of the Strong-Man, The Bodley Head, London.

Ramani, Samuel (2021): Russia and China in Africa: Prospective Partners or Asymmetric Rivals? South African Institute of International Affairs, Policy Insights 120, December 2021.

Ramani, Samuel (2022): Russia Has Big Plans for Africa. America Must Push Back—Without Getting Dragged In, https://www.foreignaffairs.com/print/node/1128391.

Reuters (2018): Factbox: Russian military cooperation deals with African countries, https://www.reuters.com/article/us-africa-russia-factbox-idUSKCN1MR0KH, 20.06.2022.

Reyntjens, Filip (2020): Respecting and Circumventing Presidential Term Limits in Sub-Saharan Africa: A Comparative Survey, *African Affairs*,119: 475, pp. 275–295.

RSF (2022): Reporters Without Borders Index, https://rsf.org/en/index.

Shinn, David (2015): Turkey's Engagement in Sub-Saharan Africa: Shifting Alliances and Strategic Diversification, Chatham House, the Royal Institute of International Affairs, Research Paper.

Siegle, Joseph (2021): Russia and Africa: Expanding, Influence and Instability, in Graeme P. Herd, ed., Russia's Global Reach: A Security and Statecraft

Assessment (Garmisch-Partenkirchen: George C. Marshall European Center for Security Studies, 2021).

Siegle, Joseph (2022b): The future of Russia-Africa relations, in *Foresight Africa*: top priorities for the continent in 2022, pp. 117–118.

Siegle, Joseph and Smith, Jeffrey (2022): Putin's World Order Would Be Devastating for Africa. Moscow is already deeply involved in destabilizing wars, https://foreignpolicy.com/2022/05/30/russia-war-africa/.

SIPRI, Stockholm International Peace Research Institute (2017): UN arms embargo on Somalia, SIPRI databases. https://www.sipri.org/databases/embargoes/un_arms_embargoes/somalia, accessed 08.11.2018.

Sikainga, Ahmad (2009): 'The World's Worst Humanitarian Crisis': Understanding the Darfur Conflict, https://origins.osu.edu/article/worlds-worst-humanitarian-crisis-understanding-darfur-conflict?language_content_entity=en, 30.05.2023.

Soyinka, Wole (2012): Of Africa, Yale University Press: New Haven.

Stearns, Jason K. (2022): Rebels without a Cause: The New Face of African War, *Foreign Affairs,* May/June 2022, pp. 143–156.

Stronski, Paul (2019): Late to the Party: Russia's Return to Africa, Carnegie Endowment for International Peace, October 2019.

Sweet, Rachel (2020): Bureaucrats at War: The Resilient State in the Congo, *African Affairs,*119: 475, pp. 224–250.

The Economist (2019b): A sub-Saharan seduction. Africa is attracting ever more interest from powers elsewhere: They are following where China led, 7 March 2019.

The Economist (2021d): China and Africa: Pomp and Circumspection, 4 December 2021, p. 33.

The Economist (2021e): Democracy in Africa: Time out, 9 January, 2021, p. 11.

The Economist (2022b): Turkey and Africa: The call of the South, April 23rd, p. 31.

The Economist (2022d): Africa's debt crisis: Debt and denial, April 30th April, p. 29.

The Economist (2022e): Geopolitics: Friends like these, April 16th April, p. 50.

The Economist (2022h): The ripples of Putin's war: Bread and Oil, 12 March, pp. 26–28.

The Economist (2022j): China and Africa: Chasing the dragon, 19 February, pp. 29–31.

The Economist (2022k): Jihadists in the Sahel: French Leave, 19 April, p. 31.

The Economist (2022l): African Economies: When you are in a hole..., 8 January 2022, pp. 24–25.

The Economist (2022m): Crossing the Mediterranean: An EU-funded horror story, 15 January, pp. 31–32.

The Economist (2022n): Russia and Africa: Wagner, worse than it sounds, 15 January, pp. 31–32.
The Economist (2022o): Unequal Partnership, Special Report: China in Africa, pp. 1–12.
The Economist (2022p): Unequal Partnership, Special Report: China in Africa, pp. 1–12.
Thussu, Daya (2016): The Scramble for Asian Soft Power in Africa, La ruée vers l'Afrique du Soft power asiatique, pp. 224–237.
Tom Wilson, David Blood, and David Pilling (2019): Congo voting data reveal huge fraud in poll to replace Kabila, https://www.ft.com/content/2b97f6e6-189d-11e9-b93e-f4351a53f1c3, accessed 15.06.2022.
Transparency International (2019): Where are Africa's billions? https://www.transparency.org/en/news/where-are-africas-billions.
Twagiramungu, Noel (2019): Re-describing transnational conflict in Africa, Journal of Modern African Studies, 57:3, pp. 377–391.
Wezeman, Pieter D. 2009: Arms transfers to Central, North and West Africa. In: SIPRI Background Paper, retrieved 21 July 2009, from http://books.sipri.org/files/misc/SIPRIBP0904b.pdf.
Wezeman, Pieter D. et al. 2019: *Trends in international arms transfers*, in SIPRI Factsheet, March 2020.
Wezeman, Pieter et al. (2022): Trends in International Arms Transfers, SIPRI Fact Sheet, March 2021.

CHAPTER 5

# The Postcolonial African Migration to the West as a Desire for Liberation

Slavery was the negation of African human beings. For over 400 years, more than 15 million Africans were the victims of the transatlantic slave trade (UN 2015; CARICOM 2023). Historians estimate that between 650 and 1900, 10–18 million sub-Saharan Africans were enslaved by Arab slave traders (Akinbode 2021; Gakunzi 2018). Black Africans were denied their liberty physically and mentally. In this chapter, I argue that the postcolonial African migration to the West must be analyzed against the background of the legacies of slavery and colonialism. The postcolonial African migration to the West is a part of the African liberation process from these historical experiences. This liberty can be achieved not through detachment from but through imitation of and coexistence with the former slaveholders and colonizers. Paradoxically, the desired liberation leads to different forms of imitation of the former slaveholders and colonizers. However, this is a denied or suppressed desire to be recognized and to be, as will be discussed in Chap. 7. In this chapter, we shall see how the postcolonial African migration to the West is a hidden or denied entreaty for liberation, recognition, and being.



B. Gebrewold, *Postcolonial African Migration to the West*, Politics of Citizenship and Migration,
https://doi.org/10.1007/978-3-031-58568-5_5

## The Denied African Humanity

The slaveholders were cynical and argued that the African slave in America was happier than in his own African civilization (James 2022: 5). Africans were just the property of whites and mere objects lying in the bottom of slave ships for days on end, their faces exposed to the tropical sun and the tropical rain, their backs in putrid water which was never bailed out; at the ports penned into trunks for the buyers to inspect while they fainted and died in the dense of putrefaction (James 2022: 6). "There has always been an intimate relationship between the name "black" and death, murder, being buried alive, along with the silence to which the thing necessarily had to be reduced—the order to be quiet and remain unseen" (Mbembe 2017: 152). Africans were just kept alive but weak so that they could not attempt escape or rebellion (Soyinka 2012: 63).

As the Tuskegee syphilis experiment (medical experiments from 1932 to 1972) on Africans in the United States showed, they were used as guinea pigs without their knowledge of medical experiments (Soyinka 2012: 74; Degruy 2017: 75). In the US, African Americans were declared as just 3/5 human as James Madison suggested in the 1787 Constitutional Convention (Anderson 2021: 28). Madison argued that slaves should be considered property for tax purposes and individuals to determine a state's population to have greater power in the soon-to-be-formed federal government (Degruy 2017). Thomas Lynch (1727–1776) of South Carolina compared the slaves in the Southern States with sheep in the North: as sheep in the North were not included in the headcount the slaves in the South could be counted either (Anderson 2021: 10). The US Constitution allowed the whites to carry weapons, whereas it was forbidden for blacks to have guns even for self-defense. When the slaves tried to defend their rights and protested against these unjust rules, they were whipped, scalded, burnt, and castrated, and their eyes or tongues extracted. The "Negro Act of 1722" spelled out how blacks would be punished if caught with arms without a special license of their master (Anderson 2021: 14). The whites were convinced that it was their constitutional right to import Africans as slaves to the United States. Blacks were considered natural slaves due to their skin color; they were "thinking property", rightless persons, predestined by the natural act of God to the position of permanent bondage as suggested by James Madison and other founding fathers (Degruy 2017: 35). For slaveholders, Africans were monsters, a mixture of two natures. Africans were transformed into human-objects, human-commodities, human-money, hated objects, humans without names, and languages, beings with inferior instincts and chaotic. "They were condemned to live and work with those who controlled them and denied them recognition as co-humans" (Mbembe 2017: 2, 6). As Fanon suggests, blacks were considered the quintessence of evil, lacking in values, insensible to ethics, the negation of values, the enemy of values, and an animal against which the white man has to be protected (Fanon 2001: 32, 2008: 146). Africans were the Shakespearean Othellos who desecrated the divinely white Desdemonas.

Blacks were not considered worthy of citizenship or other rights. Even the blacks' fighting in the War of Independence did not help them become US citizens because of their natural-born state of unfitness and dangerousness in a country where white skin determines one's worthiness (Anderson 2021: 44). As the outrage of Brabantio and Iago in Othello shows, black men were considered a threat to white women, something that had to be punished as a crime. Blacks as archetypal markers of evil were perceived as slaves at God's will and therefore did not deserve liberty, the right to vote, or the right to bear arms according to the Second Amendment. Black Africans were dehumanized as mad dogs, thugs, and inferior because of inept upbringing, and untamed rage who could make the white American society lawless encouraged by the Civil Rights movement (Anderson 2021: 140). Whiteness is the symbol of innocence, whereas blackness represents danger. A white person kills a black person because he feels threatened by the very existence of the black person. The probability that a person to be found innocent is higher if the victim was black than white, and the juries were twice as likely to convict the perpetrator of a crime against a white person than against a person of color (Anderson 2021: 152). If a white person carries a gun openly, it is not a problem; however, if a black person does it, it is a cause of public insecurity. Various studies have shown that black boys are considered older than whites, which makes them less innocent. Blacks were associated with apes, which dehumanizes and justifies violence against them. A noteworthy phenomenon in such a racist cultural, institutional, and structural mindset is that as some studies show black and white officers are each more likely to open fire on blacks they believe are armed, compared to whites who are dressed and behaving identically (Anderson 2021: 157). The blackness itself is feared, not necessarily the criminal deed. A white killer will be treated differently or kinder than a black potential killer. A white killer is often interpreted as a defender of himself, his property, businesses, public property, or security, while a black (potential) killer is a criminal and beast (Anderson 2021: 159).

Africans were lazy and irresponsible if they did not want to make the white man's land better, to fatten his mule, or save his corn (Du Bois

1903: 109). Du Bois writes about a conversation between a white Judge and his black friend,

> You know I am a friend to your people. I have helped you and your family and would have done more if you had not got the notion of going off. Now I like the colored people, and sympathize with all their reasonable aspirations; but you and I both know, John, that in this country the Negro must remain subordinate, and can never expect to be the equal of white men. In their place, your people can be honest and respectful; and God knows, I'll do what I can to help them. But when they want to reverse nature, and rule white men, and marry white women, and sit in my parlor, then, by God! We'll hold them under if we have to lynch every Nigger in the land. Now, John, the question is, are you, with your education and Northern notions, going to accept the situation and teach the darkies to be faithful servants and laborers as your fathers were,—I knew your father, John, he belonged to my brother, and he was a good Nigger. Well—well, are you going to be like him, or are you going to try to put fool ideas of rising and equality into these folks' heads, and make the discontented and unhappy? (Du Bois 1903: 166)

If the West is no longer pure white, there will be a regression, darkness, confusion, and another Middle Ages, and black races will never produce an Einstein, Stravinsky, a Greshwin, etc. (Césaire 2000: 49–51). The Philosopher Hegel was absolutely convinced that blacks were in a completely wild and untamed state and had a character that was not harmonious with humanity. Therefore, Africans are not only underdeveloped but also are unhistorical part of the world because they are in the condition of mere nature (Mbembe 2021: 7). Africa has been historically represented as a pathological case and a figure of lack (Mbembe 2021: 26).

Gershman suggests that "… Tocqueville thought that the owning of one human being by another contradicted both Christian belief and tradition and the political philosophy of the rights of man…" (Gershman 1976). However, Alexis de Tocqueville was convinced of African inferiority. For Alexis de Tocqueville, the hearts of Africans—as savages—are driven by fanaticism and greed, and like a perpetually agitated sea, where the wind does not blow from the same direction (Mbembe 2021: 93). The Negroes were depicted as unjust, cruel, barbarous, half-human, shameless, deceitful, thieves, drunkards, lazy, unclean, jealous to fury (as Othello was often presented by Shakespeare), and cowards (James 2022: 14). They had to be kept stupid and in their place as slaves. In the best case they should be taught to read but not to write so that they cannot organize and emancipate themselves (James 2022: 319).

The Arabs found that African slaves were a more lucrative business than gold and ivory and exported them to India, Persia, and Arabia. Even nowadays, Africans on their way to Europe are treated, exploited, mistreated by Arab smugglers (worse than non-black migrants), beaten up, raped, and despised as slaves in the Arab countries of North Africa with no police or legal recourse protecting them (Kingsley 2017: 70–72). Slavery was a source of cheap labor and greater national wealth for many European countries such as England, Denmark, Holland, Sweden, France, and Portugal (Dugard 2003: 65). As Jeffers describes, slaveholding white men had the fullest sexual liberty and domination over black women and girls. "No one would believe a Negro girl's accusation of a white man. And even if she was believed, no one would care" (Jeffers 2022: 407). The white "helpless" man assaults only because he was seduced by the black girls and women; the Negro women were to blame. The blacks—as the descendants of Cain—not only belonged to an inferior group but also were known purveyors of temptation for white men. They were weighed down by the curse of God and the darkest night of humanity lay over pre-colonial history concerns the whole of the African continent (Fanon 2001: 170). Therefore, black girls must feel honored and be thankful to have the sexual attention of a white man (Jeffers 2022: 407–408). There was a law created in New Orleans in 1786 that made it a criminal offense for women of "pure or mixed African blood" to give excessive attention to their dress. And if these women were too light-skinned, too European in appearance, or dressed too elegantly, it disturbed the social order (Lukasik 2021: 101). If a white man rapes a black woman, he is not committing any crime, he is only dishonoring himself (Jeffers 2022: 785).

> … Ham, the son of Noah, was responsible for the cursed color of the Negro. God had placed them on the earth to carry burdens, the Negro boys and men were ordained to exhaust themselves in brutish labor, and Negro women and girls to tolerate the weight of white men on their bodies, if it pleased God, to nurture the seed of their white masters within their wombs. That was the prescribed order of this world, and even in heaven, Negros would be expected to serve cheerfully. (Jeffers 2022: 415)

Blackness was considered so inferior that the "one-drop of blood" rule was introduced in the United States which marginalized anyone with black ancestors. The children of mixed races, free people of color were seen by the white community as a threat to their way of life and social hierarchy (Lukasik: 2021: 102). The detest of Blackness, oppression of African Americans, discrimination, low or no education, and dim job opportunities forced them to hate Blackness, hide it, or pass for white (Lukasik 2021). To pass as white means a symbol of freedom, unlimited opportunities, and all privileges that accompany being white (Lukasik 2021: 78). Whiteness enables privileges without merits or achievements (Degruy 2017: 78).

In Louisiana in 1970, what was known as a less stringent formula of race mathematics was enacted according to which, "In signifying race a person having one-thirty-second or less of Negro blood, shall not be deemed, described, or designated by any public official in the State of Louisiana as 'colored', a 'black', a negro', an Afro-American', … a 'colored person', or a 'person of color'" (quoted according to Lukasik 2021: 32).

In the white exhibition of the human skeleton and physiognomy, the blacks barely outranked a chimpanzee (Du Bois 2007: 48–9; Pitts 2005: 14; Jeffers 2022: 415; Mannoni 1990: 32). The round cheeks, high cheekbones, their somehow elevated forehead, their short, broad, flat nose, thick lips, small ears, the overall irregularity of their shape, etc. not only characterize their external appearance but also, they make the blacks ugly. They are not only ugly but also idle, treacherous, vengeful, cruel, impudent, thieves, liars, profane, nasty, ill-tempered, and therefore an unhappy race. The ugliness is not only an aesthetical but also an ethical quality. Ugly is also morally bad. Beauty or whiteness symbolizes moral integrity (Eco 2013: 4–5). Ugly is a threat to the harmony and beauty of society, like Quasimodo in Victor Hugo's novel. Who could represent this ugliness and immorality better than the black African? The "ugliness" of the Quasimodos not only confirms one's own beauty but also justifies violence against them.

For Voltaire, blacks were just mere animals, talking beasts (Mbembe 2017: 76; Memmi 1967: xxvii). Such as Joseph De Maistre, Lapouge, Jules Romains, etc. maintained that slavery was no more abnormal than the domestication of the horse or the ox (Césaire 2000: 49–50). Almost 400 years of physical, psychological, and spiritual torture, lynching, medical experimentation, redlining, disenfranchisement, being undesirable, etc. have left their mark, which Degruy analyses as post-traumatic slave syndrome. "As a result of centuries of slavery and oppression, most white Americans, in their thoughts as well as actions, believe themselves superior to blacks. Of greater import, too many African Americans unconsciously share this belief" (Degruy 2017: 100).

Victor Hugo was convinced that Africa did not have history and was a burden to the universal human life. Therefore, it was the task of the white race to make Africa fit for civilization which is actually beyond the possibility of civilization, left without culture and history (Mbembe 2017: 71–72; Du Bois 2007: 50).

As Hegel had suggested, it was the right of the whites to rule over the blacks who are without rights and do not count in the universal history of mankind, i.e. that of the whites. All objects created by blacks who are not moral subjects and with predispositions of primitive drives could only arouse scorn, terror, and disgust (Mbembe 2021: 155–56). For Hegel, Africans are beings in a limited and vegetative state, in a state of emptiness and fundamentally unrepresentable and whose moments of existence are penetrated by the negative, without language or awareness of themselves (Mbembe 2017: 6–7).

> Intractability is the distinguishing feature of the negro character. The condition in which they live is incapable of any development or culture, and their present existence is the same as it has always been. In the face of the enormous energy of sensuous arbitrariness which dominates their lives, morality has no determinate influence upon them. Anyone who wishes to study the most terrible manifestation of human nature will find them in Africa. The earliest reports concerning this continent tell us precisely the same, and it has no history in the true sense of the word.... What we understand as Africa proper is that unhistorical and undeveloped land.... (Quoted according to Mbembe 2001: 178)

For Hegel, Africa did not have a historical interest of its own; its inhabitants were barbarians and savages without any integral ingredient of culture. Africans were aimless hordes with unthinking humanity, revolting barbarity, without subjectivity. Africans were just a series of subjects who destroyed one another. Africans as animal man did not progress beyond their immediate existence (Hegel 1975: 174–190). The white supremacist Vance Muse of the Christian American Association justified segregation and discrimination and said he liked the blacks in their place. Their place for Vance Muse was the place below the whites, the place of sub-humanity. The whites who owned the blacks owned them as objects and animals against which the white race has to be protected (Jeffers 2022: 658; Lukasik 2021: 24, 112; Fanon 2008: 147). One slaveholder asks another slaveholder, "Why do you ill-treat your mule [blacks] in that way?", meaning why do you mistreat your own property (James 2022: 12).

> Between colonizer and colonized there is room only for forced labor, intimidation, the police, taxation, theft, rape, compulsory crops, contempt, mistrust, arrogance, self-complacency, swinishness, brainless elites, degraded masses. No human contact, but relations of domination and submission which turn the colonizing man into a classroom monitor, an army sergeant, a prison guard, a slave driver, and the indigenous man into an instrument of production... colonization = "' thingification'. (Césaire 2000: 42–43)

Virginias Causal Killing Act of 1705 and its Unlawful Assembly Act of 1680 made it legal to kill a slave "who raises a hand against any Christian" (Degruy 2017: 59). Africans were considered inferior because they were not European, they were not Christian, they were black, irreligious, uncivilized, therefore, they were fit to be slaves (Degruy 2017: 45–46). As Douglass argues, "As those who believe in the visibility of ghosts can easily see them, so it is always easy to see repulsive qualities in those we despise and hate." (Douglass 1881: 567).

> ... we should, when contemplating these sons of Noah, try and carry our mind back to that time when our poor elder brother Ham was cursed by his father, and condemned to be the slave of both Shem and Japheth...—a strikingly existing proof of the Holy Scriptures. ... Whilst the people of Europe and Asia were blessed by communion with God through the medium of His prophets, and obtained divine laws to regulate their ways and keep them in mind of Him who made them, the Africans were excluded from this dispensation; they think only of self-preservation in this world. Whatever, then, may be said against them for being too avaricious or too destitute of fellow-feeling, should rather reflect on ourselves, who have been so much better favored, yet have neglected to teach them, ... whilst they are sinning, know not what they are doing. To say a negro is incapable of instruction is a mere absurdity; for those few boys who have been educated in our schools have proved themselves even quicker than our own at learning; whilst, amongst themselves, the deepness of their cunning and their power of repartee are quite surprising, and are especially shown in their proficiency for telling lies

> most appropriately in preference to truth, and with an off-handed manner that makes them most amusing. (Speke 1864: xvii)

Those Africans who cooperated with the colonizers, who helped the European colonizer in the occupation and subjugation of their fellow Africans and their own countries were portrayed as possessing qualities of strength, intelligence, and beauty. The bad Africans were those who resisted foreign conquest and occupation, and they were, therefore, portrayed as ugly, weak, cowardly, and scheming (Thiong'o 2005: 92).

As Christians, the colonizers were taught the golden rule of Christianity, which is to treat others as you want to be treated and finally, you should be willing to sacrifice your comfort, your convenience, your wealth, and even your life for mankind (Du Bois 2007: 78). The colonizers were perfect and civilized human beings among themselves but barbarians to blacks. Du Bois describes this hypocrisy and double standards. The colonizer was

> ... sustained by a fine sense of Justice for himself and his Family, past and present; he is always courteous in public with 'ladies first' and precedence to 'gray hairs'; ... he is charitable, giving to the needy and deserving, to the poor proud, to inexplicable artists and to the Church. (Du Bois 2007: 79)

To black Africans, he was the persecutor, the boss, the unjust judge, tyrant. The Bible was the ideological arm of a colonialist assault, a symbol of a whole economy of colonialism, displacement, and despoilment in the name of the Christian civilization (Bedford 2018: 214). "... the unholy alliance of cross and sword taints the bible so badly that it becomes almost impossible to retrieve any liberating message from it" (Bedford 2018: 215). Blacks were kept down with poorer equipment in the schools, less money spent upon their schools, and poorer teaching than in the white schools; they were kept far away from the better-trained part of the nation, pounded back into their places every time they showed their heads above the ramparts (Du Bois 2007: 85, 101).

The assigning of place or "keeping-them-in-their-place" (Bakewell 2008) is justified by their skin color or race. As Painter argues, this racial designation based on nonwhiteness or the designation as colored, Negro, African-American, black, or Afro-American is more than a fact and is an idea that always associates the idea of blackness with slavery and inferiority (Painter 2010: ix). In the United States, blacks were relegated to the back of the bus, to separate cars and separate seating areas in restaurants, separate and poorly maintained washrooms, and drinking fountains, not allowed to marry whites, etc. to protect purity and superiority of the whites (Lukasik 2021: 60–61; Mannoni 1990: 32). Everything that is considered advanced, good, and civilized is defined and measured in white terms (Césaire 2000: 22–26). White supremacists believed that there must be something within the blacks that puts them at the bottom (Painter 2010: xi). "They are negroes—and that is enough, in the eye of this unreasoning prejudice, to justify indignity and violence." (Douglass 1881: 568).

> … dehumanization fills the air. It meets them at the workshop and factory when they apply for work. It meets them at the church, at the hotel, at the ballot box, and worst of all, it meets them in the jury box … The workshop denies him work, and the inn denies him shelter; the ballot-box a fair vote, and the jury-box a fair trial. He has ceased to be the slave of an individual but has in some sense become the slave of society. He may not now be bought and sold like a beast in the market, but he is the trammeled victim of prejudice, well calculated to repress his manly ambition, paralyze his energies, and make him a dejected and spiritless man, … Everything against the person with the hated color is promptly taken for granted; while everything in his favor is received with suspicion and doubt. A boy of this color is found in his bed tied, mutilated, and bleeding, … This same spirit, which promptly assumes everything against us, just as readily denies or explains away everything in our favor. We are not, as a race, even permitted to appropriate the virtues and achievements of our individual representatives. Manliness, capacity, learning, laudable ambition, heroic service, by any of our numbers, are easily placed to the credit of the superior race. … one drop of negro blood, though in the veins of a man of Teutonic whiteness, is enough of which to predicate all offensive and ignoble qualities. In the presence of this spirit, if a crime is committed, and the criminal is not positively known, a suspicious-looking colored man is sure to have been seen in the neighborhood. If an unarmed colored man is shot down and dies in his tracks, a jury, under the influence of this spirit, does not hesitate to find the murdered man the real criminal, and the murderer innocent. (Douglass 1881: 568–569)

Liberal thinkers like J.S. Mill had at least tacit assumptions that there was a relation between physical or biological factors and human character, according to which blacks were at an earlier stage of development and cognitively limited, enslaved to superstition, incapable of abstract thinking, and untrustworthy (Pitts 2005: 20).

The whites there—be they free or convicted—believed that "God Almighty in his most holy and wise providence hath so disposed of the condition of mankind as in all times some must be rich, some poor; some high and eminent in power and dignity; others mean and in subjection" (Painter 2010: 41). Furthermore, the political power was assumed to be a trait of white race. Color was equated with primitivism and slavery whereas whiteness was with beauty, justice, truth, and virginity, whereas the black man symbolized evil, darkness, obscurity, gloom, the labyrinth of the underworld, bad luck, war, death, sin, wretchedness, shameful feelings, base instincts, dark side of the soul, and ugliness (Fanon 2008: 157, 166–167). For the "one-drop" rule, one drop of black blood does not make you black, but it makes you colored (Lukasik 2021: 83). For example, Thomas Jefferson rejected slavery, not because of the equal dignity of all human beings but instead because he believed that slavery had an unhappy influence on the slave-owning class. After all, white society would develop the bad character of domination and tyranny if white children imitate the brutality of their white fathers (Painter 2010: 110). Even the otherwise detested and derided Irish of the nineteenth-century United States felt themselves better than blacks just because of their whiteness (Painter 2010: 133). Even the poorest and criminal white is better than a virtuous black or slave as whiteness is a blessing by itself (Jeffers 2022: 637). The fundamental thing was the distinction between the white and black man (James 2022: 28–9).

White supremacists like Johann Kaspar Lavater (1741–1801) believed that God had decreed one's outer appearance (whiteness), especially the face, to reflect one's inner state and that there is a correlation between personal beauty and human virtue, between outer looks and inner soul, and association of whiteness with separate acts of divine creation, advanced humanity, economic development, physical attractiveness, intelligence, natural superiority, and endowed with the infallible modern science and beauty (Painter 2010: 67–70, 229; Mbembe 2021: 45; Pitts 2005: 14; Mannoni 1990: 32). As Fanon puts it, gradually the blacks internalized this inferiority: "A feeling of inferiority? No, a feeling of not existing. Sin is black as virtue is white. All those white men, fingering their guns, can't be wrong. I am guilty. I do not know what of, but I know I am a wretch" (Fanon 2008: 118).

Even the despised and criminal whites did all they could to distinguish themselves from blacks. Even the free blacks aspired to be separate from the slaves (Mbembe 2017: 20). For the colonizers, Africans represented morbid and degenerate forms of man, obscure bodies waiting for help and demanding white man's aid. On the other hand, the whites were predestined to order, organize, arrange, and give form to the rest of the human herd, especially the blacks (Mbembe 2021: 148). According to the racist Gobineau, Africans were possessed by their Dionysian passion, dance, music, rhythm, lightheartedness, and sensuality, whereas the whites were endowed with the Apollonian gifts like energy, action, perseverance, rationality, and technical aptitude that make the whites racially superior (Painter 2010: 197; Pitts 2005: 14). For Gobineau, there is only a white history; there is only one worthwhile history, that is the white history (Césaire 2000: 71). Blacks cannot control their instincts, cannot think for themselves, cannot differentiate between right and wrong, are lazy and poor, do not like work, are untrustworthy, irresponsible, a liar, and do not like freedom (Mbembe 2001: 180). For the French Jean-Joseph Virey, humanity was divided into "ugly browns or blacks" on the one hand and, "beautiful whites" with their red and white complexion, blond hair, blue eyes, tall stature, and ferocious manhood on the other (Painter 2010: 90, 165). The black mentality was not capable of rational argumentation. "Only the white race possessed a will and capacity to construct life within history. The black race, in particular, had neither life, nor will, nor the energy of its own." (Mbembe 2017: 42). As Getachew argues, the colonial mindset of white superiority continued even after the end of slavery. The whites that dominated the League of Nations were convinced that the membership of even the non-colonized Africans—Ethiopia and Liberia—could only be a qualified membership due to the impossibility of black sovereignty. The colonial whites had the right of oversight to transform Africans who were considered without the capability to implement international norms. Their international rights and entitlements were limited and conditional (Getachew 2019: 53–55).

For Montesquieu,

> Sugar would be too dear if the plants which produce it were cultivated by any other than slaves. These creatures are all over black, and with such flat noses that they can scarcely be pitied. It is hardly to be believed that God, who is a wise Being, should place a soul, especially a good soul, in such a black ugly body.... It is impossible for us to suppose these creatures to be men, because, allowing them to be men, a suspicion would follow that we ourselves are not Christians. (Quoted according to Mbembe 2001: 181)

German philosopher Herder said, "... the Negro with his skin, with his ink-bubble blackness, with his lips, and hair, and turkey language, and stupidity, and laziness, is a natural brother of the apes of the same clime." (Herder 2002: 150) The "unfortunate" state of the blacks witnesses their inferiority (Herder 2002: 391). Pope Nicholas V (1447–1455) gave pontifical permission to the white and nobler human beings to turn all unbelievers into slaves (Herder 2002: 394). Philosophers like David Hume (1711–1776) and Hegel (1770–1831) were convinced of the inferiority of the Africans because they were not white, not civilized, without history, and living in the darkness (Botwe-Asamoah 2005: 7). The fact that every person shares 99.99% of the genetic material of every other human being is irrelevant for racists (Painter 2010: 391).

Blackness, slavery, and inferiority were essentially interrelated, and as a result, a black man was considered feeble, created on a lower plane than the white, and unimportant to the existing races (Painter 2010: 186). As Mbembe argues, slavery was a symbol of scission and denial of autonomy. Therefore, postcolonial freedom is about the abolition of this scission and regaining of African subjectivity (Mbembe 2021: 49). Various white scientists believed at the time that "black, the brown and red races differ anatomically so much from the white... that even with equal cerebral capacity, they could never rival the results by equal efforts" (Painter 2010: 236). It was believed that the blacks were doomed to die out, unfit as they were to live outside of slavery (Painter 2010: 251). Whites were exhorted to breed as much and fast as possible to outperform the blacks in reproduction and be able to control the land (Painter 2010: 259, 273).

Since indentured whites or convicted whites could be free by virtue of their whiteness in the United States active citizenship was opened to virtually all adult white men when the ideology of democracy gained wide acceptance between 1790 and the mid-1850s and made the United States "a white man's country" (Painter 2010: 106). In his studies, Du Bois argues that the government's movement toward abolition was very slow. The Atlantic slave trade to the United States continued until it was no longer economically beneficial. Moral outrage played less role than economic utility (Du Bois 2007: xix). Even slavery was not ended because Europeans suddenly noticed humanity in the Africans or because their moral understanding changed. Armed Africans started to become a threat to the colonial powers. As Adam Smith argues, slavery "was not abolished by humanity or improvement of manners—but as the slaves were armed by their lords and so dangerous to the king, the king abolished slavery" (Pitts 2005: 31).

It is not by chance that even today whites are just "Americans", whereas others have always a qualifying adjective, "African-Americans", Asian-Americans", Hispanic-Americans", since only white Europeans could be real Americans. Especially for the blacks, the most exploited and oppressed of all races, color had become an abiding unchangeable fact that determined their social condition.

Whereas discrimination against blacks remains, Jews, Italians, and the Irish are rarely discriminated against because of their skin color in the US. A multicultural middle class may diversify the suburbs and college campuses, but the face of poor, segregated inner cities remains black. (Painter 2010: 396). "Through Jim Crow segregation was supposed to be separate but equal, in practice, it worked to discriminate, by excluding nonwhites from public institutions, whether from libraries, schools, swimming pools, or the ballot box" (Painter 2010: 384).

Adom Getachew's book, *Worldmaking After Empire*, demonstrates how European imperialism violently inaugurated an unprecedented era of globality only in the interest of the Europeans. European conquest, colonization, native dispossession, genocide, and forced migration of more than 12 million African slaves into the Americas enduring and decisive legacies on today's Africans psychologically, economically, and politically (Getachew 2019: 3). The encounter between Europeans and Africans that started officially in 1492 resulted in an unequal, unjust, barbaric, and racialized global integration and order for the sake of dehumanization and exploitation (Getachew 2019: 9).

> Whether in everyday discourse or in ostensibly scholarly narratives, the continent is the very figure of 'the strange.' It is similar to that inaccessible 'Other with a capital 'O' … In this extremity of the Earth, reason is supposedly permanently at bay, and the unknown has supposedly attained its highest point. Africa, a headless figure threatened with madness and quite innocent of any notion of center, hierarchy, or stability, is portrayed as a vast dark cave where every benchmark and distinction come together in total confusion, and the rifts of a tragic and unhappy human history stand revealed: a mixture of the half-created and incomplete, strange signs, convulsive movements—in short, a bottomless abyss where everything is noise, yawning, and primordial chaos. (Mbembe 2001: 3)

## Colonization as "Civilization", "Regeneration", and "Humanization"

Colonization was euphemized as civilization. Etymologically, *to colonize* means to cultivate, to till; *colonus*: tiller of the soil, farmer; from 1630 to *colonize* meant to migrate to and settle in (online etymology dictionary 2023). Colonizers presented themselves as civilizers. However, it was overshadowed by physical, cultural, and psychological violence based on racism and white supremacy. The violence against Africans was systematized and institutionalized during the European scramble for Africa. The intra-European violence was transferred to Africa. When the rivalry among the Europeans started to escalate to the point of violence, the Europeans decided to settle their "African problem in a civilized way": to assemble in 1884–1885 and agree in Berlin among themselves and divide Africa as their private property. They not only exploited African resources, but also dehumanized, colonized, and exploited African labor. The Berlin Africa Conference (1884–85) itself was an intra-European pacification through African dehumanization. "... the colonialist never planned to transform the colony into the image of his homeland, nor to make the colonized in his own image! He cannot allow such an equation—it would destroy the principle of his privileges." (Memmi 1967: 69).

Colonial powers were competing to occupy as much of Africa as possible. France occupied Tunisia in 1881 to prevent Italian dominance in the region. Under the disguise of "stabilizing" the Egyptian government and its finances, Great Britain underpinned its power in 1882 in Egypt. Germany established its protectorates in Namibia, Cameroon, and Togo in 1884. However, it was during the Berlin Conference (1884–85) that colonialism was agreed upon and formalized under international law to prevent intra-European conflict on colonization (Iliffe 1997: 254). The future European demands for African territory began to be more substantial than the informal supremacy enjoyed by Great Britain. The politics of protectorates enabled Germany in 1886 to occupy Tanzania, whereas Great Britain could take up Uganda as its colonial territory in 1890. In 1886 the right to a protectorate in the Niger Delta enabled Great Britain to expand its supremacy in the region (Meredith 2005: 3). Toward the end of the nineteenth century, the French succeeded in occupying the territories at the upper course of the Niger and at Lake Chad (Shillington 2005: 308). Startled by the French ambitions in Ethiopia, Great Britain pushed the Italians to overtake the French in the competition, which caused the occupation of Eritrea by the Italians in 1889. Moreover, the Italians tried to expand their occupation into the Ethiopian Empire (Iliffe 1997: 255). In 1889 Sudan was controlled by Great Britain in competition with France. Since Great Britain was stubborn in its demand for Egypt, France was allowed to occupy Western Africa, and the Italians could march into Libya. In 1885 Great Britain declared Botswana as its protectorate. The gold deposit in southern Africa increased Great Britain's interest in expanding its colonial territory to the north. Even though the occupation of north Rhodesia (Zambia) and Nyassaland (Malawi) by Great Britain could endanger the interests of Portugal in the region, the colonial demand of both sides could be settled through an agreement on the course of the border between Mozambique and Angola (Iliffe 1997: 256).

One of the crucially positive effects of the Berlin Africa Conference for the Europeans was the intra-European pacification by means of collective violence against Africans. The intra-European violent rivalry could be averted by their collective and violent agreement against Africans. Starting with the agreement, the Europeans imitated each other in their violence against Africans.

The mimetic theory by René Girard explains the internal pacification of rivals through the externalization of violence against a surrogate victim. Girard argues that unappeased violence seeks and always finds a surrogate victim, and the original object of violence is abruptly replaced by another, which is usually vulnerable and close at hand (Girard 2016: 2; Girard 2019: 24). At the same time, the surrogate victims to which the intra-community violence is channeled are transformed beyond their real nature. Girard shows that even animals could be surrogate sacrificial victims. However, they will be upgraded to the human level to some extent to make them worthy of sacrifice. These animals are "differentiated in such a way as to create a scale of values that approximates human distinctions and represents a virtual duplicate of human society" (Girard 2016: 3). On the other hand, the human sacrificial victims would be dehumanized. The objective of this "transubstantiation" of the victims is to channel intra-community violence toward outside victims who are or are made different from the rest of the community members. That is the violence that would otherwise be vented on its own members and which is deflected upon an external and substitute victim. Girard suggests that the "substitute victim serves to protect the entire community from its own violence; it prompts the entire community to choose victims outside itself (Girard 2016: 8).

Dehumanized potential victims include prisoners of war, slaves, *pharmakos* (human scapegoat), children, king (or his fool), women, animals, etc. Girard says that all sacrificial victims, whether dehumanized human beings or "humanized" animals are "invariably distinguishable from the non-sacrificeable beings by one essential characteristic: between these victims and the community a crucial link is missing, so they can be exposed to violence without fear of reprisal. ... that sacrifice is primarily an act of violence without risk of vengeance" (Girard 2016: 14; Girard 2019: 23). Girard describes intra-community violence as impure violence that is contagious and reciprocal. Therefore, the community needs pure violence, which purifies the community because it transfers the violence toward an external victim, the "others", the "different men" (Girard 2016: 54, 59; Girard 2019: 24–25).

In the language of René Girard, we can call the actual and potential precolonial conflict within Europe a sacrificial crisis that was solved by sacrificing the Africans. "... social coexistence would be impossible if no surrogate victim existed if violence persisted beyond a certain threshold and failed to be transmuted into the culture. It is only at this point that the vicious circle of *reciprocal* violence, wholly destructive in nature, is replaced by the vicious circle of *ritual* violence [against outsiders], creative and protective in nature" (Girard 2016: 163).

Africans were externalized from the human community and humanity which was reserved for whites only. This alienation of Africans justified violence against them because no revenge on their behalf was expected or justified. As the case of Apartheid showed, white Europeans went to the land of the blacks, displaced them, took away their land, told them not to enter the cities, and forced them to work for them. During this dehumanization of Africans, the intra-European war was channeled toward Africans.

Violence signifies and promises a triumphant majesty, a condition, which can be enjoyed only and always at the expense of others (Girard 2016: 171). "The surrogate victim dies so that the entire community ... can be reborn in a new or renewed cultural order... This may well have been what Heraclitus meant when he declared: 'Dionysus and Hades are the same'" (Girard 2016: 191). Through the death of some, others enjoy community harmony through what Girard calls "generative unanimity" (Girard 2016: 302). The European civilization, capitalism, or Enlightenment flourished on African exploitation, pain, and death (Mayblin 2021: 26; Bhambra 2007). The Berlin Conference and the agreement thereof of 1885 were a political sacrificial act through which the European mimetic oppositions could be reconciled and channeled against Africa as a surrogate victim (Girard 2016: 348; Girard 2019: 30; Palaver 2003: 348). This is a collective European transfer of violent competition and its potential internal conflict on a reconciliatory colonial surrogate victim (Soyinka 2012: 34; Girard 2019: 40). Enslaved and colonized Africa was in a sense *sacred* (because it was *sacri*ficed) because it was an innocent victim which had served as a reconciling surrogate victim for the Europeans (Girard 2019: 217). The African human being is metaphorically a medical *pharmakon*—poison and remedy at the same time—for the European colonizers and slaveholders because it incorporates at the same time the despicable and redeemer (Girard 2016: 328; Girard 2019: 96; Palaver 2003: 332–333).

Girard differentiates between *acquisitive mimesis* and *conflictual mimesis.* The first one is about the mimetic relationship in which two or more individuals converge on the same object to appropriate it, whereas the second one is about the conflict, which unifies two or more individuals to converge on the same victim that all wish to strike down (Girard 2019: 25). In the first case the objective is to attain and maintain superiority and glory or to achieve recognition (the focus is on competition among equals), the second one is about the violence against the unequal innocent victim, an outsider.

What does this mean for the colonial violence on the one hand and the postcolonial African desire for liberation? The colonial violence against Africans is conflictual mimesis as the conflict among the Europeans was directed against Africans collectively. It was also an acquisitive mimesis because Europeans imitated each other in the exploitation of African resources as much as possible. The postcolonial conflict between the colonizer and colonized is also an acquisitive mimesis because the colonizer tries to retain his superiority through *othering*, *bordering*, and *ordering*, whereas the colonized tries to negate this by *de-othering, de-bordering*, and *re-ordering*, as this will be shown in Chap. 7. And also, in Chap. 8, we shall see that it is the diminishing inequality that leads the colonizer and the colonized to acquisitive mimesis.

For white supremacists, there is something of a problem with the Africans as their corruption, violence, incompetence, and indifference show (Soyinka 2012: X). Wole Soyinka reports about an incident in Europe in his presence: "Africans, you must admit, are inherently inferior. You must be, or other races would not have enslaved you for centuries. Your enslavers saw you for what you are, so you cannot blame them."

(Soyinka 2012: xii). Because of the internalization of the dehumanization and burden of social disqualification many blacks develop the feeling of not being good enough and that European descent and white skin people are superior, as a study by Nogueira in Brazil shows. Those who consider themselves Whites or are considered as such accept all the unearned privilege uncritically as self-evident and own merit. White lifestyle becomes the norm, whereas non-white is primitive or not worthy (Nogueira 2013).

In 1876, Leopold II, King of the Belgians, had declared that "To open to civilization the sole part of the globe which it has not yet penetrated, to pierce the darkness which envelops the entire population: this, I venture to say, is a crusade worthy of this century of progress." (Conrad 2008: xvii). The colonialists dehumanized, degraded, and associated Africa with supernatural evil and a place of negation out of which it will be saved by European grace. Colonialism denied human rights to human beings; Africans were made sub-humans to whom the Universal Declaration of Human Rights did not apply (Memmi 1967: xxiv).

Conrad writes that the duty of colonial expansion was the sacred civilizing work, to bring light, faith, and commerce to Africa, the darkest place on the earth. What a consolation for the colonizer and missionary to think of that! (Conrad 2008: 9). The civilizing mission of the Europeans was thought of as a European humanitarian enterprise and duty to help and protect the forsaken and inferior African beings (Mbembe 2017: 12). In her book, *A Mission to Civilize*, Alice Conklin describes how colonial France approached Africans. The French colonialists approached Africa with the conviction of French superiority and the perfectibility of the Africans who were too primitive to rule themselves but capable of being uplifted. The French were predestined by their culture, intellect, and industrial, political, and military power to civilize Africans. They had not only the right but also the duty to remake primitive African cultures along lines inspired by the cultural, political, and economic development of France guided by its moral, cultural, and social supremacy (Conklin 1997: 1–2).

> To characterize any conduct whatever towards a barbarous people as violation of the law of nations, only shows that he who so speaks has never considered the subject. A violation of great principles of morality it may easily be; but barbarians have no rights as a nation, except a right to such treatment as may, at the earliest possible period, fit them for becoming one. The only moral laws for the relation between a civilized and barbarous gov-

> ernment, are the universal rules of morality between man and man. (Quoted according to Pitts 2005: 144)

For French colonizers, the principle of civilization was just mastery of nature, the human body, and social behavior which the Africans lacked, unlike the French (Conklin 1997: 6). According to French colonizers, Africans—inferior to white Europeans—could be made to triumph over geography, climate, disease, and barbarism thanks to the ameliorating civilizing activity. Africans as colonies were subjects, not citizens; they had duties, not rights. Therefore, French democracy at home and colonialism in Africa were not irreconcilable in the civilizing measures on behalf of their African subjects (Conklin 1997: 9–10). Colonialism was an interaction between black and African barbarity and backwardness and white and European civilization (Pitts 2005: 1). The colonizers, politicians, evangelizers, and liberal intellectuals in Europe declared that they had the moral authority and obligation to civilize the barbarians at any cost. As Pitts put it, "… Europe's progressive civilization granted Europeans the authority to suspend, in their relations with non-European societies, the moral and political standards they believed applied among themselves" (Pitts 2005: 11). For John Stuart Mill, British despotism was the best government to which Africans or other colonized peoples and backward societies could aspire (Pitts 2005: 15).

According to Mannoni, who studied Malagasy, the colonized were suffering from not only dependency complexes but also, relied upon the French as their substitute father. Without the colonizer, they felt lost. They were so irrational that they started to revolt against their masters, the French. Malagasy had to be led by the French little by little to their independence and self-government lest they fail as they were incompetent in self-government (Mannoni 1990: x–xi). The colonized man shows no gratitude. Since he is so much dependent on the colonizer he does not see any need to be grateful. In the view of Mannoni, dependence excludes gratitude, and therefore, the child cannot really learn gratitude until he has attained a certain independence (Mannoni 1990: 44). "Dependence continues to give the average Malagasy a greater feeling of security while relieving him of the need to show initiative or assume responsibility" (70). For the Malagasy, the European is a supernatural being, mighty and terrible, according to Mannoni (163). The colonized African is the inverse or negative image of the colonizer (Gandhi 1998: 15).

The term “black” represents the evil, the bad, the ugly, a figure of brutality and cruelty, and distinct humanity whose very humanity was (and still is) in question (Mbembe 2017: xi). Therefore, the barbarity of slavery and colonialism were justified as part of the French civilizing mission and cannot be interpreted as injustice and do not demand repentance. On the contrary, the colonized people should be thankful to the former colonizers as civilizers who built roads, bridges, schools, and hospitals in Africa, as former President Sarkozy of France said, as the civilization could be created only with white blood (Mbembe 2021: 138, 2017: 63, 178). The colonizer’s superiority and civilizedness depended on turning or projecting own barbarity on the forcefully degraded and dehumanized Other, an act of “*thingification*” (Césaire 2000: 8–9, 35; Memmi 1967: xxvi). The slaveholder poured salt, pepper, citron, cinders, aloes, and hot ashes onto the bleeding wounds of the colonized and enslaved African (James 2022: 10). Dehumanization legitimizes the denial of human rights (Bauman 2016: 86). To use the language of Mbembe, slavery, and colonization was necropolitics through which the Western whites had reserved themselves the right and authority to dictate who may live and who must die, to kill or to allow to live, to exercise sovereignty, to exercise control over mortality and to define life as the deployment and manifestation of power (Mbembe 2003).

The French colonial mission in Africa was not only to civilize the barbarians but also motivated by the French desire to erase the history of defeat by Britain in their competition to control America, Africa, and India during the Seven Years’ War around the middle of the eighteenth-century which involved most of the European great powers (Pitts 2005: 165). France saw itself as the nation that represented the future of civilization and was charged with rescuing other peoples from tyranny and ignorance (Pitts 2005: 166). Liberal intellectuals like Tocqueville were infuriated by the anti-imperialist movements and defended even violent French conquest and rule of Algerians. The “morally superior European civilizing mission” was defended by Condorcet to eradicate the backwardness beyond Europe’s borders (Pitts 2005: 168). Condorcet, Tocqueville, Mill, etc. considered Africans as backward societies like children in need of tutelage by the morally and culturally superior Europeans (Pitts 2005: 170). The African explorer Henry Morton Stanley (1841–1904) understood that Africans were second-class citizens, under the authority of white men (Dugard 2003: 158). In this time, racism was not only condoned, but also expected, and therefore, it gave all the legitimacy and authority to the likes of Stanley to treat Africans like ignorant children. For Condorcet, liberal imperialism of benevolent, enlightened, morally superior Europe with the help of civilized European colonists would liberate Africans from barbarism and savage ignorance (Pitts 2005: 172–73). The white colonizer was convinced that "Historically, the black man steeped in the inessentiality of servitude, was set free by the master. He did not fight for his freedom." (Fanon 2008: 194). For Tocqueville, imperial European conquest and rule over Africans was a just, noble, and righteous deed because it improved colonial subjects even by violence as a necessary means. For this mission, not only violent but even iniquitous means were also permissible (Pitts 2005: 213). Even the violent civilizing mission is permissible because it is only through the cultural elements developed by the whites that it can be ensured that the people of the tropical regions can be brought to progress and higher civilization. The whites have by nature scientific, moral, and religious superiority that justify and facilitate the civilizing mission. Europe has the authority and the mission to assimilate the Africans with Europe, to turn them into Frenchmen with black skin (Césaire 2000: 57, 70, 88).

> Around the colonized there has grown a whole vocabulary of phrases, each in its own way reinforcing the dreadful secondariness of people… Thus, the status of colonized people has been fixed in zones of dependency and peripherality, stigmatized in the designation of underdeveloped, less-developed, developing states, ruled by a superior, developed, or metropolitan colonizer who was theoretically posited as a categorically antithetical overlord. (Said 1989: 207)

For Tocqueville, Europeans were such determined conquerors of colonial subjects for the sake of salvation through conquest because they were guided by Christian Europe's understanding of the unity of humanity (Pitts 2005: 224). Tocqueville suggested that freed Africans had to be trained gradually in their freedom. Otherwise, they would be overwhelmed by the emancipation and they do not know how to use or enjoy it (Pitts 2005: 228).

Various researchers have shown that the "European miracle" and Western modernity would have not been possible without slavery and exploitation. Coloniality (a colonial, racially and culturally hierarchical worldview) and modernity are inseparable (Mayblin 2021: 26; Bhambra 2007). The "colonizability" or degradability of the blacks reminds the colonizer of his superiority. "Colonial racism is built from three major ideological components: one, the gulf between the culture of the colonialist and the colonized; two, the exploitation of these differences for the benefit of the colonialists; three, the use of these supposed differences as standards of absolute fact" (Memmi 1967: 71).

The colonizer maintained it was the ethical and Christian responsibility and rationale of the colonizer to civilize the colonized (Bell 2002: 45). As Norbert Elias would say, the "civilizer" felt "civilized" in comparison to the "uncivilized", who is declared by the "civilizer" as "uncivilized" (Elias 1978). While defining other peoples as "wild", "unrestrained", or "repulsed by barbaric customs" the colonizer at the same time constructed the hallmark of "civilized people" (Goudsblom 1998: 40). The colonized were not yet human enough, but the perfectibility, i.e. the potentiality to be complete and perfect, according to standards given by the colonizer, was not always denied. Within this perfectibility mission, Christianization, commercial interests, and the civilization of savages were triumphantly allied (Maquet 1971: 242). The mission of the colonizers and missionaries was to bring to Africa a rational civilization instead of superstitious animism.

The relationship between the colonizer and the colonized is based on humiliation, oppression, and intimidation. Physically, the arms of the colonized were chopped off in the Congo; holes were made in the lips of the revolting Africans to tie them together in Angola. They were kept undernourished.

According to the colonizer, colonization is civilization, and decolonization is a graciously and generously "granted freedom". For example, regarding decolonization, in the political and historical literature of German-speaking countries, it is very common to find phrases like "set them free into independence" (*in die Unabhängigkeit entlassen*), which implies that the colonizer *sets* the colonized *free* (or *expels*, in literal translation). This means, the colonizer is even "expelling" the colonized into freedom, who is basically *unwilling* to be free. To set the colonized free, means graciousness of freedom by the colonizer and passivity of the colonized. The dehumanized Africans are—thanks to the generosity of the colonizer—elevated to the level of the human being by the colonizer. This generous attitude of the colonizer demands gratefulness from the colonized. Only the colonizer robs the freedom of the colonized and gives it back to him. In his act of re-creation or re-generation of the dehumanized, the colonizer says "There is no difference between us" (Silverman 1999: 41).

The objective of the Christian missionaries was to save the "lost souls of the African pagans" by destroying pagan worshipping sites. Because of the relapse to paganism slave trade was not only tolerated by the missionaries but was also justified and practiced by the missionaries themselves (Bitterli 1991: 106–107). The process of denigration was exacerbated when the missionaries considered evangelization as a divine mandate. Africans were declared poor, abandoned, and forsaken by God. The evangelizers declared Africans as "forsaken" and "abandoned" by God and by man, *man* means here the "civilized", the Western. Without the help of the missionaries, the colonial influence and administration in the hinterland would have not been easily established. As Shillington asserts missionaries became agents of imperialism (Shillington 2005: 291).

Cecil Rhodes said, "Only one race [white European] was destined to help in God's work and to fulfill his purpose in the world … and to bring nearer the reign of justice, liberty, and peace" (Sindima 1998: 3). Africans were told by the missionaries that they were obliged by the bible to obey their colonial rulers. When Mozambicans fought for independence they were told by Portuguese Church authorities in Maputo not to be seduced by fantasies or led astray by "evil" counselors in the hope of cultural and economic prosperity. Cultural and economic prosperity can happen only by cooperating honestly with the Portuguese authorities and obeying their orders (Sindima 1998: 3). Christian evangelizers together with colonizers were convinced of their moral, intellectual, political, economic, and technological superiority and of their right to invade the black world violently and contemptuously (Soyinka 2012: 166; Pitts 2005: 14). As David Livingston said about Africans, "All I can add in my solitude is, my Heaven's rich blessing come down on everyone, American, English, or Turk, who will help to heal this open sore of the world [Africa]" (Dugard 2003: 309).

For Daniel Comboni (1831–1881), a Catholic missionary to Africa, Africa, the poorest, the unhappiest, and most abandoned of all human beings needed salvation and regeneration through Christian evangelization. African lie buried in the shadows of death; African face had to be renovated; unfortunate and unhappy African souls were crying to Christ; unfortunate sons of Ham; black, but beautiful (Gilli 1979: 68–191). "You alone can give light to so many poor non-believers, children of the unhappy Ham, who still live in the shadow of death (Gilli 1979: 193). For Pope

Alexander VI (1492–1503), the Catholic Church had the divine mandate and legitimacy to overthrow barbarous nations, instruct them in the Catholic faith, and train them in good morals, which is a holy and praiseworthy undertaking to the happiness and glory of all Christendom. "Let no one, therefore, infringe, or with rash boldness contravene, this our recommendation, exhortation, requisition, gift, grant, assignment, constitution, deputation, decree, mandate, prohibition, and will. Should anyone presume to attempt this, be it known to him that he will incur the wrath of Almighty God." (Pope Alexander VI 1493). Comboni warned that missionaries cannot go to heaven without the company of the souls they have helped to save (Gilli 1979: 15). The Africans were unfortunate people, living in the shadow of death; they were lost sheep to be brought back to the sheepfold of salvation. Because of his missionary and civilizational zeal, Comboni had said "Africa or Death!" would be his war cry (Gilli 1979: 184). Tributes at the time of Daniel Comboni's death convey his image of Africa and his belief in saving it:

> The people of Africa—who lay unhappily in idolatry—he attracted to the love of the one true God—and with the peace given by Heaven he comforted them—who would not call him an angel? Courageously he carried—the name and the faith of Christ—among the nations—different in land, culture, and customs—who would not call him an apostle? ... From man's proud enemy—he saved many by his word and example—gently opened them heaven—doing the work of the redeemer. (Gilli 1979: 238)

In another inscription, African's feelings toward "her Bishop and Father" (Comboni) were depicted as follows:

> O Daniel—it did indeed seem that you would take away at least—the ancient curse—that weighs down on our Africa—Now that you are taken from us before your time—we prostrate ourselves at your tomb—most unhappy Africans. Not to rob of us of that was ours—not to take away our children as slaves—you went throughout the unexplored regions of Africa—but to give us—at the cost of sufferings that cannot be told—the faith of Christ and liberty. Impelled by God—to our poor Africa you vowed your great heart—and now you—fallen because of your labors but not vanquished—apostle and martyr—we exalt and venerate. You have fallen, Father—first hope of Africa—.... (Gilli 1979: 239)

According to Daniel Comboni, "the 'poor Africans' needed regeneration because they were the most unfortunate, abandoned, and unhappy souls in the world and who groan in the deep darkness of paganism buried in the darkness of unbelief and error" (Gilli 1979: 79). For David Livingstone, missionaries were members of a superior race to elevate the destructed and downtrodden race of Africa, the degraded portions of the human family (Hetherwick 1939).

This "god-forsakenness" predestines Africans for cultural and human denigration. Denigration etymologically comes from *de* + *nigrare* "to blacken", *niger* "black". "Denigrate" means "to cast aspersion on", "defame", "to deny the importance or validity of", "belittle", as the literature since 1526 suggests (Merriam Webster's Collegiate Dictionary, 10th Edition, 1993). Africans have been trying to liberate themselves from psychological enslavement and denigration. They have to negate the denigrating premises of the colonizer.

Colonialism and Christianity led to the subjugation of black Africans who had to be humanized, civilized, and baptized. The merchant, the missionary, and the military man were convinced of white supremacy (Mbuwayesango 2018: 31). The missionary religious influence provided an ideological aim for the colonization and spread of Western beliefs to the Africans guided by the alleged obligation of Christians to use all means for the conversion of the heathen through bringing the true and only God to them who lacked the knowledge of God (Mbuwayesango 2018: 33). The missionaries sought the support of the colonizers in their endeavor. "Missionaries and colonizers did not find much in African life and culture that they considered good and compatible with the Christian life. The missionaries desired and worked toward an overhaul of African life" (Mbuwayesango 2018: 35).

The objective of the colonizers and the evangelizers was to strip the belief of the Africans and replaced them with Christianity to make them governable and disciplined. Christianity was the symbol of civilization, whereas paganism or non-Christianity represented savagery (Césaire 2000: 33). The bible became a means of obedience, control, and oppression (Mbuwayesango 2018: 36). In the face of the fact that many African cultures have been patriarchally oppressive to women, through their measures against polygamy the "missionaries believed that they were liberating African women from degradation at the hands of lazy and uncivilized men. They believed that the men held on to polygamy because they did not want to work and left all work to women and children. The gospel was considered the only means powerful enough to instill in African men the ideals of obedience, hard work, and respect for authority. These were traits that the colonizers needed in African men for the sake of an adequate supply of docile and governable laborers (Mbuwayesango 2018: 37). Both the missionaries and colonizers knew how to transform the Africans to their advantage: for the missionaries, the African soul was saved; for the colonizers, they will learn how to be docile and disciplined workers (Mbuwayesango 2018: 37). As Ngwa puts it, through the bible and the gun, the evangelizers and the colonizers claimed the right and authority to critique, determine, or regulate the African character, life, existence, culture, and history have identifiable economic, political, and spiritual practices and legacies that have enduring impacts (Ngwa 2018: 43; Pitts 2005: 21).

The colonizer had all the right to own, lock up, compel the colonized to forced labor, hang him in public, kill, and abandon his corpse along the road (Mbembe 2001: 186). For the colonizers it was a self-evident right bestowed to them by the divine mandate to explore, conquer, subjugate, and save this "dark" place for its own sake with the help of a gun as a weapon of encounter (Ngwa2018: 48). Mbembe suggests that for colonizers, the colonized did not truly exist as a person or as a subject since only human beings can exist. The colonized did not exist as a living being endowed with reason nor accomplish intentional acts. They existed only as useful material beings. From the colonizer's perspective of humanity, the colonized was *nothing*. The colonizer is the creator of this nothingness and destroyer of a being. The colonized has no freedom, individuality, no history (Mbembe 2001: 188–190). The slaveholder and colonizer had two natures: the power to create and to destroy, to make a being or a nothingness out of the Africans. In the US, the Second Amendment dehumanized Africans, whereas the 15th Amendment made Negroes humans (Du Bois 2007: 4).

For the colonizers, the colonized people had inherent deficiencies deeply rooted in their cultures and social hierarchies that made them incapable of generating change or progress internally without European intervention as a civilizing force to diffuse freedom and science, political order, liberty and free trade, and Christian knowledge in regions sunken in superstitious barbarism (Pitts 2005: 40, 52–53). The Christian evangelizers wanted to liberate the Africans into the light from traditional African culture with satanic imagination, diabolism, idolatry, and darkness by transforming them into a Christian state through trade and evangelization (Mbembe 2017: 26).

> Despotism is a legitimate mode of government in dealing with barbarians, provided the end be their improvement and the means justified by actually effecting that end. Liberty, as a principle, has no application to any state of things anterior to the time when mankind has become capable of being improved by free and equal discussion. Until then, there is nothing for them but implicit obedience. (Mill 2016 [1859]: 13)

For both father and son Mills, there was a sharp dichotomy between civilized Britain and uncivilized barbarous Indians incapable of self-government handicapped by inferior mental capacities and irrationality. Therefore, a progressive colonial (British) despotism was the most appropriate form of government for them (Pitts 2005: 105, 132). For James Mill, Indians were rude, mentally debilitated, and unable to recognize their interests, and their reason was enslaved by whim or passion (Pitts 2005: 131). Colonization was a form of assistance, education, protection, emancipation, illumination, and moral treatment against African idiocy and degeneration toward regeneration (Mbembe 2017: 65).

If the blacks wanted independence from the colonizers—it was said—they could get it but they would go back to the Middle Ages. "Did the colonizer not open roads, build hospitals and schools?" (Memmi 1967: 112). The superiority complex caused King Baudouin of Belgium to utter a speech on the independence day of Congo/Zaire (30 June 1960):

> It is now up to you, gentlemen, to show that you are worthy of our confidence...The independence of Congo constitutes the culmination of the work conceived by the genius of King Léopold II, undertaken by him with a tenacious courage and continued with perseverance with Belgium... Don't compromise the future with hasty reforms, and don't replace the structures that Belgium hands over to you until you are sure you can do better. (Meredith 2005: 93)

For the King of Belgium, the colonial evils and cruelty were exaggerated, whereas slavery had protected blacks from destitution, put them to work, and taught them the way toward salvation and civilization (James 2022: 11; Degruy 2017: 39). The blacks need the thick end of the stick if they are to get out of the Middle Ages (Fanon 2001: 76, 94). Since they are deficient, they need the self-less protection of the colonizer (Memmi

1967: 82). For de Tocqueville, the blacks are personified debasement, abjection, aversion, repulsion, disgust, and a herd animal, a castrated and atrophied humanity from emanates poisoned exhalation. For the viability of the whites, according to Tocqueville, the segregation of blacks was essential because of their inherent ignominy and lowliness (Mbembe 2017: 82–84). According to Conrad,

> The negro is an example of an animal man in all his savagery and lawlessness, and if we wish to understand him at all, we must put aside all our European attitudes. We must not think either of a spiritual God or of moral laws; to comprehend him correctly, we must abstract from all reverence and morality, and from everything which call feeling. All that is foreign to man in his immediate existence, and nothing constant with humanity is to be found in his character. (Quoted according to Mbembe 2001: 176)

According to J.S. Mill, from the British perspective, the more you go to the south, the more people are backward, passive, sensuous, and excitable, i.e. the character deteriorates as one moves south and east (Pitts 2005: 136). The white world was predestined to dominate human culture and is working for the continued subordination of the blacks as nothing ever happened in the world of any importance that could not or should not be labeled "white" (Du Bois 2007: 70, 73). The colonizers took the freedom to exterminate the blacks to make the land habitable (Conrad 2008: 19). The taking away of the land and the bodies of those with different complexion or slightly flatter noses was not anything criminal or inhumane for the Christian colonizer (Conrad 2008: 120, 125). They were just insolent black limbs (Conrad 2008: 139, 178).

## The Postcolonial Man and the Lasting Impacts of Slavery and Colonization

The denial of African humanity and the civilizing mission we saw in the previous two sections have multifaceted impacts on Africans. The dehumanizing process has been transferred from generation to generation. Often it happens that in the West, Africans feel the pressure to perform error-free lest it could be interpreted that the error was because they are blacks. For example, when Gen Charles Brown Jr a history-making fighter pilot became the new Joint Chiefs of Staff chairman in 2023, he recalled the supervisors had expected less of him as an African American.

It is extensively researched that the historical happenings (slavery and colonization) have lasting impacts on Africans, not only on those who were enslaved and colonized but also on the generations that followed. Colonization, slavery, and racism have deep-going impacts on the victims. A psychological study by Forgiarini et al. shows that white observers reacted to the pain suffered by black people significantly less than to the pain of white people. This was most probably the result of personal prejudices and, more generally, racist attitudes toward the outgroup member (Forgiarini et al. 2011). There are cumulative and long-term effects of racial discrimination and other racial oppression leading to psychological injuries and scars, trauma, mental health problems, and energy loss for individuals, families, and communities who often are too tired to participate in social activities (Thompson-Miller and Feagin 2007: 111). As Thompson-Miller and Feagin argue, "the pain of, and the developed coping strategies for, racial discrimination are passed across several generations of those racially oppressed" (Thompson-Miller and Feagin 2007: 106).

> I don't think white people, generally, understand the full meaning of racist discriminatory behaviors directed toward Americans of African descent. They seem to see each act of discrimination or any act of violence as an "isolated" event. As a result, most white Americans cannot understand the strong reaction manifested by blacks when such events occur… They forget that in most cases, we live lives of quiet desperation generated by a litany of daily large and small events that, whether or not by design, remind us of our "place" in American society. [Whites] ignore the personal context of the stimulus. That is, they deny the historical impact that a negative act may have on an individual. "Nigger" to a white may simply be an epithet that should be ignored. To most blacks, the term brings into sharp and current focus all kinds of acts of racism—murder, rape, torture, denial of constitutional rights, insults, limited opportunity structure, economic problems, unequal justice under the law, and a myriad of … other racist and discriminatory acts that occur daily in the lives of most Americans of African descent. (Feagin and Sikes 1994: 23–24)

Some studies show that racism and discrimination are related to symptoms of hypertension (African Africans develop high blood pressure at an earlier age than whites and Hispanics) and post-traumatic stress disorder by contributing to events and living conditions in which poverty, crime, and violence are persistent sources of stress. The victims of racism internalize stereotypes and denigrating images in ways that adversely affect their feelings of self-worth, and undermine their physical well-being and social and psychological functioning (Omuziligbo n.y.). Studies show a relationship between chronic exposure to racism and poorer indexes of psychological and somatic health among African Americans, several stress-related diseases, including hypertension, coronary heart disease, cancer, lung ailments, accidental injuries, and cirrhosis of the liver, and psychologically, an increased levels of depression as a result of the more than 400 years of vicious and unrelenting assaults to the image, psyche, spirit, humanity, personhood, and physical body of African Americans who have been subjected to the lowest life expectancy of all U.S. racial groups (Utsey and Payne 2000: 57–59).

Intergenerational transmission of race-related traumas is often communicated and influential across networks of family and friends, even across communities. Internalization and teaching children to be obedient to Whites, distrust of Whites (for personal and family protection—distrust as a survival strategy—), and even adoration of Whites are intergenerationally transmitted.

In their desperation, the colonized and enslaved felt forsaken and outcast not only by man but also by God. According to Du Bois, the world of slavery and colonization does not yield the African man's true self-consciousness. The African man has a "double consciousness, looking at one's self through the eyes of others, of measuring one's soul by the tape of a world that looks on in amused contempt and pity. One ever feels his twoness—an American, a Negro; two souls, two thoughts, two unreconciled strivings; two warring ideals in one dark body, whose dogged strength alone keeps it from being torn asunder" (Du Bois 1903: 9). The African has been trying to "merge his double self into a better and true self, to be a Negro and an American, not to be spit upon by his fellows, without having the doors of opportunity closed roughly in his face; that the legislatures stop to reducing the Negroes to serfdom, to make them the slaves of the state (Du Bois 1903: 9). The racist slaveholder believed that the Negro failed because of his wrong education. The legacy of the sufferings of the black slaves, the clank of their chained feet, the wails of the motherless, and the muttered curses of the wretched was trivialized (Du Bois 1903: 88).

Edward Said argues that

> To have been colonized was a fate with lasting, indeed grotesquely unfair results, especially after national independence had been achieved. Poverty,

> dependency, underdevelopment, various pathologies of power and corruption, plus of course notable achievements in war, literacy, and economic development: this mix of characteristics designated the colonized people who had freed themselves on one level but who remained victims of their past on another. (Said 1989: 207)

Degruy shows that through genetic memory and environmental influences on our genes, we carry in us today the social information that was transmitted hundreds of years ago and behave accordingly. Degruy gives various examples of such behavior. For example, a poor black person upon seeing two rich people (one white and one black) becomes envious of the black person because he reminds him of his unsuccess and inferiority. If there is a piece of news on criminal offenses black people hope that it was not committed by a black person. "African Americans feel as though one bad act committed by a black person reflects upon all black people" (Degruy 2017: 111). From the Posttraumatic Slave Syndrome perspective, she explains that "… there is an assumption that the repercussions for this stranger's act would somehow negatively impact how we, who share the same skin color and/or ethnicity, will be perceived or treated by whites." (Degruy 2017: 111).

The Africans' desire to be recognized as equal human beings and to prove to the colonizer that Africans are not inferior was unstoppable. This is in my view one of the psychological liberation strategies of the Africans. Now blacks are physically free, but mentally the enslavement and colonization have long-lasting impacts. African authors and Statesmen like Léopold Sédar Senghor, Kwame Nkrumah, Julius Nyerere, etc. wrote to prove to the colonizer the peculiarity of African culture and literature. The colonizer proved to be technologically advanced and militarily powerful. However, the desire of the colonized to imitate the colonizer materially and technologically on the one hand, and the difficulty to achieve it on the other started to ignite a love-hatred relationship of fascination and frustration in the colonized. This led even at times to believe in one's own inferiority as he was made to feel inferior (Fanon 2008: 127). The acceptance and internalization of the indoctrinated inferiority is also a survival strategy in the face of unjust institutions and socio-political and economic structures (Degruy 2017: 9). Blacks are overrepresented in the criminal justice system and underrepresented in the university system (Degruy 2017: 14) Black parents are hyper-vigilant when their children are not home because they cannot be sure whether they come back safe as a white racist can take a gun and just kills blacks. The enslaved and colonized accepted and internalized the inculcated inferiority and the role assigned to them as sub-humans and uncivilized (Memmi 1967: 87–89). As Anthony Appiah in Foreword to Fanon's Black Skin White Mask suggests, black African's self-contempt manifests itself in a neurotic refusal to face up to the fact of one's own blackness (Fanon 2008: vix). Fanon suggests that a black is not a man. Blacks have a series of affective disorders from which they have to be extricated since they are locked in their blackness and internalized inferiority.

I see the postcolonial migration to the West in this context. Before it is a physical movement, migration is a mindset or mental predisposition that tries to break through the invisible and intergenerationally transmitted wall of degrading separation and a desire for belonging.

Africans were made to learn and internalize the fact that even hundreds of them had to tremble before a single white man (James 2022: xix; Fanon 2008: xi). The African novelist, Abdulrazak Gurnah, in his "*Memory of Departure*" describes how Africans adore the Whites and detest their likes, the Indians. "We do not throw the white man out. We are too afraid of him. We want him to like us. African Art, African History… we plead with them to think of us as human beings. But the Indian we persecute and chase out." (Gurnah 2021: 86).

The assertion of African identity and the desire for liberation and equality as human beings are decisive contributors to the postcolonial African migration to the West. The European collective violence against innocent Africans created a hierarchical relationship of a superior group of the slaveholder, colonizer, and evangelizer on the one hand, and an inferior group of the enslaved, colonized, and evangelized human beings on the other (Pitts 2005: 14). As Frantz Fanon describes, migration or visits to metropoles like Paris was an expression of fascination for the colonizer and frustration for one's limitations vis-à-vis the colonizer. Therefore, I see the postcolonial migration from Africa to the West primarily as a mental migration and only secondarily a spatial movement. It is both a physical and mental movement toward belonging and a desire for being.

## Postcolonial African Imitation of the West as the Desire for Liberation

One of the liberation strategies of the postcolonial African man is the attempt to prove to the colonizer that Africans have similar cultural, intellectual, or psychological qualities to the colonizer. They do this, especially by trying to imitate the colonizer. After having been denigrated for centuries, Africans attempt to prove that they are not inferior to the colonizer.

The negated (existentially denied) and colonized elites began to prove to the colonizer that they could achieve what the colonizer has achieved, that they possess culture and philosophy. They craved an elevated system of thought (Bell 2002: 23). The colonization of mind, as Frantz Fanon says, inculcated in the colonized, that the colonized is restless unless she/he has proven to the colonizer that she/he can achieve all that the colonizer has achieved materially and immaterially. Materially, the economic and industrial achievements of the colonizer were the ultimate goal of the colonized. Immaterially, the philosophy, culture, and political system of the colonizer are the benchmark of the colonized to assess the level of his humanness. Fanon says, "The Europeans have, as it is believed, attained their development goal by their own efforts. Therefore, we have to prove to the world and ourselves that we are capable of achieving the same thing" (Fanon 1966: 78). By trying to defend the blacks and their culture against the colonial prejudices and discriminations, Senghor (former president of Senegal) says the following: "Land and everything on it is a collective property which is divided under the members of the family [in Africa]. Everyone is insured, she/he has the minimum to live according to the necessities. When the harvest is ripe, it belongs to all" (Senghor 1967: 18).

The eternal attempt of the colonized is to demonstrate to the colonizer his/her civilization and culture; this means liberating oneself from immaterial colonization. The colonized want to negate the negation. In the eyes of the colonizer, the colonized African is not only black, but also a personified anti-thesis of the white world; it is not only a despised being, but also the *not-being* of the *worth-being*.

> I demand that one has to take into consideration my negating act as far as I look for something other than life; as far as I fight for the birth of a human world, for a world of mutual recognition. The one who does not recognize me opposes me. In a wild struggle, I accept the shattering death, the

irreversible dissolution, but also the possibility of the impossible. (Fanon 1991: 139)

Nkrumah suggests raising African *consciencism*. Consciencism is for Nkrumah, intellectual dispositions and intellectual force that enables blacks to digest the Western and the Islamic and the Euro-Christian elements in Africa to make them fit into the African personality (Nkrumah 1978: 79).

To survive means to demonstrate one's own identity, to defend the non-*destroyedness* and the *indestructibility*, and the capacity to prove and defend the *black's whiteness*. "It is contended that the black has invented nothing new in the areas of religion; no Dogma or morals, but just a small religiosity. But if one considers that, the fundamental point lies not in this ridiculous assertion, but instead in the matter itself. Therefore, I would like to analyze the dogma and the moral of the blacks ..." (Senghor 1967: 13). Further, Senghor tries to prove black's equality and even superiority: "We would like to illustrate those elements, which the black families have to cherish so that they can be in line with the modern humanism by enriching it at the same time ... If the African succeeds in this transition period to protect themselves from any kind of external pollution and deformation, there is no need to worry about the African future" (Senghor 1967: 16). The colonizer has maintained that the black man does not have culture. In response, the colonized man begins to explain the possible similarities between his culture and that of his colonizer or justifies the difference and emphasizes special elements of the African culture. Senghor's words try to underline the specialty of the black culture: "It is a practical but not a utilitarian art, and it is in this original sense classical. Particularly it is a spiritual art—some have unjustly called it idealistic and intellectual—since it is religious. The central function of [African] sculptors consists of presenting the dead ancestors and spirits in sculptures, which are at the same time symbols of their spirits and their place of dwelling." (Senghor 1967: 23).

In order to emphasize the peculiarity of African music, Senghor criticizes the music of the colonizer as lacking stimulation (Senghor 1967: 25). According to him African music can enrich the "deficient" European music. "It gives the lacking flavor to the impoverished Western music that is based on narrow rules... Because of the purity, power, and nobility of their sound, instruments like Xylophone are predestined to convey the style of the blacks... the contribution of the black consists of recreating with other peoples the unity of man and world; reconciling the body and spirit, human beings among each other and the stone with God" (Senghor 1967: 26–28). To overcome the inculcated nothingness, Senghor underlines the otherness: "the white reason is utilitarian and analytical, whereas the black reason is participative and intuitive" (Senghor 1967: 157). In his desire to prove to the colonizer the peculiarity of his cultural values, the colonized even constructs them, such as the Africanity:

> The capitalist attitude of mind which was introduced into Africa with the coming of colonialism is totally foreign to our own way of thinking. In the old days, the African had never aspired to the possession of personal wealth for the purpose of dominating any of his peers. She/he had never had laborers or 'factory hands' to do his work for him. But then along came the foreign capitalists. They were wealthy. They were powerful. And the Africans naturally started wanting to be wealthy too... The foundation, and the objective, of African socialism, is the extended family. The true African socialist does not look at one class of men as his brethren and another as his natural enemies. She/he does not form an alliance with the 'brethren' for the extermination of the 'non-brethren'. She/he rather regards all men as his brethren—as members of his forever extending family. (Nyerere 1968: 6; 11)

As Clapham correctly observed, "the rhetoric of unity papered over the extreme reluctance of any but a very small minority of African rulers to sacrifice any of their power in the interests of any continental grouping" (Clapham 1996: 106). Moreover, this unity is only on the political elite level, not on the societal level. In many cases, African political elites are rather rivals destabilizing each other, not allies, especially if the West is not involved in a conflict. Regional interstate mutual destabilizations in the Horn of Africa, as well as Western Africa, are good examples. "Africanity" is just a construct. In case of disputes between an African state and the West, African states pull together as Mugabe was supported by almost all Africans when he was isolated by the West.

Right after the independence statesmen like President Kenneth Kaunda of Zambia, Sekou Touré of Guinea, and Kwame Nkrumah Ghana attempted to show to the outside the peculiarity of the African culture that man is the center of everything, that the individual must put the community and society before oneself, the African socialism of extended family instead of personal wealth, etc. (Doppelfeld 1994). All these concepts emerged to prove to the colonizer (who denied African capability for cultural values) that Africa has also culture and values.

To prove African philosophy, Onyewuenyi maintains that Greek philosophy has a strong Egyptian influence.

> Little do some of us know that the first woman philosopher, Hypathia, was from Alexandria and was murdered by Christians. Names like Saint Augustine, Origen, Cyril, Tertullian are not unfamiliar to you; they are black Africans. More pertinent to our subject is the fact what today we call Greek or Western Philosophy is copied from indigenous African philosophy of the 'mystery system'. All the values of the mystery system were adopted by the Greeks and Ionians who came to Egypt to study, or studied elsewhere under Egyptian-trained teachers. These included Herodotus, Socrates, Hippocrates, Anaxagoras, Plato, Aristotle and others ... (Onyewuenyi 1998: 250)

However, Onyewuenyi forgets the fact that Egyptians do not consider themselves Africans because to be an African means to be black, which is inferior and slave.

When it seems difficult to create a similarity between Western and African cultures, the colonized intellect becomes proud of his constructed peculiarity, the difference. "We must construct an African Philosophy with categories that are typically African and where these categories do not correspond with those of either East or West; we dare to be different" (Ruch 1998: 267).

"African personality", "Negritude", and "Pan-Africanism" attempted to prove to the colonizer that Africa has culture and philosophy. There are two ways to get liberty from the legacies of degrading slavery and colonialism: either by assimilating oneself with the colonizer and adapting his values, which is the process of ingratiation; or—where this ingratiation does not seem to be possible—by an artificial construction of own superior identity by way of dissociation. Pan-Africanism had the second objective: "the homeland of African persons of African origin, solidarity among men of African descent, belief in an African personality, rehabilitation of Africa's past, pride in African culture, Africa for Africans in church and state, hope for a united and glorious future Africa" (Esedebe 1982: 3).

Some white scientists and racists began to substantiate the impossibility of ingratiation of the blacks and their innate inferiority because of two reasons: evolution and teleology. In the first case, some maintained that in the evolutionary process of the human being, culture, and civilization, modern history as the last (highest) step was inherited from the Greeks by the Romans and from the Romans by the Northern Europeans. Since the latter not only received these Greek and Roman civilizations but also created modern history, any race that did not participate in these evolutionary acts is less human. Whereas, the teleological school maintains that "God deliberately made men unequal. He gave intelligence to the whites to enable them to direct the activities of others wisely. To the non-whites, this usually meant blacks, he gave strong backs fortified with a weak mind and an obedient temper so that they might labor effectively under the supervision of the white masters" (Esedebe 1982: 19). To be means to be recognized. The dehumanized Africans hope for this recognition through imitation of the former colonizer. It is not through detachment from the former colonizer that the colonized are seeking liberty but through imitation. The best way to imitate means to live together with the colonizer who is a persecutor as well as "liberator" at the same time, a monster, a mythical creature.

## Conclusion

Africans were considered forsaken by God and the "civilized" world. They were considered inferior by nature and denied humanity. Hence, for the slaveholders and colonizers, their subjugation was legitimate and providential. However, blacks negated this from the very beginning of the contact with the Europeans. In the last section of this chapter, we saw some strategies through which they tried to negate their negation. As the colonizers were obsessed with African inferiority, Africans are obsessed with paradoxically denied desire to imitate the West: an abomination of the West as well as slavish adoration of it at the same time.

We the postcolonial Africans hate the dehumanizing West but are at the same time fascinated by its aesthetic, technological, military, economic, and cultural power that has become the global standard. Liberty for the postcolonial man means overcoming the remnants of dehumanization by becoming like the colonizer and slaveholder as much as possible through living together with him in the West and organizing his lifestyle according to the colonizer.

Against the background of dehumanizing legacy, postcolonial migration from Africa to the West is a denied desire to overcome the legacies of the dehumanization that we saw in this chapter. Successful postcolonial

African migrants to the West consider themselves to be free human beings contrary to those who have not done it. They enjoy recognition and being as we shall see in the next chapter.

## Literature


Akinbode, Ayomide (2021): The forgotten Arab slave trade, https://www.thehistoryville.com/arab-slave-trade/, 29.05.2023.

Anderson, Carol (2021): The Second: Race and guns in a fatally unequal America, Bloomsbury: London.

Bakewell, Oliver (2008): 'Keeping Them in Their Place': the ambivalent relationship between development and migration in Africa, *Third World Quarterly*, 29:7, 1341–1358.

Bauman, Zygmunt (2016): Strangers at our door, Polity: Cambridge.

Bedford, Nancy (2018): The most burning of Lavas: The Bible in Latin America, in Colonialism and the bible: contemporary reflections from the global south, Lexington books: Lanham, pp. 213–232.

Bell, Richard H. 2002. *Understanding African Philosophy: A Cross-cultural Approach to Classical and Contemporary Issues.* New York: Routledge.

Bhambra, G. K. (2007): Rethinking Modernity: Postcolonialism and the Sociological Immagination, Palgrave Macmillan: Basingstoke.

Bitterli, U. 1991. *Die Wilden und die Zivilisierten: Grundzüge einer Geistes- und Kulturgeschichte der Europäisch-Überseeischen Begegnung.* München: Beck.

Botwe-Asamoah, K. 2005. *Kwame Nkrumah's Politico-Cultural Thought and Policies: An African-Centred Paradigm for the Second Phase of the African Revolution.* New York: Routledge.

CARICOM (2023): 10-Point Reparation Plan, https://caricomreparations.org/caricom/caricoms-10-point-reparation-plan/, 07.07.2023.

Césaire, Aimé (2000): Discourse on Colonialism, Monthly Review Press: New York.

Clapham, Christopher 1996. *Africa and the International System: The Politics of State Survival.* Cambridge: Cambridge University Press.

Conklin, Alice (1997): A mission to civilize: The republican idea of empire in France and West Africa, 1895–1930, Stanford University Press: Stanford.

Conrad, Joseph (2008): Heart of Darkness and Other Tales, Oxford University Press: Oxford.

Degruy, Joy (2017): Post-Traumatic Slave Syndrome: America's Legacy of Enduring Injury and Healing, DeGruy.

Douglass, Frederick (1881): The Color Line, *The North American Review*, 132, 295, pp. 567–577.

Du Bois, W.E.B., (1903): The Souls of Black Folk, Gröls Classics.

Du Bois, W.E.B. (2007): Dusk of Dawn, Oxford University Press: Oxford.

Dugard, Martin (2003): Into Africa: the epic adventures of Stanley and Livingstone, Broadway Books: New York.
Eco, Umberto (2013): Inventing the enemy, Vintage: London.
Elias, Norbert 1978. *The Civilising Process: The History of Manners*. Oxford: Blackwell.
Esedebe, Olisanwuche P. 1982. *Pan-Africanism: The Idea and Movement 1776–1963*. Washington, DC: Howard University Press.
Fanon, Frantz (2001): The wretched of the earth, Penguin Books: London.
Fanon, Frantz (2008): Black Skin, White Masks, Grove Press: New York.
Fanon, Frantz 1966. *Die Verdammten dieser Erde*. Frankfurt am Main: Suhrkamp.
Fanon, Frantz 1991. *Black Skins, White Masks*, London: Pluto Press.
Feagin, J. R., & Sikes, M. P. (1994). Living with racism. Boston: Beacon.
Forgiarini, Matteo et al. (2011): Racism and the empathy for pain on our skin, *Frontiers in Psychology*, 2, 108, pp. 1–7, https://doi.org/10.3389/fpsyg.2011.00108.
Gakunzi, David (2018): The Arab-Muslim Slave Trade: Lifting the Taboo, *Jewish Political Studies Review*, 29: 3/4, pp. 40–42.
Gandhi, Leela (1998): Postcolonial Theory: A critical introduction, Columbia University Press: New York.
Gershman, Sally (1976): Alexis de Tocqueville and Slavery, *French Historical Studies*, 9: 3, pp. 467–483.
Getachew, Adom (2019): Worldmaking after Empire: The rise and fall of self-determination, Princeton University Press: Princeton.
Gilli, Aldo (1979) *Daniel Comboni: the man and his message; a selection from the writings of Bishop Daniel Comboni (1831–1881)*, Bologna, Editrice Missionaria Italiana.
Girard, René (2016): Violence and the Sacred, Bloomsbury Revelations: London.
Girard, René (2019): Things hidden since the foundation of the world, Bloomsbury Revelations: London.
Goudsblom, J. 1998. *The Norbert Elias Reader: A Biographical Selection*. Oxford: Blackwell.
Gurnah, Abdulrazak (2021): Memory of Departure, Bloomsbury, London.
Hegel, Georg W.F. (1975): Lectures on the philosophy of world history: Introduction, translated by H.B. Nisbet; Introduction by Duncan Forbes, Cambridge University Press: Cambridge.
Herder, Johann Gottfried von (2002): Philosophical Writings. Translated and ed. by Michael N. Forster. Cambridge: Cambridge University Press.
Hetherwick, A. 1939. *The Romance of Blantyre: How Livingstone's dream came true*. Dunfermline: Lasodine Press.
Iliffe, John 1997. *Geschichte Afrikas*. München: Beck.
James, C.L.R. (2022): The black Jacobins: Toussaint Louverture and the San Domingo revolution, Penguin: Dublin.

Jeffers, Honoreé Fanonne (2022): The love songs of W. E. B. Du Bois, 4th State: London.
Kingsley, Patrick (2017): The new Odyssey: The story of Europe's refugee crisis, The Guardian: London.
Lukasik, Gail (2021): White like her: my family's story of race and racial passing, Skyhorse Publishing: New York.
Mannoni, O. (1990): Prospero and Caliban: The psychology of colonization, Ann Arbor: Michigan University Press.
Maquet, J. 1971. *Power and Society in Africa*. London: World University Library.
Mayblin, Lucy (2021): Postcolonial perspectives on migration governance, in Emma Carmel et al. Handbook on the Governance and Politics of Migration, Edward Elgar: Cheltenham, pp. 25–35.
Meredith, M. 2005. The Fate of Africa: a History of Fifty years of Independence. New York: PublicAffairs.
Mbembe, Achille (2001): On the Postcolony, University of California Press: Berkeley.
Mbembe, Achille (2003): Necropolitics, *Public Culture*, 15, 1, pp. 11–40.
Mbembe, Achille (2017): Critique of Black Reason, Duke University Press: Durham.
Mbembe, Achille (2021): Out of the dark night: Essays on decolonization, Columbia University Press, New York.
Mbuwayesango, Dora Rudo (2018): The bible as a tool of colonization: The Zimbabwean context, in Colonialism and the bible: contemporary reflections from the global south, Lexington books: Lanham, pp. 31–42.
Memmi, Albert (1967): The colonizer and the colonized, Boston: Beacon Press.
Mill, John Stuart (1859 [2016]): On Liberty, Enhanced Media: Los Angels.
Ngwa, Kenneth (2018): Postwar hermeneutics: bible and colony-related necropolitics, in Colonialism and the bible: contemporary reflections from the global south, Lexington books: Lanham, pp. 43–74.
Nkrumah, Kwame 1978. *Consciencism: Philsophy and Ideology for De-colonization*. London: PANAF.
Nogueira, Simone Gibran (2013): Ideology of White Racial Supremacy: Colonization and De-Colonization Processes, *Psicologia & Sociedade*, 25, pp. 23–32.
Nyerere, Julius K. 1968. *Ujamaa: Essays on African Socialism*. Dar es Salam.
Online etymology dictionary (2023): Colonize, Online Etymology Dictionary, https://www.etymonline.com/search?q=colonize.
Painter, Nell Irvin (2010): The History of white people, Norton and Company: New York.
Palaver, Wolfgang (2003): René Girards mimetische Theorie, LIT: Münster.
Pitts, Jennifer (2005): A Turn to empire: the rise of imperial liberalism in Britain and France, Princeton University Press: Princeton.

Pope Alexander VI. (1493): Demarcation Bull Granting Spain Possession of Lands Discovered by Columbus, Rome, May 4, 1493.
Said, Edward (1989): Representing the Colonized: Anthropology's Interlocutors, *Critical Inquiry*, 15, 2, pp. 205–225.
Senghor, Léopold S.1967. *Négritude und Humanismus.* Düsseldorf.
Shillington, K. 2005. *History of Africa.* 2nd rev. ed., Oxford: Macmillan Education.
Silverman, Max 1999: *Facing Postmodernity.* London: Routledge.
Sindima, H. J. 1998. *Religious and Political Ethics in Africa: A moral inquiry.* Westport: Conn., Greenwood Press.
Soyinka, Wole (2012): Of Africa, Yale University Press: New Haven.
Speke, John Hanning (1864): THE DISCOVERY OF THE SOURCE OF THE NILE, https://www.gutenberg.org/files/3284/3284-h/3284-h.htm#link2HCH0014, 04.04.2023S.
Thiong'o, Ngugi wa (2005): Decolonizing the mind: the politics of language in African literature, James Curry: Nairobi.
Thompson-Miller, Ruth and Feagin, Joe R. (2007): Continuing Injuries of Racism: Counseling in a Racist Context, *The Counseling Psychologist*, 35, 1, pp. 106–115.
UN (2015): Slave Trade: International Day of Remembrance of the Victims of Slavery and the Transatlantic Slave Trade, https://www.un.org/en/observances/decade-people-african-descent/slave-trade, 29.05.2023.
Utsey, Shawn O. and Payne, Yasser (2000): Psychological Impacts of Racism In a Clinical Versus Normal Sample of African American Men, *Journal of African American Men*, 5, 3, pp. 57–72.

CHAPTER 6

# The Postcolonial African Migration to the West as a Desire for Recognition

In Chap. 2 we discussed that migration is not always caused by poverty, climate change, or conflicts, which I summarily call *material* causes of migration. In this chapter, I focus on the *relational* dimension of migration. The underlying cause of migration from postcolonial Africa to the West is relational, which is the main focus of this book. In this chapter, I discuss current African migration as a manifestation of the desire for recognition and a means for achieving social status in the origin. Since not all postcolonial Africans can migrate to the West those who have migrated and are perceived successful bring the adored West to the origin. This gives them not only a special social status at the origin but also makes them almost white, as we shall see in the following chapter. The desired equality with the West and liberty from the legacies of slavery and colonization through imitation of the West, however, will remain an illusion. The postcolonial African man will remain the Dostoevskian *Underground Man*, as we shall see in this and the following two chapters.

The African desire for recognition has at least two underlying causes: African patriarchal and autocratic cultural structures and colonial legacy. Patriarchy not only keeps up structural violence against women but also puts pressure on men to be "real men" to enjoy social status. The achievement of recognition is believed to depend on the success of overcoming the white Western borders both physically and metaphorically, to imitate successfully the white world, and to bring it to black Africa. Secondly, it is


B. Gebrewold, *Postcolonial African Migration to the West*, Politics of Citizenship and Migration,
https://doi.org/10.1007/978-3-031-58568-5_6

widespread in Africa that colonial time was the cause of almost all postcolonial African problems.

The main focus of this chapter is especially the patriarchal and autocratic cultural structures, whereas the colonial legacies will be discussed in Chaps. 7 and 8 in detail. Often, those who "successfully" migrated to the West and their families back home are not only richer than those who want to migrate but cannot, but they're also "whiter", which increases their social value and their chance for recognition.

## Recognition as Belonging

As various studies and surveys show, African emigrants say that they move to Europe because they want to escape poverty, conflicts, and the impact of climate change, in search of greener pastures in the West. From 45,823 Afrobarometer interviews completed in 34 countries between September 2016 and September 2018, on average three-fourths of potential emigrants stated that they would consider migration to search for work (44%) or to escape poverty or economic hardship (29%) (Sanny et al. 2019). In most cases, according to this survey, poverty, climate change, or conflicts were not the primary causes of the decision to migrate.

From a mimetic perspective, the culturally and locally established role of the desire for recognition is an important tool of analysis. The postcolonial African migration to the West is part of the mimetic desire for belonging or recognition. Remittances are often used for conspicuous or ostentatious consumption in the home areas, such as purchasing televisions or cars. However, migration is not only a means to accumulate material goods. Postcolonial African migration is also a means to be respected and recognized. A migrant is someone who closes the gap between the devalued origin and the idealized destination. The migrant brings the ideal Other at the destination nearer to the incomplete and longing Self at origin, which desires to become a better being. Migration is a very mimetic process because once the first one has dared to overcome the Western borders others follow like a herd in the Great Migration in Tanzania and Kenya every year despite hunger, thirst, exhaustion, and deadly crocodiles lurking in the waters to kill the wildebeests. Once the first animal entered the water others followed. The animals know it is deadly but still, they do it. The important thing is that the first animal enters the water. The rest of the animals imitate the first animal. The more animals jump in, the more animals dare to follow. Whether crocodiles and other predators have killed other animals will be ignored. What counts now is only imitation.

Similarly, in the postcolonial African migration, the machinery of migration starts to run faster and faster not always because people become suddenly poorer than before, or not only because of the increasing economic gains through remittances, but also, and probably more importantly because of the mimetic dynamic of imitation, a desire for belonging, and recognition as a better human being, as someone who has managed to overcome the physical and metaphorical border between the white West and black Africa, the border between *Being* and *Nothingness*. Once this idea has entered the migrants' minds, they are not always guided by rationality but instead by mimesis. The machinery of the desire to belong and to be recognized is at work.

In the mimetic mechanism, the most important thing is someone takes the first step. Others follow him like animals because they would like to belong to the group (Girard 2023: 176). The local pressure to be recognized is the factor pushing them from behind. The preventive information about the dangers on the way to the West and socio-economic challenges at the destination will be completely disregarded. The mimetic desire for recognition and facts-based rationality are mutually exclusive. Disinformation campaigns to discourage emigration do not avail much. Like the herd mentioned above, people know that the Mediterranean, the Sahara Desert, and the Atlantic are dangerous, but still, they start the journey because the desire for being and recognition is dependent on this only card. In patriarchal and neo-patrimonial African societies, men want to be recognized as real men; women want to be recognized as wives of real men; the "big men" want to be recognized as important chiefs (Van de Walle 2001). As Mills says,

> ... neo-patrimonial big man chieftain styles of rule, dispensing favors and using all manner of tools to bolster their rule, from traditional governance structures to kinship ties and less palpable aspects, including witchcraft and the church. The system many African leaders have preferred thrives on corruption and nepotism. (Mills 2010: 15)

Social recognition or esteem is a key component of one's identity in the African patriarchal mindset. African political elites love the one-man show, which is rooted in patriarchy and patrimonialism like in Rwanda, Eritrea, Equatorial Guinea, Cameroon, Uganda... The African "benign dictators" believe that they have cultural authorization to rule indefinitely (Mills 2010: 206). When Mobutu was in power, his name was: Mobutu Sese Seko Kuku Ngbendu Wa Za Banga: meaning "*The all-powerful warrior who, because of his endurance and inflexible will to win, goes from conquest to conquest, leaving fire in his wake.*") The desire to be recognized as a "big man" (important man) is a widespread phenomenon. Speaking from a political-institutional perspective, any opposition to the authority of the "big man" is seen as questioning the legitimacy of the authority of the person recognized socio-culturally as an important person. Political opposition groups experience difficulty in many African countries to organize themselves and thrive in such patriarchal and autocratic African cultural structures. The educated, the rich, the politicians, those who visit the countryside coming from big cities (especially capital cities), and above all migrants who live in the West and go home on vacation enjoy almost a divine adoration. This puts enormous pressure on especially young men to migrate. One's being and identity are dependent on it.

Identity is about our understanding of who we are. As Taylor argues, our identity is partly shaped by recognition or misrecognition, i.e. others mirroring back demeaning, inferiorizing, depreciatory, or contemptible picture of the subject (Taylor 1994: 25, 66). This power relation can result in internalized self-depreciation and acceptance of oppression or inferior and demeaning images as normal. These subjects believe in their inferiority and uncivilizedness. As Taylor puts it, "We define our identity always in dialogue with, sometimes in struggle against, the things our significant others want to see in us" (Taylor 1994: 32–33). Therefore, our identity is always relational and is based on recognition by others. As Taylor suggests, it depends on my dialogical relationship with others. The desire for recognition is a complex human need for acceptance and belonging (Rockefeller 1994: 97) Therefore, identity is vulnerable because it is dependent on recognition, which can be given or withheld by significant others (Taylor 1994: 36). However, the more we desire for recognition, the more we imitate, and the more we imitate our model, the more we hate him because he is the reminder of our incompleteness. I think this is what Nietzsche had in mind when he said, "My contempt and my desire increase together; the higher I climb, the more do I despise him who climbs. What do I want in the heights?" (Nietzsche 2003: 70). We search for our being through imitation from which we except our recognition. There begins the problem that Nietzsche highlights: "One man runs to his neighbor because he is looking for himself, and another because he wants to lose himself. Your bad love for yourself makes solitude a prison to you" (Nietzsche 2003: 87).

Rousseau has shown that one is dependent on others for his being, as his esteem and being are dependent on recognition or other's good opinion. Taylor argues that "the other-dependent person is a slave to opinion" (Taylor 1994: 45). Hence, all human beings are slaves, but some are made even more slaves as their humanity was completely denied. However, the human objective of liberty is not to overcome the other-dependence completely but instead to transform and transcend the manner of dependence or asymmetric other-dependence and establish equality, i.e. reciprocal recognition among equals, as Rousseau suggests (Taylor 1994: 50).

## Imitation, Migration, and the Desire for Recognition

Migration transforms society. Social transformation is causing more migration. New social values enabled by migrants are gradually undermining traditional values, gender roles, and social status. The article by Carling and Collins is one of the few publications that discuss drivers of migration in relation to aspiration and desire. Carling and Collins discuss aspiration and desire related to the subject's relation to migration possibilities (migration aspirations); the subject's relation to potential transformations in the context of migration; and another's relation to mobile or potentially mobile subjects (Carling and Collins 2018: 915). Individual attitudes toward migration reflect the values in the social context or the so-called cultures of migration, i.e. migratory achievements of others in the neighborhoods and regions and the resulting migration aspirations (Carling and Collins 2018: 916). In the same issue of the journal, Collins makes an important contribution to migration understanding by focusing on desire. He reconceptualizes migration as an ongoing process of becoming and transformation of migrants' subjectivity or migrants' own desire for becoming through migration (Collins 2018). As a study by Conrad Suso (2019) on The Gambian migration shows, actual or perceived economic motivations abroad, remittances from Europe sent home to help build family homes and subsequent prestige, social pressure to migrate (from friends, relatives, and society at large), and the possibilities to fill a social role, etc. are strong motivational factors in migration decisions and migration aspirations (Conrad Suso 2019: 118–122). This is what Jackson

(2008) analyzes as the imagination of one's social existence in images, anecdotes, and stories. Since it is about recognition, as long as the expectations in the family and the neighborhood are high, even the substantial risk of death on the way and at the destination cannot deter the would-be migrants irrespective of incorrect information or challenges on the way (Mbaye 2013). However, the question is, would there ever be "correct information" in a mimetic migration"?

In migration research, the forced migration approach is predominant. The "force dimension" focuses on poverty, conflicts, and climate change. As Erdal and Oeppen suggest, forced migration and voluntary migration are a continuum of experience, not a dichotomy. They argue that there is a considerable degree of volition in most migrants' experiences and migration decisions. "Labelling as 'forced' (or not) matters to migrants and states when asylum status is on the line" (Erdal and Oeppen 2018). But the concept of "force" entails different contents depending on the perspectives of the respective speaker. Asylum officers understand force from a legal perspective, i.e. physical safety. However, what "forces" mimetic migrants is the desire for recognition.

Economic, political, and environmental causes of migration have been extensively researched based on empirical data, as for example, the volume by Kefale and Gebresenbet (2021) shows. In this volume, Fana contends that what matters most for migration is not the economic plight of the migrants but the networks with previous migrants (Kefale and Gebresenbet 2021: 17–18). However, I argue in this chapter that even networks are facilitating factors but not the causes of migration. The desire for migration is even subconsciously preceding the abundance of networks that enable it.

There is some correlation between the "global ideas of the good life", the social-psychological and economic impact of remittances, and the transformation of the social position of youth in African societies. Remittances transform not only the economy or the society in an abstract way, they also transform the desires of the people. Youth are exposed to direct peer pressures, social stimulations to be recognized or respected, and transition to adulthood, fame, and status in the community for themselves, and for the glory of their family. The more successful people return the more migration takes place ending up in a culture of migration (Debalke 2021: 145–47).

Both economically as well as psychologically *to be* means to *migrate*. "In Hadiya [a region in southern Ethiopia], if you do not have a brother or a son in South Africa, you do not have a life" (Tufa et al. 2021: 52). Migration to the West, to the Arabian Peninsula, etc. has been perceived a guarantor of "good life". A good life means good income or wealth accumulation, "modern" houses, businesses, etc. However, this good life has an important social-psychological objective of social status, recognition, and respect. "The social recognition that the migrant obtains upon returning to the village justifies and sharpens his courage to face the wilderness, whatever the adversity maybe" (Dougnon 2013: 46). Migration enhances the youths' status in the village; migrants look smarter than those stayed back home; people look at them differently; returnees' accounts heighten the desire for new discoveries among those who stayed behind; returnees have got always something to say. "Those men and women who, unlike their peers, did not have the opportunity to migrate, consider something to be missing from their lives" (Hertrich and Lesclingand 2013: 181).

Migration is a way to achieve the desired Self which migrants want to become (Dom 2021: 100–101, 109; Debalke 2021: 141). "Success, in most areas, can be defined as having an iron sheet roof and fully furnished house, private cars, trucks, or shops. Social status and respect are also accorded to families and households with members working abroad" (Gezahegne 2021: 129; Adem 2021: 242). The study by Sarah Mahler on Salvadorian migrants in the US shows how they invent success to be recognized as successful in honor of themselves and their families back home (Mahler 1995). Unsuccessful returnees or those who do not send remittances are seen as a disgrace to themselves and their families (Gezahegne 2021: 130). Aspiration, desires, and the perception of the good life change based on what information one gets about the opportunities somewhere else, about the social status and recognition in the society they generate. As a result, to return home without success means failure and unmanliness, which puts significant pressure on those who want to return. As Dom suggests, rich young people as well as children as young as 10 years decide to migrate because they imitate those successful emigrants they, their relatives, or friends know (Dom 2021: 104; Debalke 2021: 154). As Dougnon suggests, migration is not the mere relation of individuals to property or earnings but the realization of youthful dreams of becoming a man by going through adventure, risks, and perils, and showing bravery, tenacity, and audacity. It is a kind of facing a rival or taking on a challenge. It is a social aspiration to become and to be by leaving the protection of one's own people (Dougnon 2013: 37–39, 47). Unsuccessful returning migrants lose face and will be disrespected and, therefore, they prefer to remain abroad and hide from their relatives and neighbors (Whitehouse 2013: 25). When they communicate with those back home, they misrepresent their reality at the destination, exaggerate successes, and play down the hardships they are in, which leads to misinformation about and idealization of the destination. However, even if they would tell those back home about their real situation, this would be filtered out by the recipients who do not want to believe even an accurate message about the hardships. Those intending to migrate think that migrants are sending wrong information about the risks on the way, and their lives, are hiding their wealth from those back home, are afraid of competition for jobs, are self-fish and want to keep the opportunities for themselves (Whitehouse 2013: 25–26). Therefore, would-be migrants are eclectic in the collection of information. That is, they disregard or play down any negative information. Because of this "positive information bias" negative information campaigns would have a rather limited impact as migration deterrent. Therefore, what they look for is not necessarily a piece of accurate information, but instead the information they would like to hear from sources they trust (Danish Institute for International Studies 2021).

Van Hear et al. (2018) suggest that in theorizing migration drivers, it is useful to distinguish between predisposing (economic disparities between migrants' place of origin and place of destination), proximate (economic and political emergency situations), precipitating (factors that trigger departure), and mediating (enabling, facilitating, constraining, accelerating or consolidating) drivers. However, they underline rather the potential of proximate and mediating drivers rather than the other two. The drivers are all in all economic and political (security). However, in my view this approach, though very important and helpful it is, overlooks two important issues: firstly, the predisposing issues are underrated; secondly, these predisposing drivers focus only on political and economic (*material*) causes and ignore the historical and socio-psychological (*relational*) factors.

Schöfelberger et al. (2020), Collins (2018), Collins and Carling (2018), etc. study migration from the perspective of "aspirations". For example, Schöfberger et al. understand migration desires and aspirations as individual preferences to emigrate independently of any possible limitation to do so (Schöfelberger et al. 2020: 89). Similarly, Bleck (2020) studied migration aspirations of returnees and youth in Mali. Aspirations, ambitions, friends and family members abroad, family expectations in the country of origin, information obtained from social media, that everybody was leaving, reports from former migrants and returnees, the large influence of emulation and peer-pressure, etc. are the major migration drivers, according to the Mixed Migration Centre findings (2019: 53). Often false and exaggeratedly positive information about the successes of migrants living in the West, comparatively huge earnings of earlier migrants, remittances, investment in real estates at home entices those back home to emulate them (Africa Polling Institute 2020: 31).

However, neither of these authors explains what the underlying reason behind these aspirations and desires is. Successful returnees are perceived as individuals who have gone through an experience that is different from those of the rest of the community and therefore they enjoy a better status within their community and are desired as marriageable (Gezahegne 2021: 131). The successful migrants receive a hero welcome upon returning home (Youngstedt 2013: 152).

Migration has become a rite of passage, which demonstrates one's manliness and makes desirable as a potential mate (Sintayehu 2021: 72; Dougnon 2013: 39). "Migration is thus seen as the transition for adolescents into womanhood and manhood with personal, economic and cultural autonomy, and in this way, it becomes a rite of passage" (Gezahegne 2021: 130). Migration can be analyzed as an initiatory rite that consists of traveling far from own society and country, proving the courage, moral strength, endurance, and capacity to assume and complete one's life project. The success of the enterprise guarantees several rights on return (access to the matrimonial system and to land), respect, and honor (Petit and Charbit 2018). Dougnon suggests that "migration is a combat against death for the beauties of the world. In this cultural milieu, not migrating is not living. One who does not leave the village stays at the bottom of the social ladder. He is seen as a hen that pecks the millet falling from the mortar of a lady while she pounds to prepare [food]" (Dougnon 2013: 40).

Similarly, the study by Mahler on Latin American migrants into the US shows that migration is a personal transformation process through which the migrants are reborn, assume a new identity through this rite of passage, challenge and change their old identities, and be prepared to be born again in the United States (Mahler 1995: 58). Even the dumping of passports and other personal items to erase the traces is also both literally and symbolically an attempt to get rid of the old identity and get a new one. One interviewee of Mahler described migration as a jump into emptiness hoping that there is a mattress or pillow below, which is the United States (Mahler 1995: 69). The new passage entails shedding off much of their past identities, taking new identities as debtors, passive chickens in the clutches of coyotes and predators, as illegals, wetbacks, or merely outlaws in the Promised Land where they thought the streets are dusted with euros or dollars ready to be swept into their arms after the baptism in the Mediterranean or the Rio Grande (Mahler 1995: 81). In the migrants' dreaming the whites West must be divine, the skyscrapers, the planes, the beautiful cars, the clean streets, and houses are also a divine creation in the land of marvels (Mahler 1995: 83–84).

However, once at the destination when they become aware of the disillusionment, they retake nostalgically their old identity without giving up the aspiration for the new identity. They become aware of the two irreconcilable identities in them, one real (the original) and the other (the aspired). After this disillusionment, they start to question their journey. No one questions their identity and humanity at home though poor. At their destination maybe they are not anymore poor but their humanity and identity are questioned, segregated, and suffer prejudice and hostility (Mahler 1995: 215).

From the perspective of the origin, migration increases one's status in society because working and living abroad is perceived as a privilege (Sintayehu 2021: 74). Mahler makes a very interesting analysis of Salvadoran migrants in the US. According to her findings, migrants counter their degraded social esteem in the US by enhancing their social status at home through remittances or sending Western-produced goods. The more the migrant's status is lowered in the United States, the more they feel compelled to buttress it at home. Mahler reports an incident where she was asked by the interviewees to take a picture of them in front of a new beautiful car—which did not belong to them—so that they could send their picture back home with this invented success symbolized by the car. They invent success through self-deception, they cannot entertain the prospect of failure (Mahler 1995: 86–89).

Sintayehu suggests that migrants far from home take jobs willingly that they usually would not engage in their areas of origin where these jobs are considered degrading menial jobs, or jobs traditionally considered and despised as women's jobs (Babou 2013: 241). Their recognition, self-esteem, and identity are primarily dependent on their families and neighbors, and not on those whites in the West who are considered superior and from whom one does not expect recognition. They would not tolerate disparagement by their likes than by the whites from whom the difference is much bigger. Mahler reports that the immigrants from Latin America to the US were prone to exploitation by their employers. What they hate the most is the exploitation by the immigrants or Latinos rather than by the "real" US Americans (Mahler 1995: 102).

Their relatives back home are willing to receive the money they send them home from these derided menial jobs far from their sight but they do not want to know what degrading jobs they are doing. Here, not only economic but also social-psychological factors play a role in the migration discourse (Sintayehu 2021: 78). Back home, what matters is not only how much one earns, but also what kind of job one has, i.e. how that job is regarded locally. At their destination, however, migrants are ready to take any job as long as it allows them to earn income where they are socio-culturally in oblivion. Social status is henceforth dependent on migration and remittances, the gendered conception of work is giving way to pragmatism, wealth, and income trump all traditional considerations (Babou 2013: 230). The social status will be enjoyed at home, not at the destination where the gap between the immigrants and the whites becomes insurmountable. "A man's value worth is measured by the value of the house he is able to build, the quality of life he is able to provide to his wife or wives and children, and his ability to support his parents and his in-laws" (Babou 2013: 235).

The migrant has to send back money so that he and his family can enjoy respect and recognition at the origin. At the destination, a low-status or a women's job is not a problem, whereas at the origin it is (Whitehouse 2013: 27; Babou 2013: 231). Their subjectivity is protected by their invisibility at the destination, whereas at the origin their subjectivity is exposed to evaluation: degradation or admiration depending on the wealth generated through migration. Invisibility protects from shame, but it also "implies a kind of devaluation of personhood or a denial" (Carter 2013: 65).

In such contexts, migrants see their dignity not in relation to their destination, but instead in relation to their origin. "They construct dignity as a place-bound attribute, intricately intertwined with local social hierarchies and genealogies. When someone enters a foreign land, their inherent worth is not recognized there; their dignity is not transferable from their homeland" (Whitehouse 2013: 27). "The empty-handed migrant exposes himself, on one hand to his own humiliation and on the other exposes his family to shame because they will not be able to present him to other villagers during the return ceremony... if he comes back empty-handed, his cousins mock him" (Dougnon 2013: 48).

Migration is the means to progress up the social hierarchy into adulthood or to be respected in society and be recognized as an important person and to enjoy respect and dignity (Gebresenbet 2021: 90; Dom 2021: 114; Adem 2021: 242). The risks on the way including death in the Sahara Desert or the Mediterranean are justified or "mystified" as the heavenly destiny that cannot be prevented, and it is a decision based on cost-benefit analysis (Gebresenbet 2021: 92; Dom 2021: 106). The symbols of success sent back home in the form of remittances, images of extravagant wedding ceremonies, messages, and images transmitted on social media platforms exaggerating the bright side of things not only provide them with high social status but also drive further migration from their areas (Estifanos 2021: 167). The West, its economic and technological achievements, the dollar, the euro, etc. are highly valued and coveted in the African migrants' origins. The migration researcher Sarah Mahler observed the same mindset in her work with migrants from Latin America to the US. She writes that the US is perceived as the land of milk and honey, the land of dollars whose buying power rarely diminishes vis-à-vis the home country's currency. As the study by Mahler shows, the more people migrated and sent dollars back home, the more people migrated from the neighborhoods with their heads filled with American dreams (Mahler 1995: 57).

As Adem argues, there is a selective remembering of successes and collective amnesia when it comes to risks from which future migrants draw lessons and make their decision to migrate and achieve social recognition (Adem 2021: 244). Though these factors have been increasingly playing a role in the migration decision they are not openly admitted by migrants and their families as causes (Sintayehu 2021: 75). However, analyses of various studies show that social recognition indeed plays a role besides the economic, political, or environmental causes (Debalke 2021: 156).

Fanon says that,

> Man is human only to the extent to which he tries to impose himself on another man in order to be recognized by him. As long as he has not been effectively recognized by the other, it is this other who remains the focus of his actions. His human worth and reality depend on this other and his recognition by the other. It is in this Other that the meaning of his life is condensed. (Fanon 2008: 191)

Contrary to Cummings (2015: 28) and De Haas (2007a: 833), I argue that emigration from postcolonial Africa to the West increases not because of the disequilibrium between increasing aspirations and livelihood opportunities at home. It is rather the increasing livelihood opportunities that increase the aspirations, both as mutually conditioning components of an unending vicious circle. The aspirations are not only the desire to achieve more livelihood opportunities but rather even more importantly, to achieve recognition at the origin. Improving social and economic development, increasing aspirations, and sustained or increased outmigration reinforce each other.

The fact that "middle-income countries" have the highest average levels of emigration, and that trade and investment in a source country reinforce emigration rather than reduce (de Haas 2011b, p. 562 SO-4) underlines the mimetic role of migration rather than an economic factor behind. If the gap between the model and the imitator is too big, the latter not only does not know enough about the model but also feels too inferior to imitate the model. As we shall see in Chap. 8, the decreasing gap (economically, epistemically, and culturally) between the West and the rest leads to more migration. As the mimetic theory shows, it is the increasing similarity—not difference—that intensifies mimetic rivalry (Girard 1965). More information and increasing economic and social development of the imitator will narrow the social and economic gap between the model (the West) and the imitator (postcolonial Africans). In the eyes of the imitator, the images of wealth and Western luxury are not biased. For him it is real. The imitator perceives his situation as always miserable (Girard 1965). The Senegalese perceived their situation as misery and Europe as paradise (Schapendonk and van Moppes PS-6 2007, p. 2 PS-6). Here, the belief is that salvation is possible only in the acquisition of the desired object, that is life in the West through migration.

Everything broadcast by Western media is available at the origin (Carter 2013: 65). The colonizers are designated as "advanced", therefore "advanced economies", "advanced societies", "developed countries" etc. The colonized are designated as underdeveloped, backward, least developed, developing countries, etc. Interestingly, it is not only the colonizers who call the colonized "backward"; even the colonized say to themselves that they are economically, technologically, or even culturally "backward".

In his blog titled "*migro ergo sum—I migrate, therefore I am—Social Pressure as a Driver of Economic Migration from West Africa*, Loprete suggests that there is considerable social pressure on young people, which forces them to migrate. "In many countries throughout West Africa, migration seems to have replaced the ancient rite of passage into adulthood. … The pressure comes from all angles and young Africans cannot resist, they have to migrate if they want to 'be'" (Loprete 2016).

Similarly, one of the very interesting conclusions of the 2019 UNDP study is that "Economic motivations, closely tied to self-actualization and a sense that aspiration can only be fulfilled through departure from Africa" (UNDP 2019: 5). Why is that? The title of the UNDP study is called "scaling fences". I interpret this *scaling of fences* as both physical and metaphorical. The fence is the means of bordering, othering, and ordering (Van Houtum and Van Naerssen 2002; Casas-Cortes and Cabarrubias 2019: 193). According to the UNDP study findings, the first feeling of migrants after arrival is "I made it".

Migration is an existential desire because it implies being *like* and *with* the former colonizer who negated African humanity. African ontology is dependent not on detachment from the colonizer but instead on similarity with him. The being depends on the mimetic desire to be *like* and *with* him. The *scaling of fences* is not just a search for better opportunities but also an undertaking to negate the historical and cultural rejection, negation, oppression, exploitation, and exclusion that started with slavery and colonialism. Postcolonial African migrants attempt to overcome the invisible walls erected by the imperialists like Winston Churchill who maintained that rights and freedoms are inapplicable to nonwhites and for "African aboriginals" civilization has no charms (Ibhawoh 2020: 44). Postcolonial Africans negate the negation of African humanity.

Migration is an undertaking to "be", to be recognized, to regain the identity denied through slavery and colonization, and to overcome the historical and cultural wall. The physical walls like the Sahara Desert, the Mediterranean Sea, and the walls of Ceuta and Melilla reinforce each other and keep the colonial and colonized identities apart. However, at the same and paradoxically they bring them both nearer and nearer.

Around 2005 Spain in particular was beefing up its wall in Ceuta and Melilla to keep away African irregular immigrants. However, the walls did not seem to deter them. One West African migrant said, "But they can build the fences [in Ceuta and Melilla] as high as they like, they can have as many soldiers as they like—nobody can stop us from getting through… I don't think it would be possible to go back [to my family without money] and face them with that shame. I think I would die" (BBC 2005). What a would-be migrant in West Africa said in 2019, "you become a man once you have migrated" (Africa Research Bulletin 2019a: 22783). As Loprete suggests above, from the existentialist perspective migration is a sort of initiation rite toward a new being, the freedom to become and to be.

Migrants are not ignorant of the risks they are taking. Even if migrants were informed about those who narrowly avoided death, had friends who have died attempting the same journey, followed news on social media of the Greek disaster in which up to 750 people are believed to have died in 2023, experienced boats in distress, and were returned to Libya, etc. still many have tried again and again to reach Europe. They are not deterred by the information about tragedies. Right after the above-mentioned migrant tragedy of 2023 in Greece, a 17-year-old boy from The Gambia said he had tried seven times and would continue trying. Just for one attempt paid £570. Postcolonial African migrants to the West are guided by the motto, "It's either you reach Europe or you die at sea". These are, according to them, the only two options they have. They do not go back empty-handed because "It's a disgrace" (BBC 2023). When mid-August 2023, more than 60 Senegalese migrants died while attempting to reach Spain, the brother of one of the dead said he would still attempt the trip himself as the "brother died for a dream we all have". Is it poverty or bravery, or is there something else behind this mindset?

According to a report by Human Rights Watch (2023), Saudi border guards killed summarily hundreds of Ethiopian migrants. Men were forced to rape girls. Those men who refused to rape the girls were immediately killed. There is a lot of information on such horrible deeds. But people still decide to undertake such migration journeys. To what extent does poverty explain such decisions? Isn't the world beyond Africa maybe a locus where you not only make money but also a medium that enables one's recognition, one's being?

Chimamanda Ngozi Adiche from Nigeria describes in her novel, *Americanah*, how colonialism, racism, and African migration are interconnected. She described that migration is not just about hunger, poverty, or conflict, but it is also about expanding horizons and aspirations, about looking for modernity, "civilization", and leaving behind the "backwardness." As Adichie points out, people who do not belong to or were not assigned to this "backward" or "uncivilized" group "would not understand why people who were raised well-fed and watered but mired in dissatisfaction, conditioned from birth to look towards somewhere else, eternally convinced that real lives happened in that somewhere else, were now resolved to do dangerous things, illegal things, so as to leave, none of them starving, or raped, or from burned villages, but merely hungry for choice and certainty" (Adichie 2017: 276).

The historically emaciated and colonized being tries to regain its humanity by rejecting the historically and violently assigned social psychological position. I see some similarities between the unrelenting and uncompromising will to migrate and the protest of Rosa Louis Parks of Alabama in 1955. She protested against the *allocation*, i.e. a white violent assigning of her to a seat that she did not want. She wanted the forbidden seat, reserved for the whites. Her protest was the rejection of the violently limited horizons and aspirations; it was about freedom; it was an existential revolt. The assignment of the seat was the creation of a wall. For Rosa Parks, you become a human being only when you can decide by yourself where to sit.

> The difficulty was that though one could trap them like animals, transform them in pens, work them alongside an ass or a horse and beat both with the same stick, stable them and starve them, they remained, despite their black skins and curly hair, quite invincible human beings; with the intelligence and resentments of human beings. (James 2022: 9)

We could designate Adichie's "hunger for choice and certainty" above as a desire to be recognized. The achievement of this recognition is an overwhelming task. Postcolonial African migration, as the independence struggle was, is about recognition. "Lack of recognition is not just an inconvenient existential modality suffered in silence but rather enters into the lived misfortune of people whose very survival is threatened" (Carter 2013: 71). The colonized being tries to achieve this through imitation of the colonizer's lifestyles. As Adichie describes, whereas more Europeans are now moving away from processed and frozen food, many Africans, who can afford it, are striving for it; while Europeans renovate 200 or more years old mill granaries, Africans would demolish and build something new; while the West makes a fetish of old things, Africans attempt to arrive first at the stage of modernity and *civilizedness* and overcome their "backwardness" they were told to incorporate (Adichie 2017: 436, 444).

Some forms of current migration are a rite of passage in the cultural anthropological sense to change one's status in society by entering another group and a level considered to be more advanced, and to be recognized as such. Overcoming physical hindrances such as the Sahara Desert, the Mediterranean, the walls of Ceuta and Melilla, the FRONTEX surveillance systems, etc. is part of the passage processes to achieve a new being. It is about reinventing or de- and reconstructing a certain image of the self in daily life, as Adeyanju and Oriola (2011) suggest. Adeyanju and Oriola talk about African migrants' "insatiable desire for Euro-American countries" (Adeyanju and Oriola 2011: 943):

> Africans' interests in the West are rooted in three interrelated extra-economic factors: colonial discourse, migrants' accounts or presentation of self, and the contemporary mass media. Africans' encounter with European colonialism had a lasting effect on their lifeworld. Colonialism amplified the difference between the self and the other by associating progress with European culture (religion, aesthetics, education, arts, philosophy, language, etc.) and primitivism and backwardness with the African other. But colonialism was also about material relations that existed within the larger context of global economic inequality that made the White individual look superior because of his or her race rather than because of his or her class. (Adeyanju and Oriola 2011: 943)

Adeyanju and Oriola report about a letter by two West African boys who tried to migrate to Europe: that African countries were represented by these boys as a dystopia and immiserated, while Europe is presented not only as a place of comfort but also as an ethereal rescuer, characterized by love and progress (Adeyanju and Oriola 2011: 950). "The notion that the West has a higher moral value and ethical standard than Africa has probably been recycled for generations and is hence received unquestioned. It is a conception that has become a reigning regime of truth in people's lifeworld" (Adeyanju and Oriola 2011: 960).

They also suggest that African migrants are concealing their hardships in the West from the glare of their peers back home" (Adeyanju and Oriola 2011: 944). Very often migrants lead a life of pretension and delusion. When they are in Europe they go through very tense economic and social situations, when they go home on vacation they ostentatiously display material goods. Adeyanju and Oriola designate this as a concealed backstage, which consists of narratives of shortcomings, the experience of racism, unemployment, and other disadvantages (Adeyanju and Oriola 2011: 956f). They live in a status of contradiction between a life aggrieved, hurt, and disadvantaged by racism on the one hand and a life that exists only in photos that show a fake good life having a car, being well dressed, and living in a city of skyscrapers (Adeyanju and Oriola 2011: 957). As Werbner

(2012: 218) suggests, for weddings and big traditional feast migrants go back to their countries of origin not only to celebrate these special feasts or events but also to display their newly acquired wealth and Western lifestyles back home. This is an expression of the desire for recognition.

As Bürkle et al. suggest, migration is a communication process. The findings of a research project by Bürkle et al. show that Turkish returning migrants from Germany, Australia, etc. build their houses similar to these countries even if they do not fit the local architecture or address the local climatic conditions. Even construction companies and construction workers come from abroad because they do not trust the local workers. Even the furniture is imported from such as Germany. The German flag is hoisted, and German cars are parked, etc. in Turkey. However, as long as they are in Germany, the Turkish flag is displayed in the living room. Turkish television runs all the time. In a foreign country where their identity is questioned, they solace themselves with these symbols of national identity. In Turkey, finally, they do not feel any more foreigners. However, they try to distinguish themselves from the rest of the Turks by being a bit German. The German car, flag, and architecture give the returnees an identity that secures their higher social status, better than the locals (Bürkle 2016).

## Conclusion

Even before migration—through the consumption of Western goods and through imitating the way of life, culture, language, etc. of the colonizer—the colonized would-be migrant searches for identity through the desire for recognition. First, after having migrated his delusory identity depends on the recognition by those back home. Through the scaling of fences erected to separate the mythologized colonial world from the dehumanized colonized one migration is a desire to underscore one's manliness in a patriarchal society and in the face of the disappearing cultural-anthropological initiation rite. He has to show them that he owns a car and lives in a city with skyscrapers like the former colonizer. To attain identity and recognition, he has to send back pictures of the beautiful life in the West even if he is leading a wretched life there. Secondly, he strives for recognition by society at the destination. However, he gives up his aspiration to be recognized by the former colonizer because the racism with which he is confronted every day does not allow this. He has overcome the walls of Ceuta and Melilla, the Sahara Desert, and the

Mediterranean but he cannot overcome the wall of racism. Nevertheless, as Nietzsche says, “Where I found a living creature, there I found will to power; and even in the will of the servant I found the will to be master” (Nietzsche 2003: 137). As we shall see in the following chapter, this will is the will to be equal, to negate the colonial negation, and to imitate the model, the sublime colonizer.

Fanon says that the history of slavery and colonization was a history of internalization and the epidermalization of inferiority (Fanon 2008, XV). “... the more the black Antillean assimilates the French language, the whiter he gets—i.e. the closer he becomes to becoming a true human being” (Fanon 2008: 2). He has to live, speak, and behave like a white man. “The black man entering France changes because for him the métropole is the holy of holies...” (Fanon 2008: 7). When he returns to Africa he is, according to Fanon, a demigod or deified being (Fanon 2008: 3). A less-African one is perceived or perceives himself, the whiter he is considered or the closer he feels to the white man. It is often suggested that those who are educated or informed about the white world would not behave like this. However, studies show that the inferiority of the educated is at least as strong as that of the non-educated. Through their lifestyles, the use of European furniture, the European way of social intercourse, adorning the native language with European expressions, etc. they try to feel at least a bit white (Fanon 2008: 9).

The colonized man’s identity is not dependent on detachment from the colonizer, contrarily or paradoxically, on imitation, adoration, and also abomination. Postcolonial African migration to the West is a process toward this vision or probably a delusion. The postcolonial African man—the Dostoevskian *Underground Man*—, is a divided, delusional, and self-contradictory being, as we shall in the next chapter.

## Literature


Adem, Teferi Abate (2021): The moral economy of irregular migration and remittance distribution in south Wollo, in Asnake Kefale and Fana Gebresenbet, *Youth on the move: Views from below on Ethiopian international migration*, Hurst and Company: London, pp. 221–238.

Adeyanju, Charles and Oriola, Temitope (2011): Colonialism and Contemporary African Migration: A Phenomelogical Approach, Journal of Black Studies, 42(6), 943–967.

Adichie, Chimamanda (2017): Americanah, 4th ESTATE: London.

Africa Polling Institute (2020): Deconstructing the Canada Rush: A study on the motivations for Nigerians emigrating to Canada – Report March 2020, Abuja, Nigeria.

Africa Research Bulletin, Economic, Financial and Technical Series, Volume 56 Number 6, June 16th 2019a–July 15th 2019.

Babou, Cheikh (2013): Migration as a factor of cultural change abroad and at home: Senegalese female hair braiders in the United States, in Abdoulaye Kane and Todd Leedy, African Migration: patterns and perspectives, Indiana University Press, Bloomington, pp. 230–247.

BBC (2005): Crossing deserts, scaling fences, http://news.bbc.co.uk/2/hi/africa/4332512.stm, accessed 26.03.2024.

BBC (2023): 'Europe or death' - the teenage migrants risking it all to cross the Med, https://www.bbc.com/news/world-europe-66132875, 11.07.2023.

Bleck, Aimie (2020): Migration aspirations from a youth perspective: focus groups with returnees and youth in Mali, Journal of Modern African Studies, 58:4, pp. 551–577.

Bürkle, Stefanie ed. (2016): *Migration* von Raum Architektur und Identität im Kontext türkischer Remigration, Vice Versa Verlag: Berlin.

Carling, Jørgen & Collins, Francis (2018): Aspiration, desire and drivers of migration, *Journal of Ethnic and Migration Studies*, 44:6, pp. 909–926, https://doi.org/10.1080/1369183X.2017.1384134.

Carter, Donald (2013): Navigating diaspora: The precarious depths of the Italian immigration crisis, in Abdoulaye Kane and Todd Leedy, African Migration: patterns and perspectives, Indiana University Press, Bloomington, pp. 59–77.

Casas-Cortes, Maribel and Cobarrubias, Sebastian (2019): Genealogies of contention in concentric circles: remote migration control and its Eurocentric geographical imaginaries, in Critical geographies of migration, Edward Elgar, pp. 193–205.

Collins, Francis L. (2018): Desire as a theory for migration studies: temporality, assemblage and becoming in the narratives of migrants, *Journal of Ethnic and Migration Studies*, 44:6, pp. 964–980, DOI: https://doi.org/10.1080/1369183X.2017.1384147.

Conrad Suso, Catherine (2019): Backway or bust: causes and consequences of Gambian irregular migration, *Journal of Modern African Studies*, 57:1, pp. 111–135.

Cummings, Clare et al. (2015): Why people move: understanding the drivers and trends of migration to Europe, Overseas Development Institute, Working Paper 430, 2015s.

Danish Institute for International Studies (2021): Does information save migrant's lives? Knowledge and needs of West African migrants en route to Europe, Danish Institute for International Studies, Copenhagen.

De Haas, H. (2007a.): Turning the Tide? Why Development Will Not Stop Migration. Dev. Change 38, 819–841. https://doi.org/10.1111/j.1467-7660.2007.00435.x.

De Haas, H. (2011b): "The determinants of international migration: Conceiving and measuring origin, destination and policy effects". IMI Working Paper Series. Oxford: International Migration Institute, University of Oxford.

Debalke, Mulugeta Gashaw (2021): Irregular migration among a Kambata community in Ethiopia: views from below and their implications, in Asnake Kefale and Fana Gebresenbet, *Youth on the move: Views from below on Ethiopian international migration*, Hurst and Company: London, pp. 137–158.

Dom, Catherine (2021): Migration aspiration and *glocal* ideas of the good life in two rural communities in southern and north-eastern Ethiopia: a comparative perspective, in Asnake Kefale and Fana Gebresenbet, *Youth on the move: Views from below on Ethiopian international migration*, Hurst and Company: London, pp. 97–118.

Dougnon, Isaie (2013): Migration as coping with risk and state barriers: Malian migrants' conception of being far from home, in Abdoulaye Kane and Todd Leedy, African Migration: patterns and perspectives, Indiana University Press, Bloomington, pp. 35–58.

Erdal, Marta Bivand & Oeppen, Ceri (2018): Forced to leave? The discursive and analytical significance of describing migration as forced and voluntary, Journal of Ethnic and Migration Studies, 44:6, 981–998, https://doi.org/10.1080/1369183X.2017.1384149.

Estifanos, Yordanos (2021): Gender relations in a transnational space: Ethiopian irregular migration in South Africa, in Asnake Kefale and Fana Gebresenbet, *Youth on the move: Views from below on Ethiopian international migration*, Hurst and Company: London, pp. 159–178.

Fanon, Frantz (2008): Black Skin, White Masks, Grove Press: New York.

Gebresenbet, Fana (2021): Hopelessness and future-making through irregular migration in Tigray, Ethiopia, in Asnake Kefale and Fana Gebresenbet, *Youth on the move: Views from below on Ethiopian international migration*, Hurst and Company: London, pp. 79–96.

Gezahegne, Kiya (2021): Rituals of migration: socially entrenched practices among female migrants from Amhara national regional states, in Asnake Kefale and Fana Gebresenbet, *Youth on the move: Views from below on Ethiopian international migration*, Hurst and Company: London, pp. 119–136.

Girard, René (1965): Deceit, desire and the novel: self and other in literary structure, John Hopkins University: Baltimore.

Girard, René (2023): All desire is a desire for being, Penguin Classics: London.

Hertrich, Véronique and Lesclingand, Marie (2013): Adolescent Migration in Rural Africa as a Challenge to Gender and Intergenerational Relationships: Evidence from Mali, The Annals of the American Academy, AAPSS, 648, pp. 175–188.

Houtum, Henk van & Naerssen, Ton van (2002): Bordering, Ordering and Othering, *Tijdschrift voor Economische en Sociale Geografie*, Vol. 93, No. 2, pp. 125–136.

Human Rights Watch (2023): "They Fired on Us Like Rain": Saudi Arabian Mass Killings of Ethiopian Migrants at the Yemen-Saudi Border, AUGUST 2023, ISBN: 979-8-88708-065-9.

Ibhawoh, Bonny (2020): Seeking the political kingdom: universal human rights and the anti-colonial movement in Africa, in Decolonization, Self-determination and the rise of global human rights ed. A. Dirk Moses et al. Cambridge University Press: Cambridge, pp. 35–53.

Jackson, Michael (2008): The Shock of the New: On Migrant Imaginaries and Critical Transitions, Ethnos: Journal of Anthropology, 73:1, 57–72, https://doi.org/10.1080/00141840801927533.

James, C.L.R. (2022): The black Jacobins: Toussaint Louverture and the San Domingo revolution, Penguin: Dublin.

Kefale, Asnake and Gebresenbet, Fana (2021): Introduction: Multiple transitions and irregular migration in Ethiopia: agency and assemblage from below, in Asnake Kefale and Fana Gebresenbet, *Youth on the move: Views from below on Ethiopian international migration*, Hurst and Company: London, pp. 1–22.

Loprete, Giuseppe (2016): MIGRO ERGO SUM – I Migrate, Therefore I Am – Social Pressure as a Driver of Economic Migration from West Africa, https://blogs.lse.ac.uk/africaatlse/2016/01/14/migro-ergo-sum-i-migratetherefore-i-am-social-pressure-as-a-driver-of-economic-migration-from-west-africa/, accessed 26.03.2024.

Mahler, Sarah (1995): American Dreaming, Princeton: Princeton University Press.

Mbaye, Linguère Mously (2013): Understanding Illegal Migration from Senegal, IZA DP No. 7728, Discussion Paper Series, Forschungsinstitut zur Zukunft der Arbeit Institute for the Study of Labor.

Mills, Greg (2010): Why Africa is Poor: And What Africans Can Do About It?, Penguin.

Mixed Migration Centre (2019): Navigating borderlands in the Sahel: border security governance and mixed migration in Liptako-Gourma. Available at www.mixedmigration.org, 12.06.2022.

Nietzsche, Friedrich (2003): Thus spoke Zarathustra, Penguin Classics: London.

Petit, Véronique and Charbit, Yves (2018): Economic Migrations, Households and Patriarchy In Africa And Southern Europe, HAL Id: halshs-02113339, https://halshs.archives-ouvertes.fr/halshs-02113339 (accessed 04.05.2023).

Rockefeller, Steven (1994): Comment, The politics of recognition, in Charles Taylor et al., multiculturalism, Princeton University Press, Princeton, pp. 87–98.

Sanny, Josephine Appiah-Nyamekye et al. (2019): In search of opportunity: Young and educated Africans most likely to consider moving abroad, Afrobarometer 2019, Dispatch No. 288 | 26 March 2019.

Schapendonk, J., van Moppes, D., (2007). Migration and Information: Images of Europe, migration encouraging factors and en route information sharing (No. 16), Working Papers Migration and Development Series. Radboud University, Nijmegen.

Schöfelberger, Irene et al. (2020): Migration aspirations in West and North Africa: what do we know about how they translate into migration flows to Europe? Migration in West and North Africa and across the Mediterranean Trends, risks, development and governance International Organization for Migration (IOM), Geneva, pp. 87–97.

Sintayehu, Firehwot (2021): The drivers of youth migration in Addis Ababa, in Asnake Kefale and Fana Gebresenbet, Youth on the move: Views from below on Ethiopian international migration, Hurst and Company: London, pp. 65–78.

Taylor, Charles (1994): The politics of recognition, in Charles Taylor et al., Multiculturalism, Princeton University Press, Princeton, pp. 25–74.

Tufa, Fekadu Adugna et al. (2021): Say what the government wants and do what is good for your family: facilitation of irregular migration in Ethiopia, in Asnake Kefale and Fana Gebresenbet, *Youth on the move: Views from below on Ethiopian international migration*, Hurst and Company: London, pp. 43–64.

UNDP (2019): Scaling Fences: Voices of Irregular African Migrants to Europe, UNDP.

Van de Walle, Nicolas (2001): African Economies and the Politics of Permanent Crisis, Cambridge University Press: Cambridge.

Van Hear, Nicholas et al. (2018): Push-pull plus: reconsidering the drivers of migration, Journal of Ethnic and Migration Studies, 44:6, 927–944, https://doi.org/10.1080/1369183X.2017.1384135.

Werbner, Pnina (2012): Migration and culture, in Marc Rosenblum and Daniel Tichenor, The Oxford Handbook of the politics of international migration, Oxford University Press: Oxford, pp. 215–242.

Whitehouse, Bruce (2013): Overcoming the economistic fallacy: social determinants of voluntary migration from the Sahel to the Congo basin, in Abdoulaye Kane and Todd Leedy, African Migration: patterns and perspectives, Indiana University Press, Bloomington, pp. 19–34.

Youngstedt, Scott (2013): Voluntary and involuntary homebodies: adaptations and lived experiences of Hausa "left behind" in Niamey, Niger, in Abdoulaye Kane and Todd Leedy, African Migration: patterns and perspectives, Indiana University Press, Bloomington, pp. 133–157.

CHAPTER 7

# The Postcolonial African Migration to the West as a Desire for Being

Around the world, there are more than 108 million people affected by displacement and migration in 2023. Often poverty, conflict, and climate change are behind them. At the same time, various studies have shown that people in extreme poverty cannot pay thousands of dollars for smugglers. Many people who are not affected by poverty, conflicts, or climate change have also been migrating to the West. Despite the dangers on the way, many young people are determined to venture into it. The European strict border controls, the exploitation by smugglers, the dangers of crossing the Mediterranean, the costliness of the journey, and other challenges have not deterred tens of thousands of Africans from attempting to migrate to the West. Many of them have tried to cross the Mediterranean more than seven times. This suggests that there is something more besides poverty, political insecurity, or climate change that pushes Africans to migrate to the West.

As the distinction between forced and voluntary migration is opaque (Bakewell 2021: 124), the focus on material or physical causes of migration at the expense of immaterial or psychological causes is inadequate and even misleading. This chapter goes beyond the poverty, conflict, and climate change discourse. I examine postcolonial migration from Africa to the West from a mimetic-psychoanalytical perspective, which addresses issues neglected by economic, political, and environmental approaches and focuses on the experiences and legacies of slavery and colonization.



B. Gebrewold, *Postcolonial African Migration to the West*, Politics of Citizenship and Migration,
https://doi.org/10.1007/978-3-031-58568-5_7

The *desire for being* lies behind the postcolonial migration from Africa to the West. Postcolonial migration from Africa to the West is caused primarily by the socio-psychological mindset of the desire for belonging and being, and only secondarily and epiphenomenally a spatial movement from Africa to the West. The socio-psychological causes are often disguised as material causes (poverty, conflict, and climate change). As Fanon suggests, white civilization, slavery, colonialism, and European culture have imposed an existential deviation on the black person. (Fanon 2008: xviii).

Some might criticize my theoretical approach and question why should Africans go to Europe in search of liberation from the legacies instead of distancing themselves from them. This shows a misunderstanding of the fundamental issue; it comes from the mindset of "why should one imitate and migrate to the former oppressors and at the same time seek liberation from them." As will be shown in this chapter, it is imitation, not detachment, that is perceived by postcolonial Africans as a liberation process from historical dehumanization (Cf. Appiah 1994: 154–55). Paradoxically, the desire for being depends on the hope for belonging through imitation or simulation, not through detachment.

## Belonging, Becoming, and Being Through Imitation

Mannoni was convinced that the West possessed superior power, which persuaded the natives of the overriding need to imitate it, like school children, and to obey (Mannoni 1990: 32). From a mimetic perspective, Mannoni writes: "If the 'abandoned' or 'betrayed' Malagasy continues his identification he becomes clamorous; he begins to demand equality in a way he had never before found necessary ... for every increase in equality makes the remaining differences seem the more intolerable, for they suddenly appear agonizingly irremovable. ... Thus, it is just when the Malagasy is beginning to resemble us a little that he turns roughly from us" (Mannoni 1990: 84). Similarly, Fanon suggests that,

> The look that the native turns on the settler's town is a look of lust, a look of envy; it expresses his dream of possession—all manner of possession: to sit at the settler's table, to sleep in the settler's bed, with his wife, if possible. The colonized man is an envious man. And this the settler knows very well; when their glances meet he ascertains bitterly, always on the defensive 'They

> want to take our place'. It is true, for there is no native who does not dream at least once a day of setting himself up in the settler's place. (Fanon 2001: 30)

Postcolonial behavior oscillates between detachment and imitation, the abomination of and fascination for the colonizer. On the one hand, the colonized reject the values of the colonizer, but on the other hand, they desire them. They strive for equality with the colonizers—wanting to eject them, to take their place; the settler's skin is not of any more value than a native's skin (Fanon 2001: 34–36). However, colonizers are convinced of their superiority, "If we leave, all is lost, and the country will go back to the Middle Ages" (Fanon 2001: 40). What the postcolonial blacks desire is not only the improvement of their material or economic situation but the "white man's" recognition of and respect for their dignity, their full human value, to be liberated from the western whites' notion that blacks are overcome by primitive and faulty thinking and incapable of logic (Césaire 2000: 58, 69).

The colonial aggressor (or model in the language of Girard) considers himself too far above the colonized African (disciple or imitator in the language of Girard). However, the latter's ultimate goal is to overcome the colonial border and be accepted as equal. However, this would destroy the model's superiority. At the same time, the imitator wants to be recognized as equal (the equality that can be bestowed only by the model!). The fact that the model owns this superiority, which not only he believes to possess but also the disciple himself witnesses by his attempts to be recognized as equal gives the model the certainty that he is indeed superior. Therefore, the model unconsciously encourages the disciple to imitate him. Ultimately, the model himself is imitating his own disciple because his being is dependent on being imitated. Once the disciple stops to imitate him his value and his sublimity stop to exist. Girard explains this reciprocal dynamic as follows:

> The model considers himself too far above the disciple, the disciple considers himself too far below the model, for either of them even fleetingly entertain the notion that their desires are identical—in short, that they might indeed be rivals. To make the reciprocity complete, we need only add that the disciple can also serve as a model, even to his own model. As for the model, no matter how self-sufficient he may appear, he invariably assumes the role of disciple... (Girard 2016: 165)

The colonized rejects the colonizer's superiority but admires him at the same time. Once the colonizer realizes that what he is and owns is also the object of desire of the colonized imitator he becomes convinced of his superiority. "Whenever the disciple borrows from him what he believes to be the 'true' object, he tries to possess that truth by desiring precisely what this model desires." (Girard 2016: 166). The colonized attempts recognition through rejection of subjugation and imitation of the colonizer. The mimetic relationship of the postcolonial African with the colonizer is characterized by a transformative process of becoming through imitation. Whereas being is a particular ontological presence at a particular point in time, becoming is a continuous process of transformation or a continuous moving presence of the ontological or subjective self (Natanasabapathya and Maathuis-Smith 2019).

The mimetically desired object of the colonized is to be free and equal. The colonizer tries to maintain his feelings of superiority and difference through othering, bordering, and ordering (van Houtum and van Naerssen 2002), whereas the colonized tries to negate them through the desire for equality by way of de-othering, de-bordering, and reordering. In his *Critique of Black Reason*, Mbembe describes the mimetic relation between the colonizer and the colonized.

> The fierce colonial desire to divide and classify, to create hierarchies and produce a difference, leaves behind wounds and scars. Worse, it created a fault line that lives on. Is it possible today to craft a relationship with the Black Man that is something other than that between a master and his valet? Does the Black Man not insist, still, on seeing himself through and within difference? Is he not convinced that he is inhabited by a double, a foreign entity that prevents him from knowing himself? Does he not live in a world shaped by loss and separation, cultivating a dream of returning to an entity founded on pure essentialism and therefore, often, on alterity? At what point does the project of a radical uprising in search of autonomy in the name of difference turn into a simple mimetic inversion of what was previously showered with malediction? (Mbembe 2017: 7)

This is similar to what Du Bois suggests that the colonized and enslaved do not have any true self-consciousness and therefore see themselves through the revelation of the colonizer and slaveholder and measure their soul by the tape of the oppressor who looks on in amused contempt and pity. The double identity every colonized person lives with nowadays—how they see themselves as colonized and how the colonial world sees them—keeps them in a painful self-consciousness that consists of a morbid sense of personality and a moral hesitancy, which is fatal to self-confidence. Even today the "veil" or the "color line" that separates the colonizer and the colonized persists (Du Bois 2007: xiv). It is not only the identity of the black colonized but also that of the colonizer that is determined by the color line. "The bondsman and the slave find their identity in each other's gaze: 'two-ness'" (Du Bois 2007: xv).

The "two-ness" phenomenon, expressed on both sides of the colonizing divide, means living a double life, not having the freedom to be one's authentic self (Lukasik 2021: 33). As Degruy suggests, "I am not who I think I am, and I am not who you think I am. I am who I think that you think I am" (Degruy 2017: 5). As we shall see later in this chapter, this two-ness is similar to *The Double* in which Dostoevsky shows the two-ness of the same Golyadkin, divided between the ideal or desired-self and the denied real-self,. Dostoevsky in the *Notes from Underground* and Camus in the *Fall* discuss this phenomenon. Du Bois analyzes how the colonized man sees himself from the perspective of his co-colonized black persons and also from that of the whites. Moreover, as soon as he begins to stand out from the colonized ignorant, diseased, inefficient mass because of his successes, he starts to consider himself part of the white group from which, in fact, he is excluded (Du Bois 2007: 88). He is possessed by dual personalities, his real self and the desired self. In this condition, two uncomforting facts face each other: "the ignorance, poverty, inefficiency of the darker peoples; the wealth, power and technical triumph of the whites" (Du Bois 2007: 89). Since the blacks cannot expect anything sublime from blacks, they wish they had been born white! (Du Bois 2007: 89). At the same time, the white supremacists do everything they can to avoid the physical as well as cultural vicinity of blacks. Blacks cannot escape the black company (former slavery, present poverty and ignorance, sickness and crime, social exclusion and degradation, inability to get work, discrimination in pay, the improbability of promotions, segregation), and they cannot reach that of the whites. The blacks must be among themselves as neighbors to their own people, segregated, dirty, diseased, noisy, ignorant, and dehumanized. They are barred from the privileged white neighborhood, which they aspire incessantly (Du Bois 2007: 89–92). This is a permanent flight and permanent aspiration, an agony of Sisyphus. The black man has a disdained name, a disfigured face, and mutilated humanity (Mbembe 2017: 36). He is in search of his being among the white colonizers. He has to migrate to the white Western world.

Postcolonial African migrants are disciples of the model (master), who was a colonizer, superior, powerful, politically, technologically, culturally, and economically. His ideals of beauty, lifestyles, and consumption are the model for the imitator, the colonized. The master is delighted that he is being imitated but alarmed that he is going to be equal with his imitator or disciple. "Yet if the imitation is too perfect, and the imitator threatens to surpass the model, the master will completely change his attitude and begin to display jealousy, mistrust, and hostility. He will be tempted to do everything he can to discredit and discourage the disciple" (Girard 2019: 278). This is exactly the process of othering, bordering, and ordering which the model tries to materialize. The European fences in Ceuta and Melilla, the Frontex, the Hungarian fences, and the fences demanded by different politicians in various European countries like Austria are part of this othering, bordering, and ordering process. This process is the meeting point of the two resistances colliding—which Girard would call the "relationship of opposition"—and thereby increases the value of what is considered to be European or Western, which embodies something exalted and superior (Palaver 2019: 169). The physical as well as virtual fences against migrants not only protect European or Western values but also increase their desirability. As Girard would say,

> The value of an object grows in proportion to the resistance met with in acquiring it. And the value of the model grows as the object's value grows. Even if the model has no particular prestige at the outset, even if all that 'prestige' implies … is quite unknown to the subject, the very rivalry will be quite enough to bring prestige into being. … [T]he prestige of the model, the resistance he puts up, the value of the, and the strength of the desire it arouses all reinforce each other, setting up a process of positive feedback. (Girard 2019: 282–283)

The migrant in this dynamic of resistance by the model implicitly accepts his inferiority and starts to ask himself if the prohibiting model has probably got good reasons for denying him the object of his desire, which is living together with the white and Western European.

> Once he has entered upon this vicious circle, the subject rapidly begins to credit himself with a radical inadequacy that the model has brought to light, which justifies the model's attitude toward him. The model, being closely identified with the object he jealously keeps for himself, possesses … self-sufficiency and omniscience that the subject can only dream of acquiring.

> The object is now more desired than ever. Since the model obstinately bars access to it, the possession of this object must make all the difference between the self-sufficiency of the model and the imitator's lack of sufficiency, the model's fullness of being and the imitator's nothingness. (Girard 2019: 283–284)

As Girard suggests, this misunderstood mechanism of mimetic rivalry, i.e. the dynamic relationship between the model and the imitator living in the West, is transformed into a metaphysical reality (Palaver 2003: 171). This mechanism transforms not only the reality of the relationship into a metaphysical reality, but conversely, it also transforms the model and the imitator themselves:

> If the model himself becomes more interested in the object that he designates to his imitator as a result of the latter's imitation, then he himself falls victim to his contagion. In fact, he imitates his own desire, through the intermediary of the disciple. The disciple thus becomes model to his own model, and the model, reciprocally, becomes disciple of his own disciple. In the last resort, there are no genuine differences left between the two, or ... between their desires. (Girard 2019: 287)

Image and imitation have the same etymological background in Latin. In this mimetic rivalry, the model and imitator become images of one another as they admire each other for the other's admirability (Girard 2019: 295; Palaver 2003: 163, 176). The other who is my image and whom I imitate knows how to make me fail since he possesses me, he can reduce me to nothingness and inadequacy. Postcolonial migration from Africa to the West is the desire to escape the existential inferiorization of the blacks. Africa is a place of spatial inferiority. African political, economic, and social incompetence to govern and manage their economies and societies in a better way and their lack of respect for African lives convince fellow Africans that whites are indeed superior and they escape Africa at any cost. Therefore, postcolonial African migration to the West is part of the pervasive African desire for the imitation of the white Western world.

Aesthetics and colorism are one way of postcolonial African imitation of the West. There is widespread pressure around the world to be whiter as having fair skin is considered a prerequisite for beauty. Whiteness has become synonymous with beauty and attractiveness throughout the world, leading people especially women to whiten their skin to meet this beauty expectation. There is a widespread perception that those without white skin consider themselves inferior, ugly, less attractive, and less influential; and that most black men prefer light-skinned women; whiteness is associated with cleanness, desirability, and better personality, whereas blackness is associated with reprehensibility. Colorism—the association of lighter skin with wealth, power, and beauty standards—is the legacy of slavery and colonization and is now re-strengthened by globalization through media and education. Colorism "privileges light-skinned people of color over dark in areas such as income, education, housing, and the marriage market" (Hunter 2007: 237). It is a system of inequality that grants special advantages to lighter-skinned individuals (Daftary 2023: 337). As a result, there is widespread use of skin-lightening cosmetic products, skin bleaching, and cosmetic surgery industries across the world especially in the global south. The desire for skin-lightening on the one hand and low self-esteem, body dysmorphic disorder, self-hate, anxiety, feelings of insecurity, self-doubt, and fear of being judged by others, depression, stress, shame, disordered eating, etc. on the other reinforce each other (Crimmins 2023).

Blackness is despicability. During the World Championship in Qatar, when Morocco progressed to the semi-finals by defeating many European countries, in the media it was reported that the Moroccans were representing also the rest of Africa. However, what many Europeans do not know is that many Moroccans do not consider themselves Africans because they are not "blacks." Africa represents blackness, backwardness, poverty, uncivilizedness. Racial and color-based discrimination that sub-Saharan Africans and African slave descendants face in North Africa is widespread (Menin 2018). The Moroccans who are not considered white enough by racist Europeans and are victims of different forms of racism such as in France (Duroy 2011), despise black Africans. An Article by Hahonou shows that anti-black racism is widespread in all North African countries from Mauritania to Egypt (Hahonou n.y.; King n.y.). Even black Tunisians consider themselves superior to sub-Saharan Africans, and they don't like to be compared (Bryant 2022). For example, in Sudan, those blacks who speak Arabic or whose culture is penetrated by Arab culture do not consider themselves blacks (Falola n.y.). The anti-black racism in North Africa is intuitional and systemic and prevalent in everyday practices such as wealth accumulation, income, criminal justice, employment, housing, health care, political power, and education **(Bougroug 2022). In March 2023,** Tunisian President Kais Saied said that black Africans are criminals trying to change the demographic composition of Tunisia, where 80% agree that there is racist discrimination against sub-Saharan Africans. As usual, when there is trouble the simplest thing is to blame the migrants or ethnic minorities. Blackness is considered so inferior that Egyptians complained over Netflix's depiction of Cleopatra as black (The Economist 2023). Arabs held themselves apart from Africans and viewed them with disdain, enslaved them, set fire to African houses and villages, and slaughtered innocent Africans (Dugard 2003: 209). Arabs find below themselves the blacks against whom they can feel superior. They gain a superior feeling and identity by distinguishing themselves from black Africans. As Hegel would say North Africans are considered more Europeans than Africans (Hegel 1975: 173). That is, North Africans are too good to be Africans.

Blacks are discriminated against and dehumanized not only by the white West. In an Arab Barometer survey (nearly 23,000 interviews) conducted across the Middle East and North Africa in 2021–2022, black Africans are still associated with servile status and racial inferiority. The term slave is regularly used throughout Arabic-speaking countries to refer to a black person (Hilizah 2022: 4). A study by King (2020) of the Arab Reform Initiative find out similar attitudes of contempt toward Black people in general: blacks are slaves, and servants, "Negro" and "Nigger," "monkey," "pig," "cannibal," "animal," "Blackie," etc. "Am I a Blackie?" is used when white Moroccans jest about being asked to do something unpleasant. "True Moroccans" are considered all-white (King 2020). While North Africans, as the paper by Duroy (2011) shows, are exposed to racism and discrimination in many European countries, they can console themselves with the thought that they are superior to the black Africans, the different and inferior Others. The Othered-ness they got from the Europeans they transferred onto Africans. As Sartre would say the Other is the secret of who I am (Gandhi 1998: 17).

In the Gulf countries, hundreds of thousands of African workers from Uganda, Kenya, Ethiopia, Eritrea, etc. are treated as slaves even nowadays (The Economist 2022q). In Central and Eastern Europe, just because of their skin color and even if their number is negligible, black Africans are considered a threat and treated worse than any other minority group even by those who have had no contact at all with black Africans (Bell 2022). The fact that, on 26 April 2023, the Romanian Ambassador to Kenya Dragos Tigau called Africans "monkeys" was just a symptom of widespread Central and Eastern European racism against black Africans (Booty

2023). When Russia invaded Ukraine in 2022 and people fled, black people were pushed off trains, black drivers reprimanded and stalled by Ukrainians as they tried to flee, and according to some reports, even animals were allowed on trains before Africans (Ray 2022). Anti-black racism is widespread also in Russia. Africans are not capable of understanding that they are only useful idiots for Russia. For Russian politicians, President Barack Obama and his wife were just monkeys (Snyder 2018: 254, 269). Blacks in Russia are exposed to different forms of dehumanizing racism (Zatari 2020). As the book by Avrutin shows, whiteness is an essential Russian identity characteristic (Avrutin 2022). Currently, Africans naively ignore the racist Russian attitude toward Africans and they unite with Russians against the West.

## Othering, Bordering, Ordering

The powerful erect borders because they want to be or remain different. At borders, the inside and outside will be defined. The definition of borders is something sacrificial or mimetic because all sacrificial rituals take place by excluding or scapegoating Others. Therefore, borders are "sacred" or *sacer*. From the perspective of the desiring Africans, the othering borders must be overcome to become different from what they were made by the Europeans. For postcolonial Africans, the border contains a fascinating monstrosity or a double nature: on the one hand, it is a *hindrance* on the way to the promised land, but on the other hand, it is the *beginning* point of whiteness, the sublime, and the desirable. Therefore, the postcolonial subjectivity hates it and loves it at the same time. Similarly, from the European perspective, the border is the end of the sublimity and the beginning of the despicability. Therefore, Africans must stay where they are so that the West can preserve its superiority.

Bakewell points out that the roots of the nature of this sedentary bias can be traced back to colonial practice. International development organizations cast migration as a problem that can be solved by keeping poor people "out there". Bakewell suggests that migration policy is a political attempt to keep African migrants in their place (Bakewell 2008a). Besides the sedentary bias, there is something much bigger. It is about *Othering*, *Bordering*, and Ordering (van Houtum and van Naerssen 2002). In racialized social institutions and structures, immigration policies select immigrants by race. Policies institute differential access to rights and privileges based on race through segregation and exclusion. In the racialized United

States, the only path to citizenship and thus greater social, political, and economic incorporation was whiteness (Filindra and Junn 2012: 430, 436).

The European real and virtual walls are erected to keep out the undeserving migrants and asylum seekers. From the perspectives of the immigrants, the higher the walls, the more they would try to scale them because there must be something worthwhile and desirable behind them. The worthwhileness of that something beyond the wall is directly proportional to the height of the wall. From the perspective of the destination societies, the more the migrants attempt to climb over the dangerous fences, the stronger the perception of their superiority. Therefore, the walls must be even higher and the attempts to cross them must be made more difficult.

In their article, *Bordering, Ordering* and *Othering*, van Houtum and van Naerssen (2002) argue that we relate bordering to the practices of othering through the practices of ordering and the discursive differentiation between us and them. Bordering is the immobilization of people through spatial differentiation. We regulate mobility, construct or reproduce places in space, reinforce otherness on both sides of the border, and cleanse the other that lives inside an imagined community as a competitor, enemy, and stranger (van Houtum and van Naerssen 2002: 127). Van Houtum and van Naerssen point out the paradoxicality of borders because they simultaneously reject as well as erect othering. "...borders are erected to erase territorial ambiguity and ambivalent identities in order to shape a unique and cohesive order, but thereby create new or reproduce latently existing differences in space and identity. ... The making of a unique, exclusive place goes hand in hand with governing practices of exclusion and purification. Exclusivity of a territory is always Janus-faced." (van Houtum and van Naerssen 2002: 126, 127). These Othered confined beyond the border are "...the socially '*dirty*fied' people, the 'Heimatlosen' [stateless], the 'displaced persons', the illegal immigrants, the people without papers and/or economic resources..." (van Houtum and van Naerssen 2002: 129).

In the context of slavery and exclusion and exploitation of blacks, Du Bois argues that the majority of people resent the idea of equality due to the spiritual and cultural desire to keep up one's privileged position based on the degradation, exclusion, and slavery of the others (Du Bois 2007: 68). As Nietzsche would say, those who desire to be exalted would look up, and those who are exalted look down and laugh at the tragedies of those who look up (Nietzsche 2003: 171). The othering, bordering, and ordering of the outsider is not only a political project. It is also a kind of secular sacrificial ritual that keeps the harmony of the community through exclusionary generative unanimity in the act of distinction between within and without (Girard 2016: 302–303) thanks to the existence of the sacrificial outsider.

In the face of political insecurities and weakening institutions politicians and xenophobes identify migrants as threats, problematic and dangerous outsiders, and at the end they will be dehumanized, criminalized, and securitized (represented as a security threat) as a matter of national security (Menjívar et al. 2019: 3–5; 12; Hollifield 2012: 352–53), or as Bello calls it a discursive construction of migrants and migration as threats self-fulfilled or self-evident (Bello 2017: 58). In a volatile political and social environment that looks for scapegoats for the ills of the society, migrants represent social, political, economic, cultural crisis, crime, terrorism, and security threats (Arana and McArdle 2019; Koslowski 2012). Any statistical facts-based information that disproves these allegations cannot change these convictions. The linguistic dehumanization of migrants (Pombo 2019: 490)—"wave", "avalanche", "invasion", "swarm", "tsunami", "current", "flow", "floods", "tide", "plague", "terrorists", "illegal aliens", "radical Islamists", "grave social risks", "threatening elements", a "disposable piece", the "unclean" from which the society has to be cleansed, "disfigurement of European women's beauty", etc.—justifies violence and hostility because their humanity is consciously or unconsciously denied and they are denounced as "undesirables", and inferior (Falguere 2019; Christian Aid 2007; Turner 2010; Hatton 2009; Pombo 2019: 495; Farer 2020: 23; Kingsley 2017: 43).

The linguistic dehumanization is followed by political dehumanization in which only naked survival as an animal, a scum of the earth, is provided without political rights or socio-economic rights as many asylum and refugee centers show (Agamben 1998; Ahrendt 2017: 349–352; Bello 2017: 71). These dehumanized beings are alive but do not exist because of the various types of ontological deprivation such as identity deprivation, separation from common humanity, and loss of membership in the political community (Ahrendt 2017: 351–352). Albahari shows the genealogies of European othering, bordering, and ordering taking place through what he calls genealogies of confinement, detention, pushbacks, exclusion of migrant detainees from citizens, deportation, repatriation, drownings, etc. (Albahari 2015; Wonders and Jones 2021:299–300; Bello 2017: 70–71).

As Wodak suggests, exclusion is linked to power because it is only the one who has the power who can exclude and holds an instrument of super-ordination, superiority, and contempt, of surveillance, control, and discipline (Wodak 2011: 54; Flam and Beauzamy 2011: 222). Exercise of this power results in profound emotional wounds. From the powerful *othering* perspective, Flam and Beauzamy explain "the non-physical hurt experienced by migrants: daily and routinely migrants confront different forms of rejection that can be intimidating, humiliating and incapacitating; repeated experience of such rejections causes feelings of fear, inferiority, and reserve." (Flam and Beauzamy 2011: 221). Shame, rejection, inferiority, search for identity, etc. characterize the postcolonial African personality. Migrants are personified fear, which Nyman calls the zombification of the migrant (Nyman 2020: 19). Nyman calls this situation a Manichean view of the world in which the good "we" stand in front of the zombie "other" (Nyman 2020: 20).

Borders are not only physical places neutral lines that divide or link two different geographies (Sbri 2020: 36). They are the process of exclusion and inclusion, which can be everywhere, where the "Other" is suspected to be (Sbiri et al. 2020: 2). Borders follow the weakest, the different, the outsider. Therefore, borders are omnipresent and omnipotent because they have the power to affect the migrant's perception of the self and its interaction with other Selves. Borders make migrants visible, objects of hostile discourses, anxiety, despair, and racialization. Through borders, migrants become new beings (Sbiri et al. 2020). Migrants encounter through the border their new being—the Otherness, the new identity—attributed to them by the powerful.

Migration is increasingly associated with crisis where racial and cultural differences have become pathological with migrants depicted as harbingers of social, security, cultural breakdown, crime, poverty, and source fear and hostility (Menjívar et al. 2019: 4; Richards 2019: 38; Nawyn 2019; Wonders and Jones 2021: 300). Based on ethnicity and regional or religious background the othering process constructs and essentializes the irreconcilable identities of the newcomers and the locals. This othering represents migrants as a personified disruption of national ideals and national projects of ethnic and cultural homogeneity; they become scapegoats, producing a hysterical response and target of complaint (Menjívar et al. 2019: 8; Richards 2019: 47; 52; Bello 2017: 88–92). The dehumanized migrants are seen as creatures that have brought crisis into society, a situation in which a purifying verdict or judgment has to be executed by scapegoating them for all the ills of society (Sager 2019: 591–593; Farer 2020: 13). This negative-othering (Pedersen and Hartley 2015) justifies the "State of exception" (Agamben 2005). The state of exception will be implemented for sovereignty is perceived to be under threat, the state's power will be extended, normal legal order will be suspended, and individual rights will be stripped.

However, those who oppose immigration at the destination as well as those who want to immigrate are not always interested in facts regarding the causes of as well as migration solutions. Some research shows that the fraction of international migration is only between 3% and 3.6%; that the EU gets less than half of one percent of the EU population, that most of these migrants are legal migrants with job offers, that immigration is good for the economy of the destination countries, that low-skilled immigrants generally do not hurt the wages and employment of the natives, that the influx of 2015 and 2016 was an exception, that one asylum seeker for every 2500 EU residents is negligible, that official statistics show that immigrants are not involved in more criminal acts than the natives or the whites or the Europeans, etc. (Banerjee and Duflo 2019: 10, 22–23, 109). However, despite such an abundance of facts provided by experts, many people at destination oppose migration not because they do not know these facts. Facts are often irrelevant. It is not about alternative facts but rather about life beyond facts, it is about the belief that the truth or fact does not sway their opinions (Banerjee and Duflo 2019: 11). Those who oppose immigration argue that immigrants bring economic, cultural, and security risks. However, there is no objective proof that a person sharing the same culture, or religion or having the same skin color is more benevolent than a person with a different skin color, religion, culture, or political history. The reason why many people reject immigration in spite of an abundance of information regarding immigration is not that they do not understand economics or politics; it is rather they are driven by identity politics, and perceived cultural distance between the natives and newcomers (Banerjee and Duflo 2019: 50). Societies with almost no immigrants believe that immigrants threaten the nation, the local culture and values (Banerjee and Duflo 2019: 109; Jones et al. 2015; Eatwell and Goodwin 2018: 151, 166). When the opponents of immigration in Europe are convinced that the welfare of the receiving countries is collapsing—women are raped, the nation is in decline, culture is destroyed by ethnic change by culturally incompatible Muslims and refugees, European civilization is coming to an abrupt end because of Islamization, and European history is at risk of disappearing—facts-based arguments or objective reality alone against this belief are not helpful (Eatwell and Goodwin 2018: 37–38;

147; 272). Net migration from Poland, Hungary, Romania, and Bulgaria to Western Europe is very high, and net immigration into them is extremely low. However, nationalism and anti-immigration sentiments are widespread in Eastern Europe (Eatwell and Goodwin 2018: 141–142). If someone is convinced that migration is changing one's country in ways he/she does not like, objective facts-based counterarguments do not have any value (Eatwell and Goodwin 2018: 148). Many politicians believe that if there are more jobs for locals, more economic growth, and fewer austerity measures, the voters' drift toward the right-wing, anti-migrant, and racist populists would stop. They ignore or misunderstand the fact that these objective or rational facts cannot address the irrational factors associated with immigration and rapid ethnic change (Eatwell and Goodwin 2018: 261; 274). Migrants are accused of stealing local jobs or benefitting unfairly from the welfare system or services (Walia 2021: 202; Mitsilegas 2019). Data that disprove this often do not count. Prejudice and racism are not something that can be fought against by means of more additional information and data (Du Bois 2007: 98). So, it is not about objective facts but instead about psychological and perceived facts. Some politicians compare migrants with invasive species like weeds (Walia 2021: 208).

Othering is always based on the way we perceive ourselves in relation to others and in the discourse of belonging and non-belonging, which creates the social location of insiders and outsiders and the relationship between self and other and collective identity (Jones and Krzyzanowski 2011: 52; Sbri 2020: 42). The next step of othering is bordering and ordering through which the politically desirable and normal is protected from the undesirables through fences and barbed wire which keeps the undesirables, the disruptive, dangerous, different, unworthy que jumpers outside the reach of the politically acceptable and desirable culture, way of life, and security as the Viktor Orbán's fence policy since 2015 shows (Cantat and Rajaram 2019), and their cases will be processed offshore or in a third country, as the UK has been trying to send asylum seekers to Rwanda to process their asylum claims. This externalization of the asylum process, as Australia does in its "Pacific Solution" is ordering through othering and bordering (Tazreiter 2019: 626). Whereas migrants from rich countries are desirables, those from poor countries are undesirables in Europe (Huete 2008) since human rights are enjoyed only by the rich, the "civilized" and the "regulars" (Kaya 2020: 34; Ahrendt 2017: 241ff). Gilmartin and Kuusisto-Arponen (2019) suggest that depending on their desirability, migrants are stratified, hierarchized, and classified, and the undesirables are isolated and exposed to uncertainty and hardship depending on their status, origin, and skin color. Your skin color becomes your own border, the maker of your failure, the border you cannot overcome. Perversely, your oppressor who rejects you because of your skin color has the power to save you from your own skin color. Your oppressor has the power to accept your skin color and face as the face of a human being (Bello 2017: 59). The undesirable immigrant signifies illegitimate mobility as those with greater power and resources contain the mobility of those with less power and fewer resources (Wonders and Jones 2021: 296). Migration is a benchmark against which we can measure the power of different migrant groups from the perspective of skin color, wealth and social status, countries of origin, religious and cultural background, etc. Usually, the wealthy, and privileged are more welcome in the West than others. As Wonders and Jones argue, migration *borderings* (re)produce exclusion, difference, and social inequality (Wonders and Jones 2021: 297).

As Garelli and Tazzioli, suggest there is an increasing convergence between humanitarianism and the military in which the military forces are deployed to perform humanitarian tasks and the humanitarian works are militarized in which the NGOs utilize military technologies leading to spatial containment away from Europe, prevention of the migration to Europe and keep migrants off the European shores, through war on smugglers (Garelli and Tazzioli 2019; Casas-Cortes and Cobarrubias 2019). Othering, bordering, and ordering determine who is deserving, who should be externalized, where the push factors should be eliminated, who should move and who should not move around the world, and why and how the destination countries can make the borders mobile based on their interests and definition of threats, as the "external dimension of migration policy" according to the Tampere summit or the Austrian presidency strategy paper on immigration and asylum policy of the Austrian Council Presidency from 1998 shows (Casas-Cortes and Cobarrubias 2019; Kasparek and Schmidt-Sembdner 2019).

As Papstavridis argues, the undesirables cannot be always easily rescued because "…there is no clear obligation on the part of the flag states or the coastal state to accept the rescued persons in their territory … And, even if private mariners do save people at sea, they may face the threat of sanctions imposed on them for facilitating illegal entry…, and coastal states of ten dispute the boundaries of their respective search and rescue regions…" (Papastavridis 2020: 230).

Therefore, what is being policed is not just borders but also the immigrant bodies around which securitization and militarization take place with the help of border guards, fences and uniformed troops, and surveillance hardware, such as drones, helicopters, and armored vehicles (Kuusisto-Arponen 2019: 18–21; Bello 2017: 63). Similarly, Geisen et al. (2008) show the close connection between migration and securitization in the context of European asylum and migration policies. A good migrant will be constructed, who has permission to enter and stay, who is not a trouble-maker, well-qualified, deserving-migrant, etc. (Kuusisto-Arponen 2019: 25).

Migration management is not only security *for* migrants but also security *from* migrants through the help of camps, return and repatriation policies, exclusion from the territory, containing the refugees in the "region of origin", through bilateral and multilateral agreements, secure third countries, a combination of migration policy with "war on terror", or through *governmentality* in which on the one hand the refugees or migrants *feel* involved but *not* change the actual structures of management and the society (Weima and Hindman 2019; Collyer 2019). As Raeymaekers suggests, states in the destination use tactics of spatial distancing, exclusion and segregation, illegalization, racialization, social and political differentiation, offshore detention, externalization of migration control and separation of the unwanted migrants, the establishment of legal no man's lands, and privatized surveillance as a means of border governance which separates citizens and non-citizens socially, culturally, and politically through differentiation and subordination.

Migrants serve the role of defining a line in society between an "us" and a "them" or "other" that is inferior and alien to the insider group. Therefore, foreigners can become a convenient target to deflect and redirect the attention of the public away from criticisms that could threaten the position of the dominant group. Foreigners become scapegoats for weak economic growth, unemployment or basic socio-economic inequality, and decreased welfare benefits for the locals.

"The nationalist can translate loss of relative economic position into loss of identity and status: you have always been a core member of our great nation, but foreigners, immigrants, and your own elite compatriots have been conspiring to hold you down; your country is no longer your won, and you are not respected in your own land" (Fukuyama 2018: 89). Therefore, some have argued that increasing immigration can lead to weakened national solidarity (Kymlicka 2015).

Forced marriage and honor killings will be instrumentalized to make a distinction between locals and immigrants. A pseudo-discourse on values starts: by instrumentalizing the rights violation of women or homosexuals to violate the rights of migrants / non-European foreigners. These groups will be personified as threats to economic prosperity or cultural identity by the radical right and even mainstream parties. Immigration will be securitized as personified "threats" that menace the physical well-being of host populations. "The more often the press mentions a particular issue and links it to a social ill, the more likely that issue is to be considered a 'crisis' meriting political action and resolution" (Caviedes 2015: 900). Migrants embody the collapse of "order", harbingers of bad news, threats, and risks, as Hungary's Prime Minister Victor Orbán said, "all terrorists were migrants" (Bauman 2016: 15, 31). Similarly, a study by Adeyanju and Neverson discusses, how the media in Canada represented immigrants as threats to racial identity and harbingers of health risks to the local people (Adeyanju and Neverson 2007). As Roger Cohen puts it, "...big lies produce big fears that produce big yearnings for big strongmen" (Cohen 2015).

The performing of crisis creates order by erecting borders between the desirables (insiders) and undesirable others (outsiders), what scholars call spectacular or ostentatious bordering through which the state tries to show its authority (Andreas 2009: 8; De Genova 2013: 1181; Bello 2017: 72). The ritualizing perpetuation of crisis and its utilization is what Cantat and Rajaram call a performing crisis (Cantat and Rajaram 2019), constructing official cosmic fear caused by immeasurably powerful and immeasurably great crisis (Bauman 2016: 50–51). Even though the number of refugees in Hungary was extremely low (1216 in 2017), the crisis discourse was the heart of Hungarian political life focused on the collective Hungarian identity, culture, and Christianity by producing others, erecting borders, and creating an order for own citizens as a protector of Christian Europe (Cantat and Rajaram 2019: 186–188).

The Western fear of nothingness is continued by the association of migration with criminality. Various empirical academic studies as well as data from Ministries of Internal Affairs show that the correlation between immigration and criminality is generally insignificant. In their book "Does Migration Increase Crime"? Fasani et al. explore a potential link between immigration and crime rate. These authors use various studies by other scholars who conclude that in general immigration is beneficial for the destination economy, it does not increase crime, and access to legal status reduces the number of crimes committed by immigrants in some countries (Fasani et al. 2019; Casas-Cortes and Cobarrubias 2019: 202).

Borders are not only territorial. They are omnipresent to separate the being and nothingness. They are hierarchical. It is not only that people go to the borders, but also borders exist where differences and hierarchies exist. As Walia suggests, border governance is territorially diffuse because it can take place within the state as well as outsourced beyond the state's border. It is elastic and can exist anywhere, mobile always looking for the undocumented and looking different (Walia 2021: 84). Borders are mobile because they can be externalized and deterritorialized through offshore detention centers, safe third-country agreements, and outsourcing of border control to third countries. (Walia 2021: 84–156; Bello 2017: 68; Sbri 2020: 38). The exporting of border management technologies, and the signing of readmission agreements with the countries of origin and transit itself is the mobilization of borders (Basheska and Kochenov 2015: 61; Eisele 2019).

Borders become de-territorialized, they differentiate between citizens and non-citizens, between formal and informal workers, between race, class, and gender, between regular and irregular migrants, unwanted and wanted, between blacks and whites, between man and woman, abled and disabled, homosexual and heterosexual, etc. Borders are a symbol of power and status (Johnson 2014: 80, 152–153). Borders determine being and nothingness because they decide autonomy and (un)certainty (Johnson 2014: 154). Borders weed out, select, filter, classify people, and segregate unwanted migrants (Casas-Cortes and Cobarrubias 2019). Trump's five-hundred-billion-dollar wall surrounded by a moat full of alligators and topped with electrified fencing would not be built against white Europeans, but instead against Latinos or Africans (Walia 2021: 79).

Borders are not just geographic, territorial, or political, but instead, they are also (social)psychological. We love borders because they constitute us. However, the Other constitutes us at the same time. As Balibar says, "There is no *given* identity; there is only *identification*" (Balibar 2002: 67). Our identity is constructed in relation to our status in the environment we live (Burke and Stets 2009: 19). However, we question the identity ascribed to us by our counterparts or the society because it has an essential impact on our self-esteem. Self-esteem is the function of what we are and what we want to be, i.e. the level we are on divided by the level we aspire to in order to achieve our ideal identity or social status (Burke and Stets 2009: 24; James 1890). As Balibar suggests, "Identification comes from others, and continues always to depend on others." (Balibar 2002: 67). Identities are always (re)created in specific contexts, co-constructed in interactive relationships, and identity construction always implies inclusionary and exclusionary processes by creating a *ritualized* border between "Us" and "Them" (Wodak 2015: 71; Balibar 2002: 66–67).

It is not only the "other" who is (re)constructed in his "Otherness", but also the "Self" that is the national singularity and homogeneity is imagined and constructed; members of the national community simultaneously exaggerate distinctions between themselves and others, a collective past, a collective present and future, a common culture will be remembered and reconstructed (Wodak 2015: 77; Balibar 2002: 44). It is the *border* that creates an *order* between the *other* and the Self for the protection of the latter. For Balibar, cultural characteristics based on imaginary "similarity" exhibit the individual's belonging to the community as a common, physical or spiritual, "nature" or substance', allegedly manifested in the resemblance of outward appearance, behavior, and gesture." (Balibar 2002: 68). Through bordering and othering, real or symbolic death of the bad [black] race, the inferior, the degenerate, and the abnormal makes the life of the white racist colonizer healthier, purer, and sublime (Mbembe 2021: 205).

*Bordering* and *definition* are the same etymologically and concerning content. As Balibar argues, "The idea of a simple definition of what constitutes a border is, by definition, absurd: to mark out a border is, precisely, to define a territory, to delimit it, and so to register the identity of that territory, or confer one upon it. Conversely, however, to define or identify in general is nothing other than to trace a border, to assign boundaries or borders … the very representation of the border is the precondition for any definition" (Balibar 2002: 76). Bordering as identification (attribution or construction of identity) is based on discrimination or distinction between the national and the alien, between the internal and external, between the normality of internalized national citizen-subject and *alienated* external threat. This is a ritualized, continuous, and quasi-religious process internalized by individuals and not an objective fact. Depending on whom the community wants to exclude for the sake of its own identification and existence, these borders are invisible, everywhere and nowhere, and therefore selective and conditional (Balibar 2002: 78). Balibar describes three characteristics of borders: *overdetermination*, *polysemic character*, and *heterogeneity*. In the case of overdetermination, an alien is not only an alien, but he can be defined as an enemy. Polysemic character means that borders do not apply to everyone equally because they differentiate between individuals in terms of social class, ethnic background, skin color, wealth, type of passport one owns, country of origin, etc. Regarding heterogeneity, Balibar suggests that geographically speaking borders are not situated at the borders at all and they are everywhere by selectively controlling and separating between the must-be-protected internalized nationals and alienated or externalized threats (Balibar 2002: 78–85).

One of the characteristics of standardized identities is that it *others*, *borders*, and *orders* based on the "politics of fear" (Wodak 2015). Immigrants' identities in Europe are the lowest of the low, they are the locus of exploitation, confined to ghettoized environments, and othered (Balibar 2002: 43). For the aspiring migrant, nothingness on this side of the border increases the imagination of being behind the border. The being of the colonized is the creation of the colonizer (Fanon 1991; Mbembe 2017: 151). For the postcolonial migrant, life exists only beyond the border where the colonizer is. This aspired life at the destination is more than mere physical existence (*zoe*) at origin and rather a life as being someone (*bios*) at the destination, a life through which the human being enjoys the beautiful, as Hannah Arendt would say in her *The Human Condition* (Arendt 1998: 12–13).

## Postcolonial African Migration to the West as a Mimetic Desire for Being

Speaking from an existentialist point of view, postcolonial African migrants want to make sense of their existence, to overcome the racialist and colonially imposed mere animal existence by venturing beyond the border, taking risks in order to find human fulfillment. Through their actions and choices (which were denied them during the colonial time), the postcolonial migrants reject the conditions into which they have been thrown, that they were predetermined inferior "finished being". They would like to *become*, shape their own existence, transcend their condition, and achieve a new actualization, as transcending subjects (Gosetti-Ferencei 2021: 5–7). Slavery and colonialism denied African or black men subjectivity (free and self-reflective consciousness and individuality). Therefore, postcolonial African migration is a postcolonial reordering: the materialization of one's capacity to evolve in the light of possibilities, potentiality, and subjectivity. The postcolonial migrant African tries to de-*Other* himself, draw a new border, re-*Order* through imitation of the colonizer, and ultimately to be. Decolonization is an act of refusal of the colonial and reassertion of the postcolonial subjectivity, a self-re-foundation, self-creation, and invention (Mbembe 2021: 44). The existentialist postcolonial migration from Africa is not only the recognition of the particularity of the black man's individuality but also the rejection of the predetermined path of colonial "facticity" or the "factical" situation into which he was thrown (Gosetti-Ferencei 2021: 9). From this perspective, migration is a libertarian free choice, creativity, self-determination instead of predetermination, in a word becoming, instead of "finished being" (Gosetti-Ferencei 2021: 10).

Slavery, racism, and colonialism are acts of dispossessing Africans comprehensively. Africans were people without a past and without history, located outside of time, and ontologically incapable of change and creation. Decolonization as a result means "to own oneself", a step toward the creation of new forms of fully human life, a struggle for life, a self-creation, to become one's own foundation, dis-alienation, creation of new forms of life that could genuinely be characterized as fully human, not to imitate preexisting models (Mbembe 2021: 53–55). The African was first a slave or colonized Other; then he is a foreigner-Other, an illegal migrant "Other", the figure par excellence of the intruder and the undesirable Other who has to be kept afar through the desire for borders, through reactivation of technologies of separation and selection associated with the desire to check his identities, and through the logic of expulsion because he is different and dangerous and personifies cultural and moral threats (Mbembe 2021: 134).

Postcolonial migration from Africa to the West is an existential freedom to define oneself autonomously against colonial *othering* by reestablishing the intrinsic dignity grounded in the liberty and rejection of colonial reality not by rejecting the colonizer but paradoxically by imitating. Interpreted from an existentialist perspective, the postcolonial migrant from Africa is a free "individual subject fighting against an oppressive world" by desiring recognition by the same world.

Postcolonial migration and migration policy are a theater of othering, bordering, and ordering on the one hand and de-*Othering*, debordering, and reordering on the other. The model and the imitator are trapped at this separating border for eternity, each needing recognition from the other that will never be granted (Sartre 1993; Gosetti-Ferencei 2021:

100). As the Heideggerian concept of being-in-the-world implies, the border does not necessarily separate the two sides but rather shows the inseparability of our being with the world, i.e. our relation to the world around us makes us what we are (Gosetti-Ferencei 2021: 149; Heidegger 2010).

Through his "all-things-are-in-flux," Heraclitus suggests this eternal inseparability of the model and the imitator. The model knows that the imitator wants equality, which would mean the end of the model's being. The more the imitator strives to be equal the fiercer the model has to defend his superior position by dehumanizing the model. The more he dehumanizes the disciple or the imitator the more the disciple desires to change the status quo. The existence and essence of the model depend on the admiration, adoration, and imitation by the disciple. Paradoxically, the disciple is the creator of the model.

In the *us-them* category, the Other is, on the one hand, an existential threat, the harbinger of crisis, but at the same time an essential part of the Self. There is an identity conflict between the outsiders (aliens), and others, them on one hand and insiders, us, on the other. But both categories are existentially and essentially interconnected. The Self and the Other are the creators and creations of each other simultaneously in a monstrous relationship in which one being exists in another. The Othering is an unconscious Self-procreation. This is the Heraclitan unity of the opposites and the fluidity that we are incapable of grasping. While we concentrate on the building of the separating borders, we fail to thank the existence of the "*bordered*" and "*borderable*" Other and its existential necessity. As Bauman puts it, the enemy gives an added advantage to politicians in frantic search of voters: such a call is bound to rouse the nation's self-esteem (Bauman 2016: 34). The strangers can be and must be blamed for all the grievances, uncertainties and disorientations, and a savior promises to the nation peace, prosperity, purity and unity by shutting these strangers out (Bauman 2016: 64–65). However, the Other must exist. Otherwise, the Self ceases to exist.

This unending exchange of roles is what Heraclitus called *panta rei*, "all-things-are-in-flux. Heraclitus (6-5 Century BC) suggested that everything is in eternal flux and there is no distinction between, the creator and the created; in discord is concord; bow and lyre are in harmony; all things move, the opposites are the same; all things are one" (Kirk 1975: 14–15, 65). "The unity of all things is for Heraclitus proved by the essential unity of apparent opposites ... Opposites are the same because they inevitably succeed one another: they are different degrees of the same quality, or different poles of the same continuum" (Kirk 1975: 72). Similar to Heraclitus, Eco suggests, "It seems we cannot manage without an enemy. The figure of the enemy cannot be abolished from the process of civilization. The need is second nature even to a mild man of peace … War enables a community to recognize itself as a 'nation'; a government cannot even establish its own sphere of legitimacy without the contrasting presence of war; war ensures the equilibrium between classes … Peace produces instability and delinquency among young people; war channels all disruptive forces in the best possible way, giving them a status" (Eco 2013: 17–19). Moreover, Eco says "We can recognize ourselves only in the presence of an Other, and on this the rules of coexistence and submission are based" (Eco 2013: 21). It is not the nature of a thing that gives an identity to it, but instead, the identity of a thing is dependent on the act or process of bordering or defining, by not forgetting the Other, the enemy who is ultimately my friend and my brother.

In Fragment 48 of Heraclitus' fragments we can read that *bow* and *life* have the same etymological origin, which shows the coexistence of the opposites constituting one another. In Heraclitus's own words, "For the bow the name is life, but the work is death" (Kirk 1975: 116). For Heraclitus, these extremes succeed one another and are essentially connected. In Fragment 48, he suggests that Hades and Dionysus are the same (Kirk 1975: 121). Similarly, in fragments 23 and 111, according to Heraclitus, opposites are complementary to each other and one extreme cannot be imagined without the other (Kirk 1975: 123). Similarly, Heraclitus argues in fragments 57, 88, 99, 126, etc. that opposites are essentially connected as extremes of a single process (Kirk 1975: 134). In Fragment 53, he formulates the mutually conditioning coexistence of the opposites as follows: "War is the father of all and king of all, and some he shows as gods, others as men; some he makes slaves and others free" (Kirk 1975: 245). For him, *polemos* as war, strife, or reaction between opposites—taken literally or metaphorically—signifies the existential paradox of human beings. I see the postcolonial migration in this light. "Where one side meant servitude, the other meant freedom" (Lukasik 2021: 306). The black African being is the Hugoan Quasimodo of slavery and colonialism, the personification of ugliness and evil, the absence of good and morality, and a repellent being (Eco 2013: 100). The "perfectly ugly" Quasimodo incorporated all negative qualities: he looked like a giant broken to pieces and badly cemented together, humpback, one-eyed (worse than blind!), bandy-legged, ugly monkey, wicked, the very devil, lives with cats, attends itches' Sabbath, with disagreeable face, villainous creature, all women—especially pregnant and the hoping—hiding their faces from him (Hugo 2006: 37–38). The beauty of such a society is existentially dependent on the ugliness of the Quasimodos (Eco 2013), as the Western sublimity was dependent on the subjugation of the black Africans.

As Memmi suggests, the living standards of the colonizer are high because those of the colonizer are low; the former is rich because the latter can be exploited and does not have any law that protects him (Memmi 1967: 8). The privileges and glory of the colonizer are essentially emanating from the degradation of the colonized. The colonizer has to use the darkest colors to depict the colonized, devalue him, and annihilate him. The colonizer has to demonstrate the irretrievable opposition between his glorious position and the despicable one of the colonized (Memmi 1967: 55). "The colonial situation manufactures colonialists, just as it manufactures the colonized" (Memmi 1967: 56). The "thingified" white criminal in his own country of origin becomes a dignified colonizer and superhuman when exiled to the blacks (James 2022: 28–29).

In his Souls of Black Folk, Du Bois describes the idea of double consciousness which consists of an oppressed, enslaved, and segregated person's dual relationship to herself, experiencing the self as a twoness: an inwardly free, unchained, and aspiring, however externally burdened and restricted by oppression (Gosetti-Ferencei 2021: 128; Du Bois 2009).

The irresistible life impetus of the postcolonial African being with the will-to-be persists, the will which the colonizer and the slave-holder were not able to destroy. Even if the colonizer and slaveholder wanted to crush the African being to mere animality, fatalism, and predeterminism (Flynn 2006: 42; Lukasik 2021: 112), the African will-to-be persisted and resisted it. The proper human way of existing through subjectivity, authenticity (to live a free, own, real life), and freedom to choose to be the person one wants to persist. On the one hand, the black Africans were treated as animals, however, on the other hand, the colonizer and the slaveholder knew that African humanity was indestructible. Therefore, he kept sharpening the weapons with which he could secure victory by dehumanizing Africans (Fanon 2001: 33).

From an existentialist perspective, colonialism is the denial of possibility and future, essentialization of determinism, congelation in the past, and denial of the proper human way of existing (Flynn 2006: 44; Sartre 1993; Camus 1991a). However, the real existential crisis of the postcolonial

African migrant to the West starts at the destination. For the first time, he will be aware of his nothingness because at the destination his social status, economic situation, and racism due to his skin color will show him who he is more than ever. This immaterial omnipresent and omnipotent border pursues him wherever he is and underlines his otherness and nothingness most intensively. In the face of this experience of nothingness and *invisibilization*, the postcolonial African migrant in the West fights for recognition and being (Carter 2013: 71–72). The African being becomes *hyper-visible* because of his skin color and *invisible* because of racist marginalization. His skin color is his eternal prison. The African migrant is hyper-visible not only as a refugee, asylum seeker, or alien but primarily as black. This is a trait, which hyper-*visibilizes* him in a white world. Even if the number of Africans migrating to the West is proportionally less than from other regions of the world, the migration policy debate in Europe focuses almost always on African migrants because of their *hyper-visibility*.

The African migrants attempt to escape this liminality but they end up in even more liminality or invisibility. Through migration, they want to gain a clear-cut identity, but they often achieve the opposite. "No matter their legal status, professional standing, or educational level, most immigrants never fully escape their liminality (Stoller 2013: 159). This liminality results in loneliness, alienation, and longing for home. They feel like "lost souls" belonging neither to the origin nor destination (Stoller 2013: 167–68).

Human subjectivity stands on two pillars: liberty or negative identity (absence of dehumanization, exploitation, and rejection) and positive identity or recognition through equality and justice. Colonization has compromised this subjectivity because it denied these two pillars of liberty. The colonized African, therefore, endeavors to reverse the colonial relations and to resemble the colonizer in the frank hope that he may cease to consider the colonized different from himself (Memmi 1967: 15). Therefore, we can say that the postcolonial African migrant subjectivity is *mimetic* because it is based on the desire for being through imitation.

Through its externalization agenda, the EU tries to engage countries of origin and transit to control or manage their borders (Adam and Trauner 2019). In 2022, the UK and Rwanda agreed to offer asylum seekers and migrants "safety and the opportunity to build a new life in Rwanda". Gerald Knaus in his book *Welche Grenzen brauchen wir?* (Which borders do we need?) highlights at least three arguments, especially in the chapter on African migration to Europe. Firstly, building borders between Africa and Europe is not a solution. Secondly, sub-Saharan Africa is not the main origin of migrants to Europe. Thirdly, Europe could prevent African migration through more investment in education, economy, vocational training, return agreements, etc. He suggests something similar to the EU-Turkey refugee agreement model in Africa (Knaus 2020: 200–249). Many Europeans seemed to believe that his policy suggestions would stop African migration.

However, as many researchers argued, the effects of an investment in education, the economy, etc. can have a migration-decreasing impact only in the very long term. However, Knaus presents his policy suggestions as if they would prevent African migration in a short time. Migration researchers like Knaus, or the EU, Rwanda or the UK, and other countries who aim to establish asylum processing centers in a safe third country outside of Europe, of all places in Africa, fail to understand that asylum seekers do not always seek safety in any country, but instead specifically in the West. Africa is the last place they would select because postcolonial African migration is more than a mere economic issue. Migration to the West is a mimetic desire to overcome the historical experiences of psychological boundaries and nothingness emanating from the legacies of slavery and colonialism.

The more the Africans are excluded the more they attempt to "de-border" and "reorder". The postcolonial African migration to the West is a rejection of the racist *thingification* and *invisibilization* of the African being and an imitation of the West through adoration and admiration in the hope of achieving equality and recognition. It is a transcendence of rejection of a predetermined African place in the order of human beings. In this sense, postcolonial migration is a choice and decision in one's self-becoming and protest against blind fortuity imposed by the colonizer (Gosetti-Ferencei 2021: 115; Nietzsche 2007).

Migration is a historically conditioned mindset rather than a mere spatial movement of postcolonial Africans to the West. The argument "It is the Africans' right to migrate like the white slaveholders and colonizers who went to Africa without asking any permission" is itself mimetic. "The white man arrived in Africa by sea without a visa ... We have learned to travel from the white man." (Kingsley 2017: 56).

This postcolonial migrant to the West does not intend to realize liberation and subjectivity through detachment from the colonizer but instead through imitation. The dehumanized African being tries first to

"Europeanize" or rehumanize itself and hopes to come to the same level as the colonizer. To be, the postcolonial African person has to coexist with the Europeans, imitate their lifestyles, consume the same goods as the Europeans, and even try to be fair-skinned or have a white partner (Jeffers 2022: 270). Even among siblings, those with darker skin are considered inferior (Jeffers 2022: 693). One has not only to appear white but also act white (Lukasik 2021: 203). "The closer one's skin tone was to white, the higher one's status was in the mixed-race and black communities..." (Lukasik 2021: 97; 109). The blacks were hated even by the Mulattoes because they were slaves and blacks. Many slaveholding Mulattoes even objected to the abolition of slavery (James 2022: 76, 110). James describes this situation in San Domingo:

> Black slaves and Mulattoes hated each other. Even while in words and, by their success in life, in many of their actions, Mulattoes demonstrated the falseness of the white claim to inherit superiority, yet the man of color who was nearly white despised the man of color who was only half-white, who in turn despised the man of color who was only quarter white, and so on through all the shades ... a white skin guaranteed to the owner superior abilities and entitled him to a monopoly of the best that the colony afforded. (James 2022: 36–37)

Similarly, Fanon puts it as follows: "..., the individual who climbs up into white, civilized society tends to reject his black, uncivilized family" (Fanon 2008: 128). Lukasik highlights the deep psychological dimension of passing. She shows that passing is associated with death, for example, passing on or passing away is to die. She says that passing out is a temporary death, to pass away from the mixed-race identity that keeps one in an inferiorized status, a symbolic death, and impermanence, to cross over to the other side, to pass over to the promised land of Whiteness. The Judo-Christian feast is also called the "Pass-over" or Easter for Christians. "And when a person passes, crosses over to the other side, when a part of them dies, whom do they become?" (Lukasik 2021: 80). Postcolonial migration from Africa to the West is a passing-over physically as well as metaphorically. It is a black entreaty to be recognized as an equal human being.

Wealth and beauty also implied literally or metaphorically whiteness. "black and white represent the two poles of this world, poles in perpetual conflict: a genuine Manichaean notion of the world. There, we have said it—black or white, that is the question" (Fanon 2008: 27). Similarly,

Degruy [herself being black] reports about an episode in her book: "My mother told me to never bring home anybody as black as me" (Degruy 2017: 115). From the *Posttraumatic Slave Syndrome* perspective, she writes, "I could imagine how it must have felt for her, as a black woman, growing up and believing there was something wrong with her skin color, then having a child who served almost as a curse and constant reminder of her own unworthiness." (Degruy 2017: 115).

Postcolonial migration from Africa is a manifestation of the deep-rooted struggle for liberty and indomitable hope to be oneself, liberation from the colonial identity and existence imposed by the colonizer (Mbembe 2017: 159). Postcolonial African existentialism is about hoping to be recognized as a human among other humans, being free from everything, being able to invent oneself, and not anymore to be degraded or dehumanized (Mbembe 2017: 178).

No race has suffered as much as the black race. Therefore, the impact of slavery, colonization, dehumanization, exploitation, and Western brutality on Africans is more perennial on Africans than on any other people.

> No other people under heaven, of whatever type or endowments, could have been so enslaved without failing into contempt and scorn on the part of those enslaving them. … During all the years of their bondage, the slave master had a direct interest in discrediting the personality of those he held as property … Having made him the companion of horses and mules, he naturally sought to justify himself by assuming that the negro was not much better than a mule … the old masters set themselves up as much too high as they set the manhood of the negro too low. (Douglass 1881: 572–573)

Not despite but because of this colonial subjugation the postcolonial Africans imitate the colonizer as an ideal being. White skin and Western faces are the ideals of beauty. Western economic wealth, political stability, and military power are proof of its superiority.

The postcolonial being lays claim to his own being; he wishes to recover it; and he is the project and process of the recovery of his own being (Gandhi 1998: 17). The postcolonial person is on its way toward total liberation; it refuses the privilege of the colonial "master"; rejects the primacy of Europe, resists the cultural viscosity of Europe (Gandhi 1998: 19). However, the postcolonial being on the one hand aspires to be independent of the colonial being through creative autonomy from Europe but on the other hand he imitates Europe even more. Envy and desire are two sides of the same coin of the postcolonial being: he hates his "master" but he imitates his master existentially. Migration is an expression of this mimetic desire to overcome historical and psychological subjugation.

Postcolonial migration from Africa to the West is a mimetic process to achieve liberation through (de)othering, (de)bordering, and reordering (van Houtum and van Naerssen 2002; Wonders and Jones 2021: 301). Migration policy defines not only borders but also identities. As Jones and Krzyzanowski suggest, "…identities are constructed both internally—by through our self (re)presentation and alignment with others—and externally—by the powerful 'other', such as institutional gatekeepers who can set threshold criteria for entry to groups…" Migration is a rejection of social hierarchies of mobility and rights, inequalities and disempowerment, and exclusion and demand for inclusion (Wonders and Jones 2021: 303, 308).

Mimetically, the postcolonial Africans have a pervasive desire to prove to the colonizer that they have indeed culture and philosophy. Many African philosophers have been obsessed to prove that there is an African philosophy. The African philosophy seminar in Addis Ababa in 1976 focused on this. Agblemagnon (1998) discussed the existence and nature of African philosophy, whereas Koffi and Abdou (1998) analyzed the controversy over its existence. Arthur (1998) attempts to prove that white racists were wrong in claiming that Africans were incapable of philosophy since they were incapable of logical and rational thinking. Ruch (1998) discusses whether there is an African philosophy; where it could be found; who the philosophers are; whether it is regressive or progressive; whether something pre-rational and unsystematic can be called philosophy, etc. Aren't African philosophers "scraping the bottom, of the barrel in an effort to show that Africa has a philosophy of its own…"? (Ruch 1998: 270). Cheikh Anta Diop (1998) attempts to prove the existence of African philosophy by claiming that Egyptian philosophy is an African philosophy. However, Diop forgets that Egyptians feel superior to black Africans. To prove African philosophy Onyewuenyi maintains that Greek philosophy has a strong Egyptian influence, which cannot be denied of course.

> Little do some of us know that the first woman philosopher, Hypathia, was from Alexandria and was murdered by Christians. Names like Saint Augustine, Origen, Cyril, Tertullian are not unfamiliar to you; they are black Africans. More pertinent to our subject is the fact what today we call Greek or Western Philosophy is copied from indigenous African philosophy of the

> 'mystery system'. All the values of the mystery system were adopted by the Greeks and Ionians who came to Egypt to study, or studied elsewhere under Egyptian-trained teachers. These included Herodotus, Socrates, Hippocrates, Anaxagoras, Plato, Aristotle and others... (Onyewuenyi 1998: 250)

Hallen (1998) discusses whether traditional African thoughts could be adequate to be considered as equivalent to the critical and modern philosophy. The article by Kinyongo (1998) raises various questions regarding African philosophy: "Is there an African philosophy? Can we speak of an African thought? Do we have an African philosophy? Is there an African wisdom? Is there a philosophy amongst the primitive? Can there be an African philosophy" (Kinyongo 1998: 124). The fact that white racists denied African humanity and intellect has pushed many African intellectuals to disprove racist Western scholars that Africa had not contributed anything to the field of philosophy (Onyewuenyi 1998: 245). Onyewuenyi argues that there was a concerted Western effort to hide African greatness and achievements so that slavery and the economic exploitation of Africans could be justified on the pretext that Africans were not real human beings with rationality (Onyewuenyi 1998: 246).

If the black man successfully imitates the whites, he is convinced of his equality with the whites; if not successful, it is a sign of its peculiarity. The colonized African tries to imitate the colonizer to the point of changing his skin, as Memmi would say. The colonizer is the tempting model. "The first ambition of the colonized is to become equal to that splendid model and to resemble him to the point of disappearing in him ... His habits, clothing, food, and architecture are closely copied, even if inappropriate. A mixed marriage is the extreme expression of this audacious leap" (Memmi 1967: 120–121). Memmi analyses the African mimetic attempt beautifully: "...In order to be assimilated, it is not enough to leave one's group, but one must enter another; now he meets with the colonizer's rejection. All that the colonized has done to emulate the colonizer has met with disdain from colonial masters." (Memmi 1967: 124).

If migration were only about material gains and physical security, it would end rapidly once these goals are achieved. Therefore, there are two dimensions of the causes of migration: the *material* dimension of migration, and the *relational-mimetic* dimension. The material dimension consists of poverty, climate change, and physical security, whereas the relational-mimetic constitutes imitation, desire for recognition, acceptance, fascination, envy, love-hatred relationship, admiration, and abomination. This fascination means the postcolonial beings hypnotized gaze upon the "master", which condemns him to derivative existence (Gandhi 1998: 21). Similarly, Memmi suggests that the colonized hate the colonizer and yet admire him passionately (Memmi 1967: x). The postcolonial being is overcome by a perverse and self-defeating longing for the "master" (Gandhi 1998: 22). The book by Ashis Nandy (1983), *The Intimate Enemy: Loss and Recovery of Self Under Colonialism*, explores the psychological and cultural impact of colonialism on the colonized people in India. According to him, the impact of colonialism is not only political and economic domination but also that the colonized have incorporated the culture and lifestyles of the colonizer, which they pretend to reject and despise. As a result, the colonized have simultaneously and paradoxically internalized a persistent cultural self-hatred on the one hand, and adoration and abomination of the colonizer on the other. Amartya Sen suggests that there is a postcolonial parasitical obsession with the West or the colonial powers. "Western imperialism … also created an attitudinal climate that is obsessed with the West, even though the form of that obsession may vary widely—from slavish imitation, on one side, to resolute hostility on the other. The dialects of the colonized mind include both admiration and disaffection" (Sen 2006: 85). The colonized mind is fixated on the West both in resentment and admiration (Sen 2006: 88). The colonized man acknowledges the Western superiority and therefore tries to replicate its accomplishments by imitating its skills and lifestyles. However, at the same time it denounces Western "decadence" exalts its own cultural values and identity (Sen 2006: 90–95). There is this paradoxical situation of the colonized being, which underscores its "otherness", rejects Western decadence" but passionately imitates the West. As Sen puts it, "The dialects of the captivated mind can lead to a deeply biased and parasitically reactive self-perception" (Sen 2006: 91).

In Chap. 2, we saw the continuation of migration despite increasing income and wealth in countries of origin. This phenomenon points out the relational-mimetic dimension of migration. The underlying cause of migration from postcolonial Africa is rather relational-mimetic, i.e. migration as a process of imitation of the model, the West. Fanon says,

> Man is human only to the extent to which he tries to impose himself on another man in order to be recognized by him. As long as he has not been effectively recognized by the other, it is this other who remains the focus of

> his actions. His human worth and reality depend on this other and his recognition by the other. It is in this other that the meaning of his life is condensed. (Fanon 2008: 191)

From a *material perspective*, the countries of origin cannot provide the postcolonial migrants with material or political security but they do not question their identity and subjectivity. The postcolonial migrants leave their countries but still love them. Their attachment to their countries of origin is often much stronger than to the destination countries which can give them security and wealth. In extreme cases, they love and desire to live in the destination countries but they hate them. Analyzed from the relation-mimetic perspective, the destination countries give them material and political security but they also question their identity and subjectivity. The postcolonial migrants are fascinated by the material success of the colonizer and even his power to question their subjectivity. Still, they do not aspire to gain their identity through detachment from the colonizer but through imitation and participation in the colonizer's way of life. At the same time, they hate him and his values because they remind the migrants of their inadequacies. The postcolonial Africans ask themselves, "In reality, who am I?" (Fanon 2001: 200). The postcolonial African being is an imitating being, obsessed with the desire to be as much European as possible, imitates European achievements, European techniques, and European lifestyle as its model (Fanon 2001: 252).

> The black man wants to be like the white man. For the black man, there is but one destiny. And it is white. A long time ago the black man acknowledged the undeniable superiority of the white man, and all his endeavors aim at achieving a white existence. (Fanon 2008: 202)

The postcolonial African migration to the West is caused primarily by a mimetic desire for being. On the one hand, postcolonial beings migrate to Europe because of fascination, but on the other hand, confronted with rejection, discrimination, limited socio-economic opportunities, and the economic gap between migrants and the majority of the host population, etc. they start to idealize their own culture and religion back home, reject "Western decadence", tend to religious radicalism, and in certain cases even to violence, rape, vandalism, criminality, Islamist fundamentalism, etc.

Often, migrants and their descendants live a paradoxical life in the West. They enjoy liberal democracy and human rights in the West but often they support authoritarian regimes back home. For example, 72% of Austro-Turks during the May 2023 Turkish election voted for the authoritarian Erdogan. When during the World Cup 2022 Morocco beat Belgium 2–0, dozens of Moroccan football fans smashed shop windows, threw fireworks, and torched vehicles in the center of Brussels. While watching this, one might ask oneself why are they living in Europe if Erdogan's authoritarianism is better than Austrian democracy or if they love Morocco much more than Belgium. Has anyone forced them to live in Austria or Belgium instead of in Turkey or Morocco?

Most migrants and their descendants live in an inner contradiction. On the one hand, they imitate and admire the West for its achievements which their countries of origin could not. The West asserted itself as superior economically, politically, and technologically thanks to slavery and colonialism. Even aesthetically the West has become the model of the whole world. Therefore, the West is admired and imitated. Conversely, this reality signifies at the same time the imitators' inferiority and nothingness. By its very being, the West asserts its superiority and questions existentially the non-Western subjectivity. For the non-Western immigrants in the West, even if their countries of origin do not give them physical security or economic prosperity they do not question their identity and subjectivity nor consider them inferior. The West's colonial history is abominable but its soft and hard power is fascinating and its superiority is admirable. The postcolonial immigrants in the West go through this trauma caused by the inner division of love-hate relationships toward the colonizer. The colonized admire, imitate, and hate the idol (the superior West) who reminds them of their inadequacy. It is not the Moroccans in Morocco who hate Belgium but those in Belgium who enjoy a liberal life, welfare benefits, job opportunities, education, etc. These are opportunities many Moroccans in Morocco do not have. The behavior of the Moroccan fans is mimetic: they admire Belgium/Europe but they hate it because it shows them that they are different whereas they desire to be similar. Suddenly, however, they pretend to desire to be different although their ultimate objective is to be similar. Therefore, the behavior of the Moroccan fans and their violence in Brussels during the 2022 world championship was an outcome of an inferiority complex, inner division, and increasing similarity. As Girard argues, it is an increasing similarity between the model and imitator that causes conflict and violence, not an increasing difference. While the similarity starts to increase, they emphasize their difference and start to hate each other for this difference, however, in reality, the hatred comes from the diminishing difference and increasing similarity (Girard 2023). Mimetically, it is not the lack of similarity or integration that causes tensions and conflicts between the locals and immigrants, but instead the overlooked and increasing similarity between them. "As the distance between the mediator and subject decreases, the difference diminishes, the comprehension becomes more acute and hatred intense ... Most of our ethical judgments are rooted in hatred of a mediator, a rival whom we copy." (Girard 1965: 75). The behavior of the Moroccan fans resembles that of Dostoevsky's unwanted or uninvited underground man. "The underground man is present during the preparation of the party but no one thinks inviting him. This unexpected snub—or perhaps only too expected—unleashes in him a morbid passion, a frantic desire to 'crash, conquer, and charm' these people whom he does not need and for whom, moreover, he feels genuine contempt." (Girard 1965: 68). Girard observes that "we do not understand what drives the underground man to worship, to hate, to collapse sobbing at the feet of his mediator, to send him incoherent messages full of insults mixed with endearments." (Girard 1965: 75).

According to the mimetic theory, the higher the imitator's desire for similarity with and recognition by the model, the higher the rejection. The imitator and the model desire the same object but they cannot enjoy it together. They are rivals whose being depends on exclusive ownership and recognition. But since as long as other desires the same object, there is no exclusive ownership but mimetic rivalry. In this rivalry, there is not much difference between the imitator and the model. Both become more or less the same since the imitation is not linear but instead reciprocal or triangular. Dostoevsky's *The Brothers Karamazov* underlines the mechanism of the desire for recognition and power based on rivalry, jealousy, and greed. Fyodor Karamazov and his sons imitate each other and rival. Fyodor Karamazov and Dmitry desire the same woman (Grushenka), Dimitry and Ivan desire the same woman (Katharina Ivanovna); Grushenka and Katharina Ivanovna rival against each other.

Human beings search for being and freedom in mimetic rivalry for recognition and power. As Sartre would say, in the absence of God and his status of abandonment man wants to invent himself as the future of himself (his being) as his own project (Sartre 2007: 29ff). The human project is to be considered above the rest, to dominate someone, to have the last word, to feel superior (Camus 2013: 28f). As Dostoevsky (1993: 124) would say, without power and tyranny over someone, it is impossible to live. In *Crime and Punishment*, Luzhin's main interest in his fiancée is that he can dominate her because she is destitute over whom he can have mastery later on (Dostoevsky 1991: 183f, 366f). Humans tend to make the lives of others a misery and lead them to the point where they view them as their Providence, and therefore, there cannot be justice and truth among people (Dostoevsky 1991: 378, 432). Even Raskolnikov confesses that once he had loved a girl primarily because she was poor and sick (Dostoevsky 1991: 275). In Dostoevsky's *Crime and Punishment*, the different characters of the novel try to exert power on someone by pretending to sacrifice themselves for that person: Sonya sacrifices herself for the orphaned children; Dunya for her mother; Raskolikov for Sonya; Luzhin for Dunya and her mother; Raskolikov for the girl he had loved before, etc.

Raskolnikov wanted to be an extraordinary being, not a louse like everyone else, but a real man (Dostoevsky 1991: 500). Flattery and compliments are methods of dominance (Dostoevsky 1991: 568). For Raskolnikov, it is the most natural thing that there are two groups of people: the extraordinary (dominant) and the ordinary (dominated, obedient) (Dostoevsky 1991: 308ff). Human beings are even fond of being trampled on and horsewhipped (Dostoevsky 1991: 336). However, the dominated want not only to reverse the degradation and attain recognition but also look for someone to dominate, as non-domination would mean non-being. Those we dominate constitute our being because we cannot *be* without domination or exclusion. Camus suggests that ultimately, even through our good deeds we want to be recognized or admired (Camus 2013: 53, 92). As Tolstoy says, Ivan Ilyich being on his deathbed his wife ordered doctors not for the sake of the dying man but she was doing it for herself, which implies hypocrisy and pretension (Tolstoy 2008: 204). In this duplicity (Camus 2013: 88), we desire for being through enslavement or domination of someone. In reality, we need slaves (Camus 2013: 85). Since our being is dependent on their enslavement, we are indeed their slaves in the end.

This is the duplicity or interchangeability that Shakespeare shows us in the metamorphosis or transformation of Lysander and Demetrius, Hermia, and Helena. The duplicity or inconstancy is not only the indistinguishability between the characters but also means the monstrosity of each one of them (two beings existing in one individual). Through their imitation and slavish adoration of the other, they are possessed by the other. There is no difference between sleep and walking, light and darkness, man and beast (Shakespeare 2015: lxxii, 64). The more Demetrius despises Helena, the more she loves him. Helena makes herself into an animal in her slavish adoration of Demetrius and Hermia (Shakespeare 2015: 24, 30).

The *Underground Man* of Dostoevsky one time feels the best among his "friends" and preserves his dignity as a hero, another time he sets them above himself and considers himself not even worthy to become an insect, he is just the mud, a crushed and humiliated creature (Dostoevsky 1993: 6, 45, 55). He hates and adores Zverkov (Dostoevsky 1993: 107). He is incorporated (possessed) by the enemies; he imitates them so much that he ceases to be himself. This divided person is a suffering person (Dostoevsky 1993: xix). He wants to belong and to be recognized the more, the more he is rejected or humiliated (Dostoevsky 1993: 64ff). The more his friend loved the Underground Man the more he hated him because he only wanted to have victory over him (Dostoevsky 1993: 68). "It is a burden for us even to be men—men with real, *our own* bodies and blood; we are ashamed of it, we consider it a disgrace, and keep trying to be some unprecedented omni-men" (Dostoevsky 1993: 130). The Underground Man is a symbol of the multiplicity of selves, each at war with others; all at war with everything. As Goethe says, the presence of the other makes us unhappy because we always compare ourselves with others (Goethe 1989: 73).

Werther who feels inferior to Albert and rejected by Lotte loves her the more. Werther adores his main obstacle, Albert, as someone supernatural (Goethe 1989: 56–57). But at the same he claims that Albert cannot make Lotte happy (Goethe 1989: 88). Albert even encourages Lotte to desire Werther more (Goethe 1989: 110). When Werther's passion for her becomes uncontrollable Lotte says something crucial: "I very much fear what makes the desire to possess me so attractive is its very impossibility" (Goethe 1989: 115). But her secret heart desires to keep Werther for herself, no other woman should love him (Goethe 1989: 118). In the end, By the pistols which passed through her hands, Werther kills himself.

In Stendhal's Red and Black, Mathilde considers herself most intelligent, beautiful, and rich but she looks at Julien as if she were his slave. She compares him even with Napoleon. (Stendhal 2002a: 307). Julien despises Mathilde but she is unable to despise him in return, and Julien is overwhelmed by happiness (Stendhal 2002a: 311). By despising and ignoring her Julien was convinced that he enhanced his value in her eyes (Stendhal 2002a: 315). She calls Julien—a man on the lowest level of society—her master. But she is determined that at the first weakness she detects in him, she will abandon him (Stendhal 2002a: 326). When he expressed his love and suffering, this gave her keen pleasure; when she was certain that she was loved, she despised him utterly (Stendhal 2002a: 368). When Julien shows coldness, she throws herself on her knees and beseeches him to despise and love her (Stendhal 2002a: 437). In both cases, doing the opposite of what is expected became the rule.

The Thebans once adored Oedipus almost as a god and then rejected him utterly. By solving the riddle of the Sphynx, he set them free from her bondage; he was adored, and he was their savior. "You freed us from Sphynx … you triumphed; Oedipus King, we bend to, your power, we implore you, all of us on our knees … your country calls you savior; … you have the power…" (Sophocles 1984: 161). However, then he was made responsible for the plague in Thebes. He will be designated as the curse, corruption of the land who has to be driven out of the land immediately, far from sight where he can never hear a human voice (Sophocles 1984: 245). When Oedipus is degraded, banned, blinded, and buried he becomes a savior again. "So when I am nothing—am I a man?" (Sophocles 1984: 306).

The other, the stranger Othello is a "malignant" infidel, a "circumcised dog", i.e. a monster, a seducer of a white girl with forbidden arts. Iago tells Othello that being black and marrying a white woman is against nature. Othello represents an alien and unnatural identity (of which blackness is the most conspicuous sign). Desdemona is unfortunate—Disdemona in Greek means the *Unfortunate*—because she married a black person (Shakespeare 2006: 28). Iago was upset that a black man was militarily successful and sexually triumphant with a white woman whom he desires (Shakespeare 2006: 31). Othello was a monster because he was a black being with human quality, a civilized appearance with innate savage passion, a hero and a monster (deformed creature because of blackness and racial inferiority, cursed by God as sons of Ham—son of Noah) (Shakespeare 2006: 47, 49, 122, 126). Not only Iago hates Othello externally while loving him hiddenly (sexually), but also the white society needs a black Othello to feel superior, a duplicity which we encounter in the works of Camus, Dostoevsky, and Shakespeare (Shakespeare 2006: 201).

Blackness represented sin, death, animality, and sexual depravity, deformity, proximity to beasts than humans. Black man in general and the marriage of Othello to Desdemona in particular means living beyond the proper limits of nature, a monstrosity (Shakespeare 2006: 119, 139). The divine Desdemona falls in love with what she feared to look on against all rules of nature, to the dishonor of the white humans (Shakespeare 2006: 221).

The essence of being human depends on honor, power, and glory, and the division between "us" and "them". The pleasure of Ivan Ilyich depended on the recognition and respect he enjoyed for his achievements, on how his former enemies were ashamed and licking his boots, and on how he was envied and became popular (Tolstoy 2008: xxv, 55, 57, 173, 176). As Tolstoy says, people attempt to impress others, not to be below others (Tolstoy 2008: 177).

In *The Iliad* Homer describes the rivalry between Achilles and Agamemnon, Achilles and Hector, Menelaus and Paris, even among gods themselves. When Agamemnon lost his "prize"-girl, he took the one of Achilles. Because "his" girl was taken away by Agamemnon, Achilles' heart became perverse and implacable (Homer 1987: 148). Achilles felt so much offended in his pride by the action of Agamemnon that he ignored the Argives massacre by the Trojans.

Moreover, for their respective sides, Achilles (Argives) and Hector (Trojans) were the symbols of pride, glory, and honor. As Achilles says, it is not just that human misery is a fate given to humans by the gods, but both the Argives and the Trojans were convinced that the mutual slaughter was a divine deed guided by the gods themselves (Homer 1987: 401). Glory, pride, and honor take center stage in *The Iliad*. All warriors believe that gods are on their side. Therefore, human beings massacre their opponents without any pity. With gods on their side, they cannot be wrong. This is the plague that Dostoevsky is talking about, that each person thought he alone possessed the truth and suffered agony as he looked at the others (Dostoevsky 1991: 652).

Achilles and Hector were almost gods themselves. Furthermore, they imitated each other both in their violence and their quasi-divinity. Hector even imitated Achilles by wearing the latter's "beautiful armor" (Homer 1987: 279). Mimetic violence, desire for glory, pride, and vanity were not much different among gods. How pleased Athene was when Menelaus prayed first to her before all the other gods! Athene, Hera, and Poseidon were on the side of the Argives, whereas Apollo, Ares, Artemis, and Aphrodite were on the side of the Trojans exhorting the respective side to do more violence, and Zeus sometimes switched sides.

As Camus and Dostoevsky show us, to be means to dominate someone. In such a relationship, humans suffer from the anxiety of having no one to dominate and, therefore not to be. The dominated always attempt to reverse it. As Dostoevsky would say, human nature itself suffers from this plague of honor, glory, rivalry, and pride (Dostoevsky 1991: xxi). Dostoevsky designates the whole human race in general as a villain (Dostoevsky 1991: 34). Similarly, Homer for that matter Zeus says that human beings are the most miserable creatures among the living beings (Homer 1987: 286).

The pleasure and self-satisfaction of Zverkov depend on the slavish adoration he gets from the Underground Man. The more Zverkov despises and rejects the Underground Man, the more the latter adores and desires him and his friends. Zerkov considers the Underground Man so base that when the latter attempts to ask him for forgiveness, Zverkov says, "Y-y-you? Offend m-m-me? I will have you know, my dear sir, that you could never under any circumstances offend me!" (Dostoevsky 1993: 81). For the Underground Man, *to be* means to belong to the adored group. To strive for essence instead of mere existence, as Sartre would say.

The postcolonial African migrant to the West is like the Dostoevskian Underground Man who desperately implores Zerkov to be included in the circle of his friends but who will be rejected the more. The more the postcolonial African migrants adore the West and try to get there, the more they are confronted with physical and metaphysical fences. The more they are rejected the more they will try to overcome these fences. The higher and the more prevalent the walls, the more postcolonial migrants try to overcome them. There must be something desirable behind the walls.

The existential or desperate attempt to overcome the walls is already an indicator of increasing equality and decreasing gap. The narrower the resource gap, knowledge gap, and liberty gap, the more intense the imitation. The more the imitation increases, the bigger the awareness of the difference. The more the experiences of rejection and hopelessness in achieving acceptance at the destination, the higher the imitators underscore their own "peculiar identity" and culture.

The desire to migrate to and live in the West on the one hand and hatred toward the West on the other is poorly analyzed in migration studies. The adoration of the colonizer has been extensively internalized in postcolonial Africa. There is almost a "geneticized" transmission of colonial legacies is a pervasive fascination for the colonizer in the form of simultaneous adoration and abomination of the West. As the Nigerian writer Achebe in his *Arrow of God* writes, "The white man has the power which comes from the true God and it burns like fire. This is the God about Whom we preach every eight days..." (Achebe 2017: 256). During the colonial period, many Africans believed that white men could not be sick like ordinary people (Achebe 2017: 341). It was considered to be foolish to defy a white man; white men do not take bribes (Achebe 2017: 368, 447). The white man is unbeatable physically and intellectually (Achebe 2017: 472–473). It is this belief Fanon had in mind when he says, "A black man behaves differently with a white man than he does with another black man" (Fanon 2008: 1). Similarly, Achebe observed that no decent restaurant would serve a Nigerian food because nobody wants a Nigerian food, instead a western food (Achebe 2017: 447). This is a "genericized" legacy of slavery, colonization, and evangelization.

Even if often it seems to have failed, the postcolonial migration from Africa to the West is an attempt or hope to be free and equal. The African liberation process against slavery, colonization, and dehumanization consists of at least three components. Firstly, intellectual arguments and academic writings by Africans rejecting racist ideologies, defending the equality of all human beings, and highlighting the peculiarity of African culture, literature, and tradition as Julius Nyerere, Kwame Nkrumah, and Léopold Sédar Senghor, etc. did. Secondly, political reaction or revolt against colonialism in all colonized African countries, as the establishment of the Organization of African Unity had shown. Thirdly, a social-psychological reaction to the legacies of slavery and colonization, the hope to set oneself free from mental slavery, and the desire for being. As postcolonial theories by such as Frantz Fanon, Aimé Cesaire, Albert Memmi, etc. show, there has been a certain identity crisis in the colonized, which ultimately has led to his admiration and fascination for the colonizer and therefore the desire to be recognized by the colonizer to achieve his being by imitating the colonizer and his ways of life and ideally by living together with him.

The colonized Africans learn from the West what is desirable. In the language of René Girard, the West is the mediator or the model from whom the colonized learn what is desirable. "The mediator is immobile and the hero turns around him like a planet around the sun" (Girard 1965: 46). We can observe a lot of similarities between the postcolonial Africans and the heroes in Dostoevsky's *Eternal Husband*, *Notes from Underground*, or Cervantes' *Don Quixote*. The colonizer is a god with a human face because he is the one who heralds paradise. By imitating this colonial god, the postcolonial Africans hope to overcome their nothingness and become someone.

Don Quixote represents human beings inhabiting a fictional realm, leading an imaginary life and often guided by a futile human desire. Does Don Quixote know that what he believes is just bogus, or doesn't he know it? In either case, it is disturbing. His only focus is on honors, social status, and recognition as a knight-errant. The dreaming Don Quixote, the questioning Sancho ("What giants?"), and the lying Sancho to get out of the difficult situation are all the traits of one and the same person (Cervantes 2003: xiv–xvii). Don Quixote and Sancho symbolize not only the distorted version of utopias but also show us how we end up utilizing our freedom in another unfreedom and slavery. Through the pathological desire for social status, recognition, and power, we end up in slavery because what we possess ultimately and paradoxically possesses and enslaves us. Sancho tries to convince Don Quixote that chivalries about which the latter tells him, the winning of kingdoms and empires, giving away of islands, and doing other favors and great deeds, etc. are just empty lies and self-delusion (Cervantes 2003: 209). Don Quixote convinced himself that all his setbacks, beatings, and injuries almost to death were all the works of wicked enchanters jealous of his coming glory (Cervantes 2003: 867–8). Don Quixote loves more than anything else on earth a woman (Dulcinea del Toboso) who even did not exist. He calls her the most beautiful on earth. Sancho lies that he knows her. Don Quixote betrays his own nothingness when he says he lives in her and breathes and has his being (Cervantes 2003: 276). Don Quixote sends Sancho into the village to learn what great things people are talking about him.

I see no difference between Don Quixote and postcolonial African migrants some of whom have tried more than seven times to enter Europe but ultimately end up in poverty as begging in European streets despite the information they had got that the journey to and life in Europe is dangerous and hard. As Don Quixote idealized the life of the Knight Errant, migrants idealize the life of the civilized and developed colonizer. The deluded Don Quixote believes that he is the best Knight Errant the world has ever seen who punishes the wicked and rewards the righteous; he believes himself to be the strongest man ever, even when he is defeated almost all the time by his enemies. He despises his squire Sancho Pansa who eats and drinks like an ordinary human being. He, however, as a Knight Errant does not care about hunger and thirst because these are just the things of the flesh and hence ephemeral. Whereas Don Quixote pursues fame incessantly, reality pulls him down all the time. As Don Quixote enslaves himself to his imagination of Amadis of Gaul, postcolonial African migrants enslave themselves to their imagination of the West.

Dostoevsky's *The Double*, the tragic hero lives in an inner conflict between his real and ideal self (Dostoevsky 2013, Xii; 58–59). One of the most important things in Dostoevsky's *Double* is the desire to become and to be someone else. The main character Golyadkin senior is convinced of his perfectness, whereas his *Double* incorporates all vices with his utterly despicable character. But Dostoevsky underlines that both Golyadkins are completely similar. The despicable other Golyadkin is just a construction of the first Golyadkin. Therefore, in different places, Dostoevsky emphasized the utter similarities between both Golyadkins. Similarly, a second Don Quixote emerges identical to the first one but antithetical in action (Cervantes 2003: 969). Both Golyadkins of Dostoevsky and Don Quixote of Cervantes are at the same time completely different and completely the same.

Dostoevsky and Cervantes point out the human desire to be someone else. The white colonizer and slaveholder projects all bad qualities and despicability on black Africans. The existence of such dehumanized and despicable black Africans gives the white slaveholder and colonizer the certainty of his superiority. Even the victim himself internalizes this and starts to believe in his inferiority and his persecutor's superiority. Fanon reports the following:

> We know a lot of girls from Martinique, students in France, who confess in lily-white innocence that they would never marry a black man. (Choose to go back there once you have escaped? No, thank you.) Besides, they add, it is not that we want to downplay the credentials of the black man, but you know it is better to be white. (Fanon 2008: 30)

What we learn from Cervantes and Dostoevsky is that social and historical power relations that cause feelings of wretchedness and despicability create a gap between the idealized social values and standards on one hand and one's factual situation on the other. The perception of this gap and the desire to overcome it and to be recognized as one of the idealized social members is the goal of the postcolonial African migrant. As Fanon would say, for the postcolonial man there is only one way out, and it leads out to the white world where he hopes for white approval (Fanon 2008: 33). The desire to become someone else after having migrated or having overcome the material and immaterial borders between the former colonizer and the colonized are shown in the *migro ergo sum* case discussed above.

## Conclusion

Postcolonial migration from Africa to the West is a desire for being by negating colonial and neo-colonial bordering, othering, and ordering and through coexistence with the mystified colonizer. There is a strong desire to reject the status quo, to negate the *negatedness*, and attain being through consumption, migration, and imitation of the colonizer's lifestyle. However, the postcolonial African migrant in the West gives up his aspiration to be recognized because the racism with which he is confronted every day does not allow this. He has overcome the walls of Ceuta and Melilla, the Sahara Desert, and the Mediterranean but he cannot overcome the wall of racism. Nevertheless, as Nietzsche says, "Where I found a living creature, there I found will to power; and even in the will of the servant I found the will to be master" (Nietzsche 2003: 137). However, the wall cannot be surmounted even if he is successful in achieving the desired education or income at the destination. Racism becomes the dead-end, which nullifies all his aspirations to be equal. He has only two options: either he ends up—like in Dostoevsky's tragedy of Mr. Golyadkin—in self-destruction by trying to be someone else incessantly. Alternatively, he accepts himself—as Don Quixote did in the end when he became the *shepherd* Quixotiz.

Liberation, information, and education, as Fanon would say metaphorically, awaken the desire of the black man to sleep with the white woman, which would elevate him to the white man's level (Fanon 2008: xvii, 54). According to Fanon, liberation means to set free the black man from the arsenal of complexes that germinated in a colonial situation (Fanon 2008: xvii 14). The colonized attempts (in the mode of Sisyphus) permanently to demystify or to overcome the metaphorical and physical wall between Africa and the West. However, the walls become higher; the borders become ubiquitous to keep the undesirable and undeserving migrants outside and far away. Quixotically, the colonial powers believed in their superiority despite their colonial brutality and barbarity.

In the colonial and "civilizing" relationships, the idea or construction of Africanity is dependent on the construction of Europeanity. However, Europeanity does not exist without the construction of Africanity. Hence, Africanity and Europeanity are not only constructs but they also essentially belong together. Postcolonial African migrants to the West are encouraged by the decreasing multifaceted gap between Africanity and Europeanity and they are striving to decrease this gap further in their desire for being as we shall this in the following chapter.

## Literature


Achebe, Chinua (2017): Things Fall Apart; Arrow of God; No Longer at Ease: The African Trilogy: Penguin Classics: New York.

Adam, Ilke and Trauner, Florian (2019): Ghana and EU Migration policy: studying an African response to the EU's externalization agenda, in Sergio Carrera et al. (eds.), Constitutionalizing the external dimensions of EU migration policies in times of crisis, Edward Elgar: Cheltenham, pp. 257–271.

Adeyanju, Charles T. and Neverson, Nicole (2007): "There Will Be a Next Time": Media Discourse about an "Apocalyptic" Vision of Immigration, Racial Diversity, and Health Risks, *Canadian Ethnic Studies*, Volume 39, Number 1, 2007, pp. 79–105.

Agamben, Giorgio (1998): Homo Sacer: Sovereign Power and Bare Life, Stanford: Stanford University Press.

Agamben, Giorgio (2005): State of Exception, Chicago University Press: Chicago.

Agblemagnon, N'sougan (1998): The existence and nature of African philsopphy in the past and in the present, In: Claude Sumner (ed.) African philosophy, Addis Abeba: Addis Abeba University Press, pp. 1–10.

Ahrendt, Hannah (2017): The Origins of Totalitarianism, Penguin Classics.

Albahari, Maurizio (2015): Crimes of Peace: Mediterranean Migration at the World's Deadliest Border, University of Pennsylvania Press: Philadelphia.

Andreas, Peter (2009): Border Games: Policing the US-Mexico Divide, Ithaca, NY: Cornell University Press.

Appiah, Anthony (1994): Identity, authenticity, survival: multicultural society and social reproduction, The politics of recognition, in Charles Taylor et al., Multiculturalism, Princeton University Press, Princeton, pp. 149–164.

Arana, Arantza Gomez and McArdle, Scarlett (2019): The EU and migration crisis: reinforcing a security-based approach to migration? in Sergio Carrera et al. (eds.), Constitutionalizing the external dimensions of EU migration policies in times of crisis, Edward Elgar: Cheltenham, pp. 272–289.

Arendt, Hannah (1998): The Human Condition, Chicago: The University of Chicago Press.

Arthur, J.K.M. (1998): African philosophers and African philosophy, In: Claude Sumner (ed.) African philosophy, Addis Abeba: Addis Abeba University Press, pp. 11–20.

Avrutin, Eugene M. (2022): Racism in Modern Russia: From the Romanovs to Putin, Bloomsbury: London.

Bakewell, Oliver (2008a): 'Keeping Them in Their Place': the ambivalent relationship between development and migration in Africa, *Third World Quarterly*, 29:7, 1341–1358.

Bakewell, Oliver (2021) Unsettling the boundaries between forced and voluntary migration: in Emma Carmel et al. Handbook on the Governance and Politics of Migration, Edward Elgar: Cheltenham, pp. 124–136.

Balibar, Ètienne (2002): Politics and the Other Scene, Verso: London.

Banerjee, V. Abhijt and Duflo, Esther (2019): Good Economics for Hard Times, Public Affairs: New York.

Basheska, Elena and Kochenov, Dimitry (2015): EuroMed, migration and frenemy-ship: pretending to deepen cooperation across the Mediterranean, in Francesca Ippolito and Seline Trevisanut, Migration in the Mediterranean: Mechanisms of International Cooperation, Cambridge: Cambridge University Press, pp. 43–67.

Bauman, Zygmunt (2016): Strangers at our door, Polity: Cambridge.

Bell, David A. (2022): The Importance of Skin Colour in Central Eastern Europe: A Comparative Analysis of Racist Attitudes in Hungary, Poland and the Czech Republic, *Central and Eastern European Migration Review*, Vol. 11, No. 1, 2022, pp. 5–25, https://doi.org/10.54667/ceemr.2022.03.

Bello, Valeria (2017): International Migration and international security: why prejudice is a global security threat, Routledge: New York.

Booty, Natasha (2023): Dragos Tigau: Romania recalls Kenya ambassador over racist monkey slur, https://www.bbc.com/news/world-africa-65867104, 07.07.2023.

Bougroug, Anwar (2022): Anti-Blackness in the MENA-Region: time to reflect on racial bias, intersectionality, and paths to collective liberation, https://www.mykalimag.com/en/2020/07/23/anti-blackness-in-the-mena-region-time-to-reflect-on-racial-bias-intersectionality-and-paths-to-collective-liberation/, 25.05.2023.

Bryant, Lisa (2022): Being Black in Tunisia, https://www.voanews.com/a/being-black-in-tunisia/6838272.html, 25.05.2023.

Burke, Peter and Stets, Jan (2009): Oxford: Oxford University Press.

Camus, Albert (1991a): The myth of Sisyphus and other essays, New York: Vintage.

Camus, Albert (2013): The Fall, Penguin Classics, London: Penguin Books.

Carter, Donald (2013): Navigating diaspora: The precarious depths of the Italian immigration crisis, in Abdoulaye Kane and Todd Leedy, African Migration: patterns and perspectives, Indiana University Press, Bloomington, pp. 59–77.

Casas-Cortes, Maribel and Cobarrubias, Sebastian (2019): Genealogies of contention in concentric circles: remote migration control and its Eurocentric geographical imaginaries, in Critical geographies of migration, Edward Elgar, pp. 193–205.

Cantat, Céline and Rajaram, Prem (2019): The politics of the Refugee Crisis, in The Oxford Handbook of Migration Crisis, Oxford University Press: Oxford, pp. 181–195.

Caviedes, Alexander (2015): An Emerging 'European' News Portrayal of Immigration?, Journal of Ethnic and Migration Studies, 41:6, 897–917.
Cervantes, Miguel de (2003): Don Quixote, Penguin Classics, London.
Césaire, Aimé (2000): Discourse on Colonialism, Monthly Review Press: New York.
Christian Aid (2007): Human tide: the real migration crisis, London.
Cohen, Roger (2015): America's Bountiful Churn, https://www.nytimes.com/2015/12/31/opinion/americas-bountiful-churn.html, accessed, 21.04.2023.
Collyer, Michael (2019): From Preventive to repressive: the changing use of development and humanitariansm to control migration, in Critical geographies of migration, Edward Elgar, pp. 170–181.
Crimmins, John (2023): The Psychological Impact of Skin Lightening, https://healthnews.com/mental-health/self-care-and-therapy/the-psychological-impact-of-skin-lightening/.
Daftary, Karishma (2023): Uncovering the roots of skin bleaching: Colorism and its detrimental effects, *Journal of Cosmetic Dermatology*, 22: 1, pp. 337–338.
De Genova, Nicholas (2013): Spectacles of migrant 'illegality': the scene of exclusion, the obscene of inclusion, *Ethnic and Racial Studies*, 36 (7), 1180–1198.
Degruy, Joy (2017): Post-Traumatic Slave Syndrome: America's Legacy of Enduring Injury and Healing, DeGruy.
Diop, Cheikh Anta (1998): Is there an African philosophy? In: Claude Sumner (ed.) African philosophy, Addis Abeba: Addis Abeba University Press, pp. 21–32.
Dostoevsky, Fyodor (1991): Crime and Punishment: Penguin Classics, London.
Dostoevsky, Fyodor (1993): Notes from Underground, Vintage Classics, New York: Vintage Books.
Dostoevsky, Fyodor (2013): The Double, Alma Classics: Richmond.
Douglass, Frederick (1881): The Color Line, *The North American Review*, 132, 295, pp. 567–577.
Du Bois, W.E.B (2009): Souls of Black Folk, Oxford University Press: Oxford.
Du Bois, W.E.B. (2007): Dusk of Dawn, Oxford University Press: Oxford.
Dugard, Martin (2003): Into Africa: the epic adventures of Stanley and Livingstone, Broadway Books: New York.
Duroy, Quentin (2011): North African Identity and Racial Discrimination in France: A Social Economic Analysis of Capability Deprivation, *Review of Social Economy*, 69: 3, pp. 307–332.
Eatwell, Roger and Goodwin, Matthew (2018): National Populism: The Revolt Against Liberal Democracy, Pelican.
Eco, Umberto (2013): Inventing the enemy, Vintage: London.
Eisele, Katharina (2019): The EU's readmission policy: of agreements and arrangements, in Sergio Carrera et al. (eds.), Constitutionalizing the external dimen-

sions of EU migration policies in times of crisis, Edward Elgar: Cheltenham, pp. 135–154.

Falguere, X. Casademont (2019): Refugee and Romani immigrant populations in Barcelona, in The Oxford Handbook of Migration Crisis, Oxford University Press: Oxford, pp. 111–126.

Fanon, Frantz (2001): The wretched of the earth, Penguin Books: London.

Fanon, Frantz (2008): Black Skin, White Masks, Grove Press: New York.

Fanon, Frantz 1991. *Black Skins, White Masks*, London: Pluto Press.

Farer, Tom (2020): Migration and Integration: The case for liberalism with borders, Cambridge University Press: Cambridge.

Fasani, Francesco et al. (2019): Does Immigration increase crime? Migration policy and the creation of the criminal immigrant, Cambridge University Press: Cambridge.

Filindra, Alexandra and Junn, Jane (2012): Aliens and people of color: the multidimensional relationship of immigration policy abd radical classification in the United States, in Marc Rosenblum and Daniel Tichenor, The Oxford Handbook of the politics of international migration, Oxford University Press: Oxford, pp. 429–455.

Flam, Helena and Beauzamy, Brigitte (2011): Symbolic Violence, in Gerard Delanty et al. (eds.), Identity, Belonging and Migration, Liverpool University Press: Liverpool, p. 221–240.

Flynn, Thomas (2006): Existentialism: A very short introduction, Oxford: Oxford University Press.

Fukuyama, Francis (2018): Identity: Contemporary identity politics and the struggle for recognition, Profile Books: London.

Gandhi, Leela (1998): Postcolonial Theory: A critical introduction, Columbia University Press: New York.

Garelli, Gelnda and Tazzioli, Martina (2019): Military-humanitarianism, in Critical geographies of migration, Edward Elgar, pp. 182–191.

Geisen, Thomas et al. (2008): (B)ordering and othering migrants by the European Union, Migration in a New Europe: People, Borders and Trajectories, Société Geografica Italiana: Rome.

Gilmartin, Mary and Kuusisto-Arponen, Anna-Kaisa (2019): Borders and Bodies: siting critical geographies of migration, in Critical geographies of migration, Edward Elgar, pp. 18–29.

Girard, René (1965): Deceit, desire and the novel: self and other in literary structure, John Hopkins University: Baltimore.

Girard, René (2016): Violence and the Sacred, Bloomsbury Revelations: London.

Girard, René (2019): Things hidden since the foundation of the world, Bloomsbury Revelations: London.

Girard, René (2023): All desire is a desire for being, Penguin Classics: London.

Goethe, Johann W. (1989): The Sorrows of Young Werther, Penguin Classics, London: Penguin Books.

Gosetti-Ferencei, Jennifer Anna (2021): On Being and Becoming: An Existentialist Approach to Life, Oxford University Press: Oxford.

Hallen, Barry (1998): Phenomenology and the exposition of African traditional thought, In: Claude Sumner (ed.) African philosophy, Addis Abeba: Addis Abeba University Press, pp. 47–66.

Hatton, Timothy (2009): The Rise and Fall of Asylum: What Happened and Why? *The Economic Journal*, Volume 119, Issue 535, 1 February 2009, Pages F183–F213.

Hegel, Georg W.F. (1975): Lectures on the philosophy of world history: Introduction, translated by H.B. Nisbet; Introduction by Duncan Forbes, Cambridge University Press: Cambridge.

Heidegger, Martin (2010): Being and Time, State University of New York.

Hilizah, Nisirine (2022): Racial Discrimination and Anti-Blackness in the Middle East and North Africa, Arab Barometer – Wave VII Racism Report, in https://www.arabbarometer.org/, accessed 02.05.2023.

Hollifield, James (2012): Migration and International Relations, in Marc Rosenblum and Daniel Tichenor, The Oxford Handbook of the politics of international migration, Oxford University Press: Oxford, pp. 345–381.

Homer (1987): The Iliad, Penguin Classics, London: Penguin Books.

Houtum, Henk van & Naerssen, Ton van (2002): Bordering, Ordering and Othering, *Tijdschrift voor Economische en Sociale Geografie*, Vol. 93, No. 2, pp. 125–136.

Huete, R.; (2008) "Analysing the Social Perception of Residential Tourism Development" in Costa, C. and Cravo, P., Advances in Tourism Research. Aveiro: IASK. pp. 153–161.

Hugo, Victor (2006): The hunchback of Notre Dame, Dover: New York.

Hunter, Margaret (2007): The Persistent Problem of Colorism: Skin Tone, Status, and Inequality, Sociology Compass 1:1, pp. 237–254.

James, C.L.R. (2022): The black Jacobins: Toussaint Louverture and the San Domingo revolution, Penguin: Dublin.

James, William (1890): Principles of Psychology. New York: Holt Rinehart and Winston.

Jeffers, Honoreé Fanonne (2022): The love songs of W. E. B. Du Bois, 4th State: London.

Johnson, Heather (2014): Borders, Asylum and Global Non-Citizenship: The other side of the fence, Cambridge University Press: Cambridge.

Jones, Paul and Krzyzanowski, Michal (2011): Identity, belonging and migration: beyond constructing 'Others', in Gerard Delanty et al. (eds.), Identity, Belonging and Migration, Liverpool University Press: Liverpool, pp. 38–53.

Jones, Robert P. et al. (2015): How Americans view Immigrants, and what they want from Immigration Reform, Public Religion Research Institute, Washington DC.
Kasparek, Bernd and Schmidt-Sembdner, Matthias (2019): Renationalization and spaces of migration: the European border regime after 2015, in Critical geographies of migration, Edward Elgar, pp. 206–219.
Kaya, Hülya (2020): The EU-Turkey Statement on Refugees: Assessing its impact on Fundamental Rights, Cheltenham: Edward Edgar.
King, Stephen J. (2020): Ending Denial: Anti-Black Racism in Morocco, Arab Reform Initiative, Bawader, 28 August 2020.
Kingsley, Patrick (2017): The new Odyssey: The story of Europe's refugee crisis, The Guardian: London.
Kinyongo, J. (1998): From discursivity to philosophical discourse in Africa, In: Claude Sumner (ed.) African philosophy, Addis Abeba: Addis Abeba University Press, pp. 123–138.
Kirk, G. S. (1975): Heraclitus: The cosmic fragments, Cambridge University Press: Cambridge.
Knaus, Gerald (2020): Welche Grenzen brauchen wir: zwischen Empathie und Angst – Flucht, Migration und die Zukunft von Asyl, Piper: München.
Koffi, Niamkey and Abdou, Toure (1998): Controversy on the existence of an African philosophy, In: Claude Sumner (ed.) African philosophy, Addis Abeba: Addis Abeba University Press, pp. 157–174.
Koslowski, Rey (2012): Immigration, crime and terrorism, in Marc Rosenblum and Daniel Tichenor, The Oxford Handbook of the politics of international migration, Oxford University Press: Oxford, pp. 511–531.
Kuusisto-Arponen and Gilmartin, Mary (2019): Embodied immigration and the geographies of care: the worlds of unaccompanied refugee minors, in Critical geographies of migration, Edward Elgar, pp. 80–91.
Kymlicka, Will (2015): Solidarity in diverse societies: beyond neoliberal multiculturalism and welfare chauvinism, Comparative Migration Studies, 3, 17, pp. 1–19.
Lukasik, Gail (2021): White like her: my family's story of race and racial passing, Skyhorse Publishing: New York.
Mannoni, O. (1990): Prospero and Caliban: The psychology of colonization, Ann Arbor: Michigan University Press.
Mbembe, Achille (2017): Critique of Black Reason, Duke University Press: Durham.
Mbembe, Achille (2021): Out of the dark night: Essays on decolonization, Columbia University Press, New York.
Memmi, Albert (1967): The colonizer and the colonized, Boston: Beacon Press.

Menin, Laura (2018): Being 'black' in North Africa and the Middle East, https://www.opendemocracy.net/en/beyond-trafficking-and-slavery/being-black-in-north-africa-and-middle-east/, 25.05.2023.

Menjivar, Cecilia et al. (2019): Migration Crisis: Definitions, Critiques, and Global Contexts, in The Oxford Handbook of Migration Crisis, Oxford University Press: Oxford, pp. 1–20.

Mitsilegas, Valsamis (2019): Extraterritorial immigration control, preventive justice and the rule of law in turbulent times: lessons from the anti-smuggling crusade, in Sergio Carrera et al. (eds.), Constitutionalizing the external dimensions of EU migration policies in times of crisis, Edward Elgar: Cheltenham, pp. 290–308.

Natanasabapathy, Puvanambihai & Maathuis-Smith, Sandra (2019): Philosophy of being and becoming: A transformative learning approach using threshold concepts, *Educational Philosophy and Theory*, 51:4, pp. 369–379, https://doi.org/10.1080/00131857.2018.1464439.

Nawyn, Stephanie (2019: Refugees in the United States and Politics of Crisis, in The Oxford Handbook of Migration Crisis, Oxford University Press: Oxford, pp. 163–179.

Nietzsche, Frederick (2007): Ecce Homo, Oxford: Oxford University Press.

Nietzsche, Friedrich (2003): Thus spoke Zarathustra, Penguin Classics: London.

Nyman, Jopi (2020): Narratives of contemporary Im/mobility: writing forced migration in the borderscape, in Sbiri et al., Mobile identities: race, ethnicity, and borders in contemporary literature and culture, Cambridge Scholars Publishing: Newcastle, pp. 15–34.

Onyewuenyi, Innocent Chilaka 1998. *Is There an African philosophy*, in: Claude Sumner (ed.) African philosophy. Addis Abeba: Addis Abeba University Press, pp. 243–260.

Palaver, Wolfgang (2003): René Girards mimetische Theorie, LIT: Münster.

Palaver, Wolfgang (2019): Populism and religion: On the politics of fear, Dialog, 58: pp. 22–29, https://doi.org/10.1111/dial.12450

Papastavridis, Efthymios (2020): Search and rescue at sea: shared responsibilities in the Mediterranean See, in Francesca Ippolito, et al., Bilateral Relations in the Mediterranean: Prospects for migration issues, Edward Elgar: Cheltenham, pp. 229–249.

Pedersen, Anne and Hartley L. (2015): Can We Make a Difference? Prejudice Towards Asylum Seekers in Australia and the Effectiveness of Antiprejudice Interventions, *Journal of Pacific Rim Psychology*, 9, 1–14.

Pombo, Maria Dolores (2019): Violence at the US-Mexican border, in The Oxford Handbook of Migration Crisis, Oxford University Press: Oxford, pp. 485–500.

Ray, Rashwan (2022): The Russian invasion of Ukraine shows racism has no boundaries, https://www.brookings.edu/articles/the-russian-invasion-of-ukraine-shows-racism-has-no-boundaries/, 07.07.2023.

Richards, Eric (2019): Migrants in Crisis in Nineteenth-Century Britain, *in The Oxford Handbook of Migration Crisis, Oxford University Press: Oxford*, pp. 37–56.

Ruch, E.A. 1998. *African Philosophy.* Regressive or Progressive? In: Claude Sumner (ed.) African philosophy, Addis Abeba: Addis Abeba University Press, pp. 261–276.

Sager, Alex (2019): Ethics and Migration Crisis, in The Oxford Handbook of Migration Crisis, Oxford University Press: Oxford, pp. 589–602.

Sartre, Jean-Paul (1993): No Exit, Vintage: New York.

Sartre, Jean-Paul (2007): Existentialism is a Humanism, New Haven: Yale University Press.

Sbiri, Kamal et al. (2020): Introduction, in Sbiri et al., Mobile identities: race, ethnicity, and borders in contemporary literature and culture, Cambridge Scholars Publishing: Newcastle, pp. 1–14.

Sbri, Kamal (2020): Black transits: Transition, recognition, and bordering in Marie Ndiaye's *Three strong women*, in Sbiri et al., Mobile identities: race, ethnicity, and borders in contemporary literature and culture, Cambridge Scholars Publishing: Newcastle, pp. 35–56.

Sen, Amartya (2006): Identity and violence: the illusion of destiny, Norton: New York.

Shakespeare, William (2006): Othello, Oxford World Classics, Oxford: Oxford University Press.

Shakespeare, William (2015): A Midsummer Night's Dream, Penguin Classics, London: Penguin Books.

Snyder, Timothy (2018): The Road to Unfreedom: Russia, Europe, and America, Tim Duggan Books: New York.

Sophocles (1984): The Three Theban Plays: Antigone, Oedipus the King, Oedipus at Colonos, Penguin Classics, London: Penguin Books.

Stendhal (2002a): The Red and the Black, Penguin Classics, London: Penguin Books.

Stoller, Paul (2013): Strangers are like the mist: language in the push and pull of the African diaspora, in Abdoulaye Kane and Todd Leedy, African Migration: patterns and perspectives, Indiana University Press, Bloomington, pp. 158–172.

Tazreiter, Claudia (2019): Narratives of Crisis Migration and Power of Visual Culture, in The Oxford Handbook of Migration Crisis, Oxford University Press: Oxford, pp. 619–614.

The Economist (2022q): Echoes of war, A resurgence of regional rivalries imperils eastern Congo: Meddling neighbors add to the mayhem, 30 June 2022, https://www.economist.com/middle-east-and-africa/2022/06/30/a-resurgence-of-regional-rivalries-imperils-eastern-congo.

The Economist (2023): Myths and mummies: A new docu-drama about Cleopatra has riled officials in Cairo, https://www.economist.com/culture/2023/05/10/a-new-docu-drama-about-cleopatra-has-riled-officials-in-cairo, 07.07.2023.

Tolstoy, Leo (2008): The Death of Ivan Ilyich and Other Stories, Penguin Classics, London: Penguin Books.

Turner, Bryan S. (2010): Enclosures, Enclaves, and Entrapment, *European Journal of Social* Theory 10(2):287–303.

Walia, Harsha (2021): Border and Rule: Global migration, capitalism, and the rise of racist nationalism, Haymarket Books: Chicago.

Weima, Yolanda and Hindman, Jennifer (2019): Managing displacement: negotiating transnationalism, encampment and return, in Critical geographies of migration, Edward Elgar, pp. 30–44.

Wodak, Ruth (2011): 'Us' and 'Them': Inclusion and Exclusion – Discrimination via Discourse, in Gerard Delanty et al. (eds.), Identity, Belonging and Migration, Liverpool University Press: Liverpool, p. 54–77.

Wodak, Ruth (2015): The politics of fear: what right-wing populist discourses mean, Sage: Los Angeles.

Wonders, Nancy and Jones, Lynn (2021): Challenging the borders of difference and inequality: power in migration as a social movement for global justice, in Handbook of Migration and Global Justice, Leanne Weber and Claudia Tazreiter, eds., Edward Elgar, Cheltenham, pp. 296–313.

Zatari, Amalia (2020): Racism in Russia: Stories of prejudice, https://www.bbc.com/news/world-europe-53055857, 07.07.2020.

CHAPTER 8

# The Postcolonial Migration from Africa to the West as the Desire for Equality and Negation of Difference

## Color Obsession and the Narcissism of Minor Differences

From the mimetic perspective, what the colonizer hated was not the *difference* between himself (model) and the colonized (imitator) but instead the *similarity*. He did not hate the color black as such. He hated only black people. His hatred is implicitly the acceptance of the latter's humanity and equality. He did not hate black objects, black animals, etc. but only black people. This has been analyzed beautifully by Douglas:

> There must be something in the color of itself to kindle rage and inflame hate, and render the white man generally uncomfortable. If the white man were really so constituted that color were, in itself, a torment to him, this grand old earth of ours would be no place for him. Colored objects confront him here at every point of the compass. If he should shrink and shudder every time he sees anything dark, he would have little time for anything else. He would require a colorless world to live in—a world where flowers, fields, and floods should all be of snowy whiteness; where rivers, lakes, and oceans should all be white; where islands, capes, and continents should all be white; where all the men, and women, and children should be white; where all the fish of the sea, all the birds of the air, all the 'cattle upon a thousand hills,' should be white; where the heavens above and the earth beneath should be



B. Gebrewold, *Postcolonial African Migration to the West*, Politics of Citizenship and Migration,
https://doi.org/10.1007/978-3-031-58568-5_8

> white, and where day and night should not be divided by light and darkness, but the world should be one eternal scene of light. In such a white world, the entrance of a black man would be hailed with joy by the inhabitants. Anybody or anything would be welcome that would break the oppressive and tormenting monotony of the all-prevailing white. In the abstract, there is no prejudice against color. No man shrinks from another because he is clothed in a suit of black, nor offended with his boots because they are black. (Douglass 1881: 574–75)

The colonizer does not have a problem with black objects but with black humans. Since the ontological difference between black objects and white human beings is so clear-cut the colonizer was not worried about his identity. However, since he implicitly admitted that black humans are still humans he had to differentiate himself from them by all means. It is the admittance of the similarity, not the lack thereof that caused the colonizer to hate the colonized. For the colonizer, the black color which is not a problem at all in the case of material objects became an existential problem in the case of human beings. To be, the colonizer had to differentiate himself from something similar (black human beings), not from completely different (black material objects). Therefore, racism is not based on absolute differences but instead on relative differences. Whereas the colonizer tried to keep up this relative difference postcolonial African migrants have been trying to close this relative difference or gap.

The end of slavery, colonial segregation, dehumanization, and hierarchization brought the dawn of a new African subjectivity. It has gradually undermined physical and mental boundaries erected by Western racists. Through increasing political, economic, technological, and social capabilities the horizons and aspirations of the postcolonial Africans have been expanding. The temporal and spatial coexistence between the colonized and the colonizer in the "advanced, civilized, and idealized West"—in the promised land—has become common because of postcolonial African migration to the West: similarity is increasing and difference is decreasing.

Globalization, communication technology, economic growth, institutionalized cultural capital (Pinxten and Lievens 2014: 1099), etc. have been playing a significant role in the migration processes and the speeding up of similarity and decreasing difference or gap between the model and the imitator. As various studies show, postcolonial social transformation promotes postcolonial migration from Africa to the West by decreasing the gap. In this chapter, I briefly discuss what social transformation is and then analyze postcolonial migration from Africa to the West from the perspective of increasing similarity and decreasing differences or gaps.

While colonization was the negation of African humanity, postcolonial African migration to the West is the negation of this African dehumanization. Postcolonial African social transformation is a driving factor behind the strive for equality and the negation of difference. It is a process of protest against the physical and metaphorical *sedentarization* of the Africans. As Rosa Parks in Alabama did in 1955, it is a way of struggle to choose one's place freely and the unwillingness to accept being assigned somewhere. Therefore, postcolonial African migration to the West is an existentialist struggle for liberty that is facilitated by social transformation, which manifests itself through the decreasing gaps between the colonizer and the colonized: liberty gap, resource gap, and knowledge gap.

The underlying decision-making process of migrants' is influenced by one's education level, transferable skills, financial and physical assets, levels of income, a family or community history of migration, networks, availability of information, facilitating factors, etc. These variables are the causes as well as consequences of social transformation (UNCTAD 2018: 3). As De Haas suggests, migration transforms the broader social, cultural, and economic contexts in sending and receiving communities and societies (De Haas 2010: 1591). Migration is also an outcome of social transformation: increasing equality and similarity and decreasing differences and gaps between the origin and destination. Urbanization has popularized global youth culture almost in all corners of African countries (UN DESA 2022). Urban lifestyles reflect a portion of lifestyles and opportunities in the West. Migrants become conduits of information from the West as go-betweens tying regions and countries of origin with the destination and enabling global membership (Landau 2013: 102). In the following, first I will discuss what social transformation is. Secondly, I will explore how social transformation (increasing similarity and equality and decreasing differences and gaps) has been contributing to the postcolonial African migration to the West.

## What Is Social Transformation?

Transformation is a way of formation by adopting a new shape, or form, changing in shape, or going beyond the current shape. A society gets a new form by adopting new values or shaping itself according to them. The globalization of consumption, drinks, food, information, lifestyles, modern technological equipment, etc. has been transforming societies around the world (de Andrade 2021). The media and communication industries such as television, music, video, film, and the internet have transformative powers by spreading consumerist lifestyles, facilitating overall globalization, and breaking down cultural and other barriers (Siochrú 2004).

As Bourdieu would say, various factors dispel existing doxa and transform society: the pursuit of social justice; urbanization; capitalist modernization; economic and political revolutions; social crises such as war, famine, and recession, and especially the intensified precarity of the educated, etc. which could lead the subordinate classes to question the taken-for-granted order of things and to orchestrate their resistance; and symbolic revolution by artists and scientists; the state formation itself and its laws etc. (Bourdieu 2014; Bourdieu 2016; Fowler 2020: 439).

Castles suggests that social transformation implies an underlying notion of the way society and culture change in response to such factors as economic growth, war, or political upheavals. Moreover, global interconnectedness has regional, national, and local effects (Castles 2001: 15). Polanyi has shown that industrialization and modernization have brought about "great transformation" (Polanyi 2001).

Conflicts and environmental disasters that displace millions around the world—(by 31 December 2020, 55 million IDPs, and by the end of 2023, more than 108 million)—have been transforming the world. A study by the World Bank estimates that by 2025, 216 million will be internally displaced worldwide due to water scarcity, lower crop productivity, seal level rise, and storm surge, heat stress, extreme events such as fires, land losses (Internal Displacement Monitoring Center 2021; Clement et al. 2021; IOM 2020: 253–270; IPCC 2022; Rigaud et al. 2018).

Various researchers have been warning against the over-simplification of complex and diverse patterns of migration behavior (Massey et al. 1998; Portes 1997; Portes and DeWind 2004; Castles 2008). What are the constituent elements of this complexity? Very often, however, this complexity is reduced only to economic (poverty), political (security), and environmental (climate change) factors as drivers of migration. These factors are approached from basic needs perspectives. Social transformation goes beyond these basic and material needs perspectives and expands the scope to the immaterial or psychological needs as transformations are taking place on different levels. These psychological needs are a desire for recognition, enjoyment of a new social status, etc.

Castle suggests that "social transformation could provide the basis for a new understanding of the links between human mobility and global change. Castles understands social transformation as a fundamental economic, political, etc. shift in the way society is organized, and all existing social patterns are questioned and reconfigured (Castles 2010: 1578). Therefore, according to Castles, migration is an integral and essential part of social transformation processes (Castles 2010: 1579). Similarly, according to Zelinsky, the modernization process has contributed to the movement of people from the countryside displacing them from traditional livelihoods and past ways of life, making them internal migrants, and leading others to international migration (Zelinsky 1971).

There are increasingly new studies that show that "migration out of Africa seems rather to be driven by processes of development and social transformation which have increased Africans' capabilities and aspirations to migrate, a trend which is likely to continue in the future" (Flahaux and De Haas 2016: 1). Without denying the relevance of conflict as a cause of (forced) migration, Bakewell & Bonfiglio (2013: 4, 10) underline the importance of social processes in search for a better life in the city that drive mobility (Zelinsky 1971). Such transformative social processes include demographic and economic transitions or "development".

I see postcolonial social transformation as the decreasing social gaps between the colonizer and the colonized which consists of a liberty gap, resource gap, and knowledge gap. The social transformation and the decreasing social gaps are driven by the postcolonial African desire for similarity and equality with the colonizer. In the postcolonial longing of the colonized to be similar and equal to the colonizer, we can observe a postcolonial mimetic desire. As Girard says, "We learn what to want from each other, and our desires spread contagiously because we copy each other, hoping the people we admire hold the magic key to success and happiness. The more we imitate each other, the more we become the same—and it is our sameness, not our differences, that make us fight" (Girard 2023: xi).

Postcolonial African migrants hope to be more human by imitating the former colonizer because they admire him and his success. This admiration happens not in spite of but exactly because the West was the colonizer. The decreasing liberty gap, resource gap, and knowledge gap between the colonizer and the colonized are both causes and consequences of postcolonial social transformation that accelerate and intensify the mimetic process. Postcolonial African migration to the West is, therefore, the result of the pervasive postcolonial African desire for similarity, equality, and de-hierarchization as a result of this social transformation, as we shall see in the following sections.

## The Decreasing Liberty Gap

For Castles, migration is one part of the process of structural and institutional transformation, which arises through major changes in global political, economic, and social relationships (Castles 2010: 1566). Externally, postcolonial migration from Africa to the West is driven by the decreasing gap of liberty and equality between the former colonizer and the colonized due to the end of institutionalized segregation, dehumanization, and discrimination. Internally, because of liberty, the gap between the power elites and the rest of the society is also decreasing leading to emigration.

Isaiah Berlin differentiates between two concepts of liberty or freedom: negative and positive liberty or freedom (Berlin 2016: 169). Negative liberty means the absence of coercion, obstruction, interference with one's activity, or non-prevention from attaining a goal by human beings. Positive liberty, on the other hand, means the individual's ability to be his/her own master; one's life and decisions depend on oneself, not on external forces of whatever kind, to be subject, not an object; to be not nobody; to be a doer, decider, not being decided for; self-directed and not acted upon by external nature or by other men as if one were a thing, an animal, a slave incapable of playing a human role (Berlin 2016: 178).

For Amartya Sen, liberty is about the ability to decide to live as we would like, to achieve what we value, not being forced into some state because of constraints imposed by others, to be able to determine what we want, what we value and ultimately what we decide to choose and to promote the ends that we may want to advance (Sen 2009: 228, 232). For Sen, freedom has an intrinsic as well as an instrumental value. In the first case, it is a value that has to be pursued for its own sake. However, essentially related to this, freedom has also an instrumental value because the enjoyment of one aspect of freedom enhances its other aspects. That is why Sen focuses in his analysis on political freedom, economic opportunities, social opportunities like education, transparency guarantees, and protective security. For Sen, poverty is not only about inadequate income or wealth; it is also about capability deprivation, which is the inability to be and to lead the life one has a reason to value (Sen 2000; Nussbaum 2000).

Therefore, postcolonial migration is highly dependent on decreasing institutional unfreedom. The right and liberty to leave depend on this. This is where liberty, capability, and migration overlap. Postcolonial migration from Africa to the West is one of the manifestations of liberty or social transformation since the end of institutionalized separation and racism.

Social transformation changes collective and individual identities and aspirations (Todd 2005). Contrary to the colonial time in which only whites and colonizers used to enjoy absolute freedom, the postcolonial period and international law have democratized the global free movement. This is an important global social transformation contributing to postcolonial migration to the West. The institutionalized limitations of liberty like during colonialism, apartheid, or slavery do not exist anymore. The number of countries that require exit visas for their citizens has substantially decreased and contributed to the liberty to migrate. To be able to leave one's place or country is a negative freedom because the state authority does not prevent someone from leaving his country. Ilya Somin (2020) in his book *Free to Move* shows how migration is a symptom of political freedom. People can move, broaden their opportunities, and choose where they can live. It is not only that international foot voting promotes freedom, but it is already a symbol or an outcome of a certain amount of political freedom (Somin 2020: 64). The more people can move, the freer the society. As Enzensberger would say, "The more freedom and equality people gain, the more they expect" (Enzensberger 1994: 39).

Africans are increasingly moving because they enjoy much more political liberty than during colonialism. Bakewell suggests that instead of discussing African migration as an exceptional problem and a symptom of failure among the poor, and understanding mobility as something normal for the wealthy, international elite, we have to acknowledge that there is a complex relationship between migration and social transformations as mobility affects changes for better and worse in any society (Bakewell 2008: 1355–6).

African societies are increasingly choosing cultural-religious values freely instead of the old ones. Traditional or big religions are no longer the dominant ones in Africa. People are choosing new religions and cultural values coming from abroad. Youth liberty is increasing vis-à-vis gerontocracy in Africa as education is expanding, as we shall in a later section. Therefore, in the postcolonial African migration study, it is important to discuss the individual agency of the youth, and the changes in family and social expectations of the youth and young adults (Kefale and Gebresenbet 2021: 6). This transformation has become possible because of the increasing enjoyment of liberty.

Gebresenbet discusses a case of social transformation in a northern Ethiopian region given youth's increasing independence not only in their decisions to migrate but also in their determination of values they perceive as important for themselves. Whereas government officials try to depict migration through the help of traffickers or smugglers as irrational and dangerous the youth sees it positively (Gebresenbet 2021: 79). Peer pressure is also a cause of migration because many friends are leaving (Campbell 2019: 232).

The increasing demand for liberty and democracy has been transforming African societies (Bentley et al. 2015; Sanny and Selormey 2021). According to an Afrobarometer 2016 survey, the majority of Africans preferred democracy to authoritarian regimes, presidential dictatorship, military rule, or one-party government (Mattes and Bratton 2016: 2). In 2014 and 2015, Afrobarometer interviews with 53,935 citizens in 36 countries were conducted. For two-thirds of all respondents (67%) democracy is always preferable whereas only 11% believe that a non-democratic regime can be preferable in some instances. Moreover, 78% of the respondents rejected presidential dictatorship and one-party rule, and 73% rejected military rule (Mattes and Bratton 2016: 4–5).

**Preference for democracy and rejection of authoritarian alternatives have continued**. Increasing independence and agency empowers and emboldens social groups and individuals to defy repressive cultural values and political norms. The inhibition level that separates individuals from the West and which used to be perceived as unsurmountable is lowering because of the increasing self-confidence. People are increasingly emboldened because the mythologized white West as something unreachable and superior is increasingly questioned.

On the national level, political activism, defiance, and protests against repression, authoritarianism, male dominance, and human rights violations have been increasing in many African countries. For example, even if migration can expose women to new vulnerabilities as the result of sexual exploitation and oppressive gender relations, migration can provide new opportunities to improve women's lives and change such as income, greater autonomy, self-confidence, and social status (Jolly and Reeves 2005: 1). A World Bank study result shows that not only migration can improve the autonomy, human capital, self-esteem, authority, rights, access to resources, dignity, and worth of women in their families and communities, it is also the result of a gradual but a steady social transformation from a patriarchal to an equal society (Fleury 2016). Therefore, women's agency (emancipation and liberty) related to education is ultimately leading to increasing women's emigration.

In Chinua Achebe's *Things Fall Apart* one can see how the cultural mindset is determined by patriarchal values that assign women to the weakest and lowest social position in the Nigerian or generally in African societies. Man is the symbol of strength and intelligence. A man who does not show these qualities is despised as a "woman" (Achebe 2017: 22). Women are often not allowed to participate in important meetings. Boys are trained to be worthy heirs who can feed their families and control the women. As Achebe shows, controlling women is even more important than economic success or wealth accumulation. Even violence and bloodshed are glorified for the sake of manliness (Achebe 2017: 41). If a woman is intelligent, independent, and strong one regrets that she is not a man; if a man is not strong, determined, or intelligent he is despised as a "woman". Achebe describes how an inadvertent murder is considered as not worthy of a man, therefore female manslaughter (Achebe 2017: 94, 99). Achebe, in his novel, writes about a dialogue between two characters: "Can you tell me, Okonkwo, why it is that one of the commonest names we give our children is Nneka, or 'Mother is Supreme'? We all know that a man is the head of the family and his wives do his bidding. A child belongs to his father and his family and not to his mother and her family. A man belongs to his fatherland and not to his motherland" (Achebe 2017: 102). The message here is that the father and man symbolize strength and success, whereas the mother and woman are refuge and comfort in case of sorrow and distress. In normal life, the father is the ideal of one's manliness and the mother is the symbol of weakness; a father who has a good daughter wishes she were a boy. A man does not ask a woman for forgiveness or beg for a big favor; a man can never be wrong in his own house; what belongs to her belongs to him "because the owner of a person is also the owner of whatever that person has" (Achebe 2017: 351).

The ideal woman has to bear a boy, many boys (Achebe 2017: 132–133, 172). In such a cultural mindset woman is a human being without subjectivity. A good wife is an obedient woman, who gives birth to boys. What constitutes her identity is not what she is instead who she belongs to. Achebe raises in his writings this issues many times ironically. "... When strangers see him they no longer ask *Whose son is he*? but *Who is he*? Of his wife they will no longer say *Whose daughter*? But *Whose wife*?" (Achebe 2017: 266).

Even if there is still a very long way to go, there is steady progress in women's liberty in African societies. A study by Carruth and Smith in Ethiopia shows that, because of an increasing self-assertiveness, women see migration as a chance to escape patriarchy. Carruth and Smith (2022: 85) suggest that women do not migrate only for financial opportunities, but also to escape combinations of domestic, political, and structural violence. Birhanu argues that for women in a patriarchal society like Ethiopia, migration is becoming a way for women to express their dissatisfaction with rural life and with patriarchal and oppressive gender roles and values imposed on women. One important form of social transformation is that women have started to become independent not only in their decision to migrate but also through financial self-sufficiency and self-ownership thanks to their work abroad if they come back home with money (Birhanu 2021: 205). African women have become more individuals than mere subordinate community members. Through this transformation, they have been negotiating their gender roles and questioning the established role of wives who stay at home, earn less than their husbands, and care for the latter and their children. As a consequence, men are adopting the traditional roles of women. Still, this social transformation has caused also conflicts in families. For many men this new situation is difficult to accept as it challenges their traditional authority and decreases their masculinity (Kane and Leedy 2013: 12). The migration-related women empowerment is undermining traditional masculinity (Babou 2013: 237). Despite widespread sexual exploitation and violence against migrant women, they have been increasingly opting for migration. Women have increasing access to the information that women affected by domestic and sexual violence have a good chance of getting asylum status in the West (Freedman 2013: 213–216). Carruth and Smith show that "migration and violence against women are linked" (Carruth and Smith 2022: 98).

Young women see migration as an escape from financial disadvantages and violence against women in patriarchal societies (Carruth and Smith 2022: 85). "Women also experience high levels of economic precarity: compared with men, they face lower educational attainment, lower wages, higher unemployment in formal sectors, and higher prevalence of low wages and informal contractual work in domestic and agricultural sectors" (Carruth and Smith 2022: 97). This study on Ethiopia shows women migrants agency searching for economic autonomy and the attempt to "build their own house" and to claim independence from their families (Carruth and Smith 2022: 103). Swai deconstructs the widespread idea that trafficked women are moved without their knowledge and against their will. Swai argues that there is a social transformation going on in Africa so that women are little by little setting themselves free from the patriarchal African cultural values that confine women at home, keep them obedient daughter or woman that needs to be punished or excommunicated if she does not comply with these values (Swai 2009: 148). Swai argues that through migration African women's subjectivities are being developed (Swai 2009: 152).

To sum up, despite available information about torture, rape, suffering, the dangers of the Sahara Desert and the Mediterranean, and killings, migrants decide to move on (Kingsley 2017: 40). There is very often a consensual deal between smugglers and migrants who pursue their dreams (Kingsley 2017: 73). This is because migration is becoming affordable for many young people because of the decreasing resource gap.

## The Decreasing Resource Gap

Various studies suggest "The poorest tend to migrate within national borders, and often within rural areas or to small towns, remaining invisible in most statistics." (Haan and Yaqub 2009). In an unequal world, poverty is a material gap between the poor and the rich, the colonized and the colonizer, the powerless and power elite, and the imitator and the model. From the mimetic perspective, the smaller the gap between these two poles, the stronger the rivalry or the mimetic similarity and competition (Girard 1965, 2023). The reason why people with increasing wealth migrate is that they have the knowledge, the resources, and the capability to question the ascribed or imagined superiority of the powerful whereas the poor lack them. The available resources enable this movement. Therefore, postcolonial migration from Africa to the West is a sign of decreasing resource gap. The colonizer's superiority, othering, bordering, and ordering are increasingly questioned through simulation, imitation, and spatial coexistence.

Various studies suggest that there is a link between global change, economic development, local social transformation, improvements in living standards, and migration (Castles 2009: 12, 15, 22; Massey et al. 1998: 223; UNDP 2009). The findings by a World Bank study show that the poor tend to either migrate less or migrate to low-return destinations whereas those who can overcome difficulties in access to remunerative migration opportunities and the high costs associated with migrating would migrate more and easily especially to the West (Murrugarra et al. 2011). Therefore, migration from poor and developing countries to the West is a sign of an improving capability or a decreasing resource gap.

On the one hand, there is a high number of Africans living in extreme poverty (460 million in 2022). Due to high gender inequality on average, around 32% of females were less likely to have the same opportunities as males in 2022 (Kamer 2022c). Africa is highly affected by extreme income inequality (average incomes of the top 10% are about 30 times higher than those of the bottom 50%) (Chancel et al. 2019: 2). On the other hand, even if the starting line of this economic growth is very low, there is a significant economic transformation. According to the IMF, many economies in sub-Saharan Africa grew at a record pace before the pandemic. Though it was not clear whether the gains in economic growth have been shared equally across regions within countries, countries like Rwanda, Ethiopia, etc. grew an average of more than 7.5% per year over the two decades before the pandemic (Fuje and Yao 2022). According to FDI-Intelligence globally, Ghana's growth was 8.79% in 2019, followed by South Sudan at 8.78%, Rwanda at 7.8%, Ethiopia at 7.7%, and Côte d'Ivoire at 7.4% besides Bangladesh at 7.28%, and followed by India at 7.25% (Mitchell 2019). Ethiopian economy grew on average 10.3% a year from 2006/07 to 2016/17—compared to a regional average of 5.4%—and was projected to grow at 8.1% during 2018–2021 if the pandemic and the civil war had not interfered. The Rwandan economy grew by 7.5% on average between 2007 and 2017 (Bajpai 2019). Between 2000 and 2016, Africa was the second fastest-growing region in the world with an average annual GDP growth of 4.6% (ODI 2018). The African Continental Free Trade Area (AfCFTA) is going to accelerate this social transformation by accelerating productive transformation by developing regional production networks, reducing intra-African trade costs, etc. (African Union 2022).

One of the main characteristics of socio-economic transformation in Africa is that consumption patterns have been changing. An increasing number of people even in the countryside have been buying modern electronic goods such as TVs, smartphones, and rechargeable lamps which was unthinkable 20 or 30 years ago. Where there is no electricity, many people are trying to purchase photovoltaic facilities to run TVs or charge their smartphones (Birhanu 2021: 210).

Consumption is also a way of communication. It is more than our relations to objects. It is a medium through which we communicate and relate with others (Poster 1988: 21, 131). "People in society will not understand life and will not enjoy it without internalizing the styles and methods of consumption. Consumption is a discourse, a discourse about contemporary society, the way society talks to itself" (Jansiz 2014: 78). Consumption constructs identity, forms relationships, and creates the need to have an identity. Therefore, consumption and personal identity are fundamentally social (Lunt and Livingstone 1992: 24).

There have been significant changes that have been taking in many countries thanks to globalization. "... consumption patterns such as dressing and eating, habits and lifestyles, preferred music, movies, and TV series, etc. briefly, feelings, thoughts and perceptions in all areas of life, preferences and tastes, needs, desires, and demands are becoming more and more homogenized day by day" (Ayaz 2021: 1209). As a result, rapid transformations and advances in media and communication are spreading the values of certain individuals and societies (Ayaz 2021: 1211). The perceptions of modernism have significant impacts on the consumption culture or behavior and accordingly on lifestyles and consumption activities. This behavior ultimately determines how one wants to satisfy his or her needs, desires, or will through lifestyle activities such as consumption patterns, spare-time activities, eating and drinking habits, home improvement, personal relationships, and activities performed with others. The consumption culture offers the individual the chance to be different from those it considers inferior, equalizes with those it considers superior, and bestows him a new identity (Şentürk and Tarhan 2017: 102–103). The lifestyles, consumer culture, and perception of needs are being significantly shaped in Africa because of this cultural globalization.

The culture of migration has penetrated regions that usually have not had a tradition of migration. Often migrants spend a lot of money to finance their journey to the West by paying a significant amount of money to smugglers. However, the smugglers are depicted as unscrupulous exploiters of vulnerable migrants. Many studies have shown that migrants are not always poor victims exploited by evil smugglers (Tufa et al. 2021: 43). Smugglers and migrants are agents acting freely. Migrants as agents, not as victims, see migration as a capability-enhancing act, an expression of development, not of a failure (Gebresenbet 2021: 93). Migrants challenge the implicit assumption that mobility is normal for the wealthy international elite and a symptom of failure for the poor (Bakewell 2008: 1355–56).

The right to be mobile used to be class-specific and selective: for the rich, western, or whites (Castles 2010: 1567). National border controls for people with less economic resources or political rights are much stricter than for the well-off. Poverty, inequality, discrimination, and (hidden) racism limit the freedom to move. However, migration is becoming increasingly "democratized" or "liberalized", and affordable. Regions that barely experienced migration have become emigration hotspots and can afford it (Kefale and Gebresenbet 2021). In the region where I grew up in Ethiopia, people are very creative in collecting or saving money to migrate and they are less willing to invest it in local businesses. At the beginning of the 1990s, almost nobody talked about migration in my neighborhood. Now almost everybody talks all the time about migration instead of education, local business, etc.

Many households are increasingly in competition to send as many children as possible abroad to get remittances and consequentially enjoy social respect in their neighborhood. At the same time, through migration, economic as well as status inequality is also increasing leading more people to migrate from other households to close the inequality gap. This leads again to more investment in migration to close the inequality gap and the social status gap.

Flahaux and De Haas suggest that people will only migrate if they have the resources and ambitions to make this happen. Moreover, they see migration as a function of people's aspirations and capabilities to migrate (De Haas 2011, 2014). The social transformation that leads to more mobility includes the economic development processes that expand people's access to material resources, improved social networks thanks to modern technology and social media, abundant and precise information about the risks and opportunities on the transit and at the destination, improvements in infrastructure and transportation, etc. (Flahaux and De Haas 2016: 4). Bakewell suggests that migrants see improved quality of life is related to new opportunities, which may include moving and establishing a new "home" elsewhere (Bakewell 2008: 1351).

As Flahaux and De Haas argue migration aspirations depend on people's more general life aspirations and their perceptions of the extent to which these aspirations can be fulfilled "here" and "there". The underlying causes are the continuously changing aspirations and the perceptions of inequality. Lifestyles and people's perceptions of the "good life" change continuously alongside increasing material aspirations and a growing appetite for consumer goods leading to more aspirations and capabilities to migrate (Flahaux and De Haas 2016: 4). These changing aspirations imply the comparison of oneself with those richer than oneself at home as well as abroad. Migration is hence not only a means to accumulate wealth but also a means to close the material and lifestyle gap.

Migration is also an expression of the desire to feel equal locally as well as with the former colonizer, the model. The more financial (economic) capital the people in the origin achieve, the higher they raise their aspirations to migrate and their hope to close the gap between them and their model, which is the West. As long as this capital is still low, that means they are very poor and their level of information remains low, the less people migrate to the West. The mimetic aspect of migration here is that the amount of resources and desire for migration are directly proportional. As Flahaux and De Haas suggest, "materially poor populations of the least developed countries have fewer capabilities to move, and when they move, they tend to move over shorter distances, either internally, or to other African countries", whereas the more the urbanization increases and the GDP per capita grows, the further they migrate (Flahaux and De Haas 2016: 17). Urbanization in Africa rose from 15% in the 1960s to 43% in 2018, and is projected to grow to 50% by 2030 (UNECA 2017) and cities being places of origin, transit, destination, and return (Fumagalli and Schaefer 2019: 41). By 2050, about 1.3 billion Africans will live in cities, an increase of almost 1 billion between 2022 and 2050 (Adegboye 2022).

As there is a positive correlation between migration and remittances, increased remittances lead to more migration (Mangalu 2012: 166). Remittances have been transforming not only the lives of those who migrated and their families but also have revealed to the neighbors, friends, and relatives the transformative power of migration and leading to the relative decrease of the gap between themselves and the West. Migrants' financial transfers have a significant impact on the decision to migrate and on the communities living in sending countries (Natali and Isaacs 2019: 117). Despite some exceptions, remittances to sub-Saharan Africa have been steadily increasing in the past years (in US$, billions): 2015 (42), 2016 (39), 2017 (42), 2018 (49), 2019 (47), 2020 (43), 2021 (49), 2022 (53), and 2023 (55) (World Bank Group 2022: 2). The official development assistance to the region by the end of 2020 was almost US$ 68 billion (The World Bank 2022).

People have become aware that migration can transform their economic lives and bring respect from family members and neighborhoods. Because of remittances, the decreasing resource gap is enabling more migration. As De Haas argues, migration transforms the broader social, cultural, and economic contexts and generates social stratification, economic growth, entrepreneurship, and cultural change (De Haas 2010: 1591). As the book by Kane and Leedy elaborates,

> "The increasing global flow of goods, money, ideas, and images is transforming the twenty-first-century world in unpredictable forms. Globalization has connected remote areas in Africa to global cities in a variety of ways. The cable/satellite news media, in a matter of seconds, bring news from one corner of the world to the most secluded places in the globe… The revolution in technologies of communication and mass media have made connections with home more viable, bringing migrants together on a daily basis with their families and friends left behind." (Kane and Leedy 2013: 8–9)

Various studies have shown that socio-economic development in source countries will continue to enable migration and not the poorest instead those who have sufficient resources embark on their journey. De Haas suggests that the combination of modest levels of economic development, relative poverty, and education enables and inspires people to migrate (De Haas 2011: S62). Walkey and Mbokazi (2019: 133) argue that migration is both the result of development and a means to achieve it, that is both are the result and enablers of the furthering decrease of the resource gap. Because of the decreasing resources gap, the availability of information about the outside world increased and improved the taste of imported modern goods. The perception of needs has changed substantially, the needed consumer goods are imported, and migration has become indispensable to meet these needs (De Haas 2010: 1595). From the social transformation perspective, we can observe that financial and social remittances increase aspirations and facilitate capabilities to migrate because of the decreasing resource gap. The feelings of relative deprivation indeed increase but the preferences expand as well due to decreasing resource gap.

As Sen would say, poverty is a capability deprivation. This deprivation is not just a shortage of material resources, but also a structural limitation of one's potential. It is not uncommon that in poor countries many poor people believe that their poverty is destiny or it is their own fault. However, through financial and social remittances the poor start to shake this belief in their inferiority and start to become aware of their potential to overcome their status, and become aware of the feasibility of meeting their needs through migration. Because of the decreasing gap, migrants form themselves anew, questioning the existing beliefs about themselves and the social structure. The migration threshold levels get lower not only because of cost- and risk-reducing network effects but also because of the rising self-consciousness of the migrants who start to question the gap between themselves and the outside world. Migration, from a social transformation perspective, is a fundamental questioning of the social hierarchy economically and psychologically as the individual aspirations and capabilities to migrate increase because of the decreasing resource gap. Young people in many African cultures have been "liberating" themselves through migration from culturally imposed values such as social pressure for redistribution (Kane and Leedy 2013: 4). The village as a safe place and the outside as a dangerous wilderness have been transformed or challenged.

De Haas hypothesizes "that, as long as people's aspirations increase faster than increases in local opportunities, migration aspirations will continue to increase, while internal dynamics of migration processes can increase people's capability to migrate" (De Haas 2010: 1609). The question is why do the aspirations increase? In De Haas' analysis, comparatively lower local opportunities are presented as the causes of migration. I argue that the decreasing knowledge gap and the socio-psychological gap between origin and destination facilitated by the decreasing resource gap increase the aspiration to migrate. Aspirations are not just material or constant. They are psychological and elastic. Therefore, they are mimetic. The cause is "positive" (the existence of increasing similarity and decreasing gap between origin and destination) not (wholly) "negative" (lack of local opportunities) that pushes the postcolonial people to migrate. The above hypothesis by De Haas makes the facilitating causes (the economic factor or local opportunities) the primordial causes. The globally interconnected migrants are not always comparing themselves with the people at the local level but also with those abroad who are usually considered as living better and more modern lives than in origin. The consumer goods and lifestyles from outside are of higher value than the local ones. They are associated with higher social status and recognition. Life in the West brings more satisfaction (even much better than consuming Western goods at origin) because it is assumed that it brings the real socio-psychological gap decrease between the postcolonial man and the colonizer. Migrants who succeed in returning with wealth gain status in their communities in a way parallel to ritual initiates who emerge after accomplishing their rite of passage. They are accorded personal favors, social capital, and notoriety (Kane and Leedy 2013: 5).

Migrants spend a lot of money to organize their migration journey. "Migrants claim that to cross the Agadez region they pay between 300 to 700 euros... Roughly... In addition, thousands of euros per individual are spent from their savings and loans in order to reach Niger, lodge in Libya, and cross over the Mediterranean." (Loprete 2016). The migrants know that they could start businesses with that money in their areas of origin. Migrants also know that relatives and friends are more willing to lend money to migrants than to those who would invest it locally. Moreover, migrants and their families enjoy a better social standing than those who do not migrate. Loprete writes in his article "One migrant told us during an interview in the transit center "I'm here because when my older brother died in the Mediterranean my mother said that now it is my turn to go." (Loprete 2016). Can the financial aspect only explain this phenomenon? European financial incentives or job offers at origin could not convince asylum seekers in Europe to return to, for example, African countries. This is why European readmission agreements with countries of origin failed. Therefore, postcolonial migration from Africa to the West is more than material. This is not peculiarly an African phenomenon. On 23 October 2019, 39 Vietnamese migrants including two sisters died in a lorry of people smugglers in Essex in eastern England. The family of the sisters paid almost €88,000 to the smugglers (Coen and Henk 2020: 14).

It is important to raise the question of why people invest so much money in migration despite such high risks, instead of investing it at home. The desire to escape poverty alone does not help us much to understand this phenomenon. The desire to decrease the difference or the gap between the postcolonial man and the colonizer through migration to the West is the primordial desire, which is facilitated by the decreasing knowledge gap, which we shall see in the following section.

## The Decreasing Knowledge Gap

Through the help of education, the youth acquires information about the benefits, routes, dangers, and opportunities on the way and at the destination. However, education is increasingly perceived as a waste of time in many African countries because of joblessness even after university graduation. There is a growing focus shift from education toward migration-based business as education is increasingly considered useless and a waste of time (Birhanu 2021: 211–2). Migration is perceived as more rational than the continuation of schooling or education which are widely believed not to provide employment opportunities at home (Sintayehu 2021: 74, 78).

Still, it is education that is enabling migration because it generates knowledge. Migrants use their skills to migrate; they collect and select useful information. Education facilitates migration itself and opens opportunities at the destination. Education creates new aspirations and new abilities (Mangalu 2012: 169). It generates information about the perceived possibilities abroad and increases aspirations for a good life (Gebresenbet 2021: 89; Debalke 2021:142). It is not despite education instead because of education people are moving to the West. Education decreases implicitly the knowledge gap between the postcolonial man and the colonizer and challenges the perceived superiority of the former even if the migration itself takes place because of admiration of and fascination for the colonizer. As the findings by Mangalu show, the more educated and well-off the migrants are, the more they tend to migrate to Europe or North America (Mangalu 2012: 157). Education creates new aspirations and new abilities (Mangalu 2012: 169). Education decreases the knowledge gap horizontally (locally and nationally) and vertically (between the societies in origin and at destination) and, therefore, those relatively better educated are highly likely to migrate (Mixed Migration Centre 2019: 51).

Those ethnic groups within which there is a widespread "diffusion" of migration (that is less knowledge gap between origin and destination) monopolize access to international migration. However, eventually, through steady connection and exchange of ideas and goods between various ethnic groups in the sending countries even those ethnic groups with less migration diffusion will get more information about the benefits of migration. Contrary to what De Haas suggests, the level of local opportunities is not decisive for the increase of migration aspirations which are elastic and evolving. More education and information about the outside world are transforming local culture not only by modernizing the local situation but also by decreasing the knowledge gap between origin and destination and by increasing migration aspirations. The more the perception of relative deprivation and the social prestige attached to migration increases the more the aspirations to migrate increase. If it were only about the economic or financial factors, possible local opportunities would have addressed the aspirations. However, the aspirations are specifically directed at the West. Therefore, the decreasing knowledge gap between postcolonial man and the colonizer spurns the desire to migrate to the West and expands the aspirations that are thought to create more similarity and equality with the West.

Information and communication technology play a decisive role in the reduction of the knowledge gap. From a technological perspective, the use of social media and mobile telephones has expanded substantially in Africa. Faster global transportation systems, the internet, mobile phones, different forms of social media, etc. facilitate networks and collection and exchange of information with would-be migrants, smugglers, and those who already migrated. These tools facilitate information collection on the political, economic, and security situation at transit and destination.

Modernization is mimetic. Mobile phones not only facilitate the mobility of migrants, but they also have symbolic value by closing the gap between the origin and destination as a symbol of modernity. Those who own smartphones in Africa enjoy a higher social status. As Mbembe suggests, they are also a medium of self-stylization and self-singularization, modernity, and autonomy, a symbol of the transgression of the boundary between the African and the idealized Western world, a similarity with the West (Mbembe 2021: 216–217). Mobile phones are not only economically important by strengthening and expanding transnational connections and securing remittances, but they also reduce the psychological distance between the postcolonial man and the colonizer (Youngstedt 2013: 148).

The number of Africans using the Internet jumped from 4 million (2000) to over 197.6 million (2012), which is 18.6% of the population. The number of people having access to broadband coverage increased from 0.1% (2005) to 16% (2012), probably reaching 99% by 2060. Similarly, the internet contribution to overall GDP is expected to increase from 1.1% around 2015 and is expected to grow to at least 5% or 6% by 2025, which is the same level as the world's leading economies (African Development Bank 2014: 55). Mobile technologies and services contributed over $130 billion to the African economy, which is about 8% GDP in 2021 (FurtherAfrica 2022). As far as the mobile phone market is concerned, Africa is the fastest growing and second largest in the world with more than 8 in 10 Africans having a mobile phone and an average penetration rising from 37% in 2010 to 80% in 2013 and growing at 4.2% annually (African Development Bank 2014: 55). The number of smartphone subscriptions in sub-Saharan Africa increased from 29.56 million in 2011 to 564.13 million in 2022 and is estimated to reach 798.06 million in

2027 (Statista 2022). With its massive and increasingly tech-savvy generation of young Africans Africa is to become one of the world's major consumers of media and entertainment (Guttmann 2022). Share of web traffic in Africa as of May 2022 shows that the devices used for internet access constituted 1.5% tablets, 25.1% laptops, and desktops, and 73.4% used their mobile phones (Kamer 2022b). The computer penetration rate among households in Africa increased from 3.7% in 2005 to 7.7% in 2019 (Alsop 2021). Though Africa—with its around 43% internet penetration rate in 2021—is below a global average of 66%, the use of the internet has evolved rapidly in Africa as the number more than doubled to 570 million internet users in 2022 compared to 2015 (Saleh 2022).

Another area of rapid social transformation and the decrease of the knowledge gap in Africa is the expansion of physical infrastructure. Even if the current trade between African countries represents only 12% of total economic activity, the trend to integrate African countries' infrastructures within countries and across borders has been increasing since the early 2000s (Lakmeeharan 2020). By 2020, though around 590 million people in sub-Saharan Africa lacked access to electricity, between 2014 and 2018 the number of people without access declined and the electrification efforts surpassed population growth (Aevarsdottir 2018: 49). Between 2014 and 2019, few countries increased the accessibility to electrical energy, improving the overall conditions in sub-Saharan Africa before the COVID-pandemic started to jeopardize this trend (Kamer 2022a). More electrification means more information.

The lack of electricity access is one of the main constraints to television penetration in many regions of African countries. Nevertheless, the number of TV households in sub-Saharan Africa has been steadily increasing. As data shows, the number of TV households increased from 42.49 million in 2010 to a projected 74.76 million in 2021 (Statista Research Department 2016). Sub-Saharan Africa is forecasted to be the world's fastest-growing television and video market, with a growth rate of 35.7% between 2018 and 2023 (EUTELSAT 2022).

For example, African internal migration to urban areas is caused by perceived or real job opportunities, availability of better infrastructure, better schools, and electricity. Urban areas are a symbol of modernity for those leaving the "backward" rural areas and agricultural and subsistence economies. At the same time, rapid urbanization, or the increasing dominance of the urban areas in the national economy and society is the first structural transformation attracting more migrants from the countryside

(Rao and Vakulabharanam 2019: 271). The more people migrate from the countryside, the more information flows about the opportunities in the urban areas. More urbanization means more information about the opportunities. When more networks emerge migration will be easier and cheaper leading to the increased reduction of the knowledge gap.

The IOM World Migration Report 2020 suggests that migration is in large part related to the broader global economic, social, political, and technological transformations that have been shaping our lives (IOM 2020: 1). As this report underlines, technology is an enabler and a game-changer related to migration. The transfer of technology and knowledge, the ability to use ICT by migrants to gather information and advice in real-time during migration journeys, for clandestine border crossings, etc. modern technology is an essential social transformative tool. Social media platforms connect geographically dispersed groups with common interests, facilitate irregular migration, and enable migrants to avoid abusive and exploitative migrant smugglers and human traffickers (IOM 2020: 8).

Information brings new ideas, exposes new lifestyles, and increases aspirations. Not only the life of the migrants is transformed, but also of their community for better or worse (De Haas 2010: 1593). De Haas suggests that income inequality, relative deprivation, and network effects can easily reinforce each other and increase migration aspirations (De Haas 2010: 1593). As the term itself suggests, income inequality and relative deprivation are, however, *relative*. They can lead to migration only after the reduction of the knowledge gap starts to gain a foothold. While the absolute poor still consider migration for the educated and wealthy, the relatively poor increasingly migrate as societies rapidly modernize and transform because of the decrease in the knowledge gap between origin and destination. Cultural change through "social remittances" is also part of the reduction of the knowledge gap because it forms new ideas, behaviors, identities, and awareness of opportunities and lifestyles (Lipton 1980: 12). The awareness of possibilities of social and material success makes migration increasingly a norm rather than the exception, and staying home can become associated with absolute poverty, failure, a mindset which increasingly leads to a "culture of migration" (Massey et al. 1993) and migration obsession (De Haas 2010: 1595).

Migratory journeys would be virtually impossible without cheap mobile communications and modern technological equipment, such as mobile phones and email access (Collyer 2007: 674). The unprecedented accessibility of contacts equipped with mobile phones enables geographically expansive networks (Schaub 2012: 126; Cummings et al. 2015: 31). In my village, for example, people who do not have electricity to charge their mobile phones walk for hours to the next town and charge them. Many people are buying solar panels and fixing them on their roofs primarily to charge their phones, as they are not connected to the electricity grid.

As a consequence of the modernization-induced social transformation, the social-psychological gap between both the relatively poor and relatively well-off people at the origin as well as between would-be migrants and the societies at the destination is decreasing. There is more knowledge about risks and opportunities. The imagined inaccessibility of the outside world has been demystified. Growing connectedness to the outside world through exposure to new information, education, media, networks, and modernizing culture are transforming local culture by opening it up, questioning traditional values, giving young men and women more autonomy, "expanding imaginative and geographical horizons, and leading to migration" (De Haas 2010: 1610).

De Haas suggests that migration-related transformative effects can be "dampened or even reversed at later stages of migration diffusion processes when migration selectivity, remittance-propelled inequality, and the social prestige attached to migration may decrease and eventually become negative" (De Haas 2010: 1611). Similarly, El Nour et al. (2019:199) suggest that as Africa empowered by the use of technology and networking would overcome barriers and create new opportunities, its economy will grow and the drive and desire to migrate outside the continent will decrease, African foreign-born and foreign-trained talents would return to Africa whereas the West slows down. According to these authors, Africa will become a key migration and investment destination including many Western migrants looking for access to growing African markets and low production costs; elderly Westerners would relocate to Africa due to low living costs and improved quality of life.

However, I argue that this could be the case if we see postcolonial African migration to the West only from an economic perspective. The imitated lifestyles are more psychological and relational than material. The idealization of Western lifestyles is rather psychological than economic. When this idealization and imitation of the West start to decrease nobody knows and it is very difficult to predict as postcolonial African migration to the West is a mimetic desire for being. It is a search for equality and negation of difference. The question is, are postcolonial African migrants to the West going to achieve this because of successful wealth accumulation? I highly doubt this.

In 2019, UNDP researchers interviewed 3069 adult irregular African migrants who had traveled from 43 African countries of origin across 13 European countries. Only 2% of respondents said that greater awareness of the risks of irregular migration would have caused them not to migrate (UNDP 2019: 5, 16). Moreover, those who traveled were relatively better off than their peers (UNDP 2019: 5). There is some kind of "information defiance" or information selectivity in the postcolonial African migration to the West. Those who decide to migrate to the West defy any migration sensitization campaigns or migration deterrence programs that show the dangers on the way and at the destination (Carruth and Smith 2022: 96). This suggests that physical security threats on the way, achieved economic success, job opportunities back home, etc. would not necessarily decrease postcolonial African migration to the West.

An Afrobarometer survey shows that on average, more than half (51%) of all respondents with post-secondary educational qualifications say they have given at least "a little" consideration to emigrating, whereas 43% and 29% of respondents with secondary and primary schooling, respectively, and only 24% of those with no formal education make up potential emigrants (Sanny et al. 2019: 6). Just lack of opportunities as a cause of migration cannot explain this phenomenon as the educated ones have more opportunities at home than those with less or no education at all. This survey shows some substantial differences in motivations to migrate between those highly affected and those not affected by poverty. Individuals from the first group are more than twice as likely to indicate that harsh economic conditions are the main reason they would consider emigrating (39%) vs. the second group (18%), while those who experienced no economic hardship ("no lived poverty") are substantially more likely to cite a search for work as a main motivator (47% vs. 40%) (Sanny et al. 2019: 18). It is interesting to see that those who do not experience poverty at all are more likely to migrate to search for work than the poor ones. Moreover, according to the UNDP study discussed above, a majority of Africans who migrate internationally are better educated and economically more successful than their peers back home (UNDP 2019: 31f; 35).

As Czaika argues, those who intend to migrate continuously collect information and assess general economic prospects, including the labor market situation, at home and abroad to form reference points and updates for their migration-related expectations (Czaika 2015). After having gained information about the possibilities of net income gain through migration, the household's sense of relative deprivation increases (Myrdal 1957). The technological social transformation facilitates the ties and networks with relatives and friends at the destination and helps decrease monetary, psychological, and other similar costs and migration risks (Brettell and Hollifield 2000, 106–118).

## Conclusion

The social transformation raises the perception of the imbalance between the desired "standard of living" and the actual "scale of living" (Piore 1979; Massey et al. 2006: 36; Todaro 1976). This perception is already an indicator of the decreasing gap. The decreasing liberty, resource, and knowledge gaps horizontally and vertically facilitate and speed up the postcolonial African imitation of the West. Thanks to globalization, the consumption patterns of the colonizer are prevalent everywhere. Ordinary citizens, intellectuals, religious conservatives, or fundamentalists in the South who usually glorify their own cultural values and who attack "Western decadence" and neocolonialism are fascinated at the same time by the lifestyles of the colonizer and adoring it. Globalization and global social transformations have contributed to the decrease of the gaps between the colonizer and the colonized. The three decreasing gaps imply the intensification of imitation. From the mimetic perspective, this is the migration paradox of liberation through imitation and increasing similarity, not through difference and detachment. All people try to become a bit Western even if they pretend to hate it.

To the dehumanized African being, Western goods are hoped to bestow identity and full humanity. The luxury goods imported by many African elites from the West tell a lot. For the postcolonial African man, the becoming and being are dependent on the successful imitation of the colonizer through the consumption of Western products and migration to the West. The colonized Africans desire to be "full humans" by becoming at least a bit Western or "whiter". I hope it will not end like the Dostoevskian tragedy of the *Double*.

## Literature

Achebe, Chinua (2017): Things Fall Apart; Arrow of God; No Longer at Ease: The African Trilogy: Penguin Classics: New York.

Adegboye, Emmanuel(2022): Youth innovation can help shape the future of African cities, https://www.chathamhouse.org/2022/08/youth-innovation-can-help-shape-future-african-cities?gclid=CjwKCAiA7IGcBhA8EiwAFfUDsbPIUA0juJeEdAHTomGMS4cTCJZqCI0d15KtdZjsHFwqp3eXl6E-nxoC9nAQAvD_BwE.

Aevarsdottir, Anna (2018): Electrifying Africa: The Oxford Institute for Energy Studies, September 2018, Issue 115.

African Development Bank (2014): African Economic Outlook 2013, Addis Abeba.

African Union (2022): Africa's Development Dynamics 2022: Regional Value Chains for a Sustainable Recovery, AUC/OECD.

Alsop, Thomas (2021): Computer penetration rate among households in Africa 2005–2019, https://www.statista.com/statistics/748549/africa-households-with-computer/.

Ayaz, Selin (2021): Globalization phenomenon and its effects. *KAUJEASF*, 12(24), 1199–1217.

Babou, Cheikh (2013): Migration as a factor of cultural change abroad and at home: Senegalese female hair braiders in the United States, in Abdoulaye Kane and Todd Leedy, African Migration: patterns and perspectives, Indiana University Press, Bloomington, pp. 230–247.

Bajpai, Prableen (2019): The 5 Fastest Growing Economies in the World, https://www.nasdaq.com/articles/the-5-fastest-growing-economies-in-the-world-2019-06-27.

Bakewell, Oliver (2008): 'Keeping Them in Their Place': the ambivalent relationship between development and migration in Africa, *Third World Quarterly*, 29:7, 1341–1358.

Bakewell, O. & Bonfiglio, A. (2013). Moving Beyond Conflict: Re-framing mobility in the African Great Lakes region. Working paper for the African Great Lakes Mobility Project (Vol. IMI Working paper 71). Oxford: International Migration Institutes, University of Oxford.

Bentley, Thomas et al. (2015): African democracy update: Democratic satisfaction remains elusive for many, Afrobarometer Dispatch No. 45.

Berlin, Isaiah (2016): Liberty, Oxford University Press: Oxford.

Birhanu, Kiros (2021): Heterogeneity in outcomes of migration: the case of irregular migration to Saudi Arabia in Harresaw, Eastern Tigray, in Asnake Kefale and Fana Gebresenbet, *Youth on the move: Views from below on Ethiopian international migration*, Hurst and Company: London, pp. 201–220.

Bourdieu, P. (2014). On the State (1989–1992). Cambridge: Polity.

Bourdieu, P. (2016). Sociologie générale (Vol. II (1983–6)). Paris: Raisons d'Agir, Seuil.

Brettell, Caroline B. and Hollifield James F. (2000): Migration Theory. Talking across Disciplines, (Routledge, New York).

Campbell, John (2019): Conflicting perspectives on the "migrant crisis" in the Horn of Africa, in The Oxford Handbook of Migration Crisis, Oxford University Press: Oxford, pp. 229–243.

Carruth, Lauren and Smith, Lahra (2022): Building one's own house: power and escape for Ethiopian women through international migration, Journal of Modern African Studies, 60 (1), pp. 85–109.

Castles, Stephen (2001): Studying Social Transformation, International Political Science Review, Vol. 22, No. 1, pp. 13–32.

Castles, S. (2008) Development and Migration*Migration and Development: What Comes First? New York: Social Science Research Council, http://essays.ssrc.org/developmentpapers/wp-content/uploads/2009/08/2Castles.pdf.

Castles, Stephen (2009): Development and Migration—Migration and Development: What Comes First? Global Perspective and African Experiences, *Theoria: A Journal of Social and Political Theory*, December 2009, Vol. 56, No. 121, pp. 1–31.

Castles, Stephen (2010) Understanding Global Migration: A Social Transformation Perspective, *Journal of Ethnic and Migration Studies*, 36:10, pp. 1565–1586.

Chancel, Lucas et al. (2019): Income inequality in Africa, 1990–2017, WID.world Issue Brief 2019/6, September 2019.

Clement, Viviane et al. (2021): GROUNDSWELL: Acting on Internal Climate Migration, PART II, International Bank for Reconstruction and Development / The World Bank.

Coen, A. and Henk, M. (2020): Bete für mich, in Die ZEIT, 7 May, 2020, N.20, p.14.

Collyer, M. (2007). In-between places: Trans-Saharan transit migrants in Morocco and the fragmented journey to Europe. Antipode, 39(4), 668–690. https://doi.org/10.1111/j.1467-8330.2007.00546.x.

Cummings, Clare et al. (2015): *Why people move: understanding the drivers and trends of migration to Europe*, Overseas Development Institute, Working Paper 430, 2015s.

Czaika, Mathias (2015): Migration and Economic Prospects, Journal of Ethnic and Migration Studies, 41:1, 58–82.

De Andrade, Pedro (2021): Globalization of Consumption, Lifestyles and 'Viral Society', E-*Revista de Estudos Interculturais do CEI–ISCAP*, N.° 9, maio de 2021.

De Haan, Arjan and Yaqub, Shahin (2009): Migration and Poverty Linkages, Knowledge Gaps and Policy Implications, Social Policy and Development Programme Paper Number 40, United Nations Research Institute for Social Development (UNRISD).

De Haas, H. (2010): The Internal Dynamics of Migration Processes: A Theoretical Inquiry, Journal of Ethnic and Migration Studies, 36:10, pp. 1587–1617.

De Haas, H. (2011). The Determinants of International Migration, DEMIG Working Paper 2. International Migration Institute, University of Oxford.

De Haas, H. (2014). Migration theory: Quo vadis? DEMIG, Working Paper, International Migration Institute, University of Oxford.

Debalke, Mulugeta Gashaw (2021): Irregular migration among a Kambata community in Ethiopia: views from below and their implications, in Asnake Kefale and Fana Gebresenbet, *Youth on the move: Views from below on Ethiopian international migration*, Hurst and Company: London, pp. 137–158.
Douglass, Frederick (1881): The Color Line, *The North American Review*, 132, 295, pp. 567–577.
El Nour, Ashraf et al. (2019): Migration Futures for Africa, in *Africa Migration Report*, International Organization for Migration, pp. 199–211.
Enzensberger, Hans Magnus (1994): Civil wars: From L.A. to Bosnia, New York: The New Press.
EUTELSAT (2022): TV Viewing Trends in Sub-Sahara Africa, https://www.eutelsat.com/en/blog/6-TV-viewing-trends-in-sub-saharan-africa.html.
Flahaux, Marie-Laurence and De Haas, Hein (2016): African migration: trends, patterns, drivers; *Comparative Migration Studies* 4:1, pp. 1–25.
Fleury, Anjali (2016): Understanding Women and Migration: A Literature Review, KNOMAD Working Paper Series, World Bank.
Fowler, Bridget (2020): Pierre Bourdieu on social transformation, with particular reference to political and symbolic revolutions, *Theory and Society* (2020) 49:439–463.
Freedman, Jane (2013): The feminization of asylum migration from Africa: problems and perspectives, in Abdoulaye Kane and Todd Leedy, African Migration: patterns and perspectives, Indiana University Press, Bloomington, pp. 211–229.
Fuje, Habtamu and Yao, Jiaxiong (2022): Africa's Rapid Economic Growth Hasn't Fully Closed Income Gaps, https://www.imf.org/en/Blogs/Articles/2022/09/20/africas-rapid-economic-growth-hasnt-fully-closed-income-gaps.
Fumagalli, Corrado and Schaefer, Katja (2019): Migration and Urbanization in Africa, in Africa Migration Report, International Organization for Migration, pp. 41–51.
FurtherAfrica (2022): African countries with the highest number of mobile phones, https://furtherafrica.com/2022/07/19/african-countries-with-the-highest-number-of-mobile-phones/.
Gebresenbet, Fana (2021): Hopelessness and future-making through irregular migration in Tigray, Ethiopia, in Asnake Kefale and Fana Gebresenbet, *Youth on the move: Views from below on Ethiopian international migration*, Hurst and Company: London, pp. 79–96.
Girard, René (1965): Deceit, desire and the novel: self and other in literary structure, John Hopkins University: Baltimore.
Girard, René (2023): All desire is a desire for being, Penguin Classics: London.
Guttmann, A. (2022): Media in Africa – Statistics & Facts, https://www.statista.com/topics/5032/media-in-africa/#dossierKeyfigures.
Internal Displacement Monitoring Center (2021): Global Report on Internal Displacement, Norwegian Refugee Council.

IOM (2020): The World Migration Report 2020, International Organization for Migration, Geneva.

IPCC (2022): Climate Change 2022: Impacts, Adaptation and Vulnerability. Contribution of Working Group II to the Sixth Assessment Report of the Intergovernmental Panel on Climate Change, Cambridge University Press, Cambridge, UK.

Jansiz, Ahmad (2014): The Ideology of Consumption: The Challenges Facing a Consumerist Society, *Journal of Politics and Law*; Vol. 7, No. 1.

Jolly, Susie and Reeves, Hazel (2005): Gender and Migration, Overview Report, Institute of Development Studies, BRIDGE.

Kamer, Lars (2022a): African population without electricity 2000–2021, https://www.statista.com/statistics/1221698/population-without-access-to-electricity-in-africa/.

Kamer, Lars (2022b): Share of web traffic in Africa as of May 2022, by device, https://www.statista.com/statistics/1176693/web-traffic-by-device-in-africa/.

Kamer, Lars (2022c): Gender gap index in Sub-Saharan Africa 2022, by country, https://www.statista.com/statistics/1220485/gender-gap-index-in-sub-saharan-africa-by-country/.

Kane, Abdoulaye and Leedy, Todd (2013): African patterns of migration in a global era: new perspectives, in Abdoulaye Kane and Todd Leedy, African Migration: patterns and perspectives, Indiana University Press, Bloomington, pp. 1–18.

Kefale, Asnake and Gebresenbet, Fana (2021): Introduction: Multiple transitions and irregular migration in Ethiopia: agency and assemblage from below, in Asnake Kefale and Fana Gebresenbet, *Youth on the move: Views from below on Ethiopian international migration*, Hurst and Company: London, pp. 1–22.

Kingsley, Patrick (2017): The new Odyssey: The story of Europe's refugee crisis, The Guardian: London.

Lakmeeharan, Kannan (2020): Solving Africa's infrastructure paradox, McKinsey & Company.

Landau, Loren (2013): Belonging amid shifting sands: insertion, self-exclusion, and the remaking of African urbanism, in Abdoulaye Kane and Todd Leedy, African Migration: patterns and perspectives, Indiana University Press, Bloomington, pp. 93–112.

Lipton, M. (1980). "Migration from Rural Areas of Poor Countries: The Impact on Rural Productivity and Income Distribution." World Development 8 (1): 1–24.

Loprete, Giuseppe (2016): MIGRO ERGO SUM – I Migrate, Therefore I Am – Social Pressure as a Driver of Economic Migration from West Africa, https://blogs.lse.ac.uk/africaatlse/2016/01/14/migro-ergo-sum-i-migratetherefore-i-am-social-pressure-as-a-driver-of-economic-migration-from-west-africa/, accessed 26.03.2024.

Lunt, Peter Kenneth and Livingstone, Sonia (1992): Mass consumption and personal identity: everyday economic experience, Buckingham : Open University Press.

Mangalu, Agbada (2012): Congolese migrants towards Africa, Europe, America, and Asia, in Mohamed Berriane and Hein de Haas, African migration research, Africa World Press: Trenton, pp. 147–176.

Massey, D.S., Arango, J., Hugo, G., Kouaouci, A., Pellegrino, A. and Taylor, J.E. (1993) 'Theories of international migration: a review and appraisal', Population and Development Review, 19(3): 431–66.

Massey, D.S., Arango, J., Hugo, G., Kouaouci, A., Pellegrino, A. and Taylor, J.E. (1998) *Worlds in Motion, Understanding International Migration at the End of the Millenium*. Oxford: Clarendon Press.

Massey, Douglas S., Joaquín Arango, Graeme Hugo, Ali Kouaouci, Adela Pellegrino and J Edward Taylor (2006) Theories of International Migration: A Review and Appraisal. In: Messina Anthony and Lahav Gallya (Eds.) The Migration Reader. Exploring Politics and Policies (Lynne Rienner, Boulder), 34–62.

Mattes, Robert and Bratton, Michael (2016): Do Africans still want democracy? Afrobarometer Policy Paper No. 36 | November 2016.

Mbembe, Achille (2021): Out of the dark night: Essays on decolonization, Columbia University Press, New York.

Mitchell, Jason (2019): IMF: African economies are the world's fastest growing, https://www.fdiintelligence.com/content/news/imf-african-economies-are-the-worlds-fastest-growing-75841.

Mixed Migration Centre (2019): Navigating borderlands in the Sahel: border security governance and mixed migration in Liptako-Gourma. Available at www.mixedmigration.org, 12.06.2022.

Murrugarra, Edmundo et al. (2011): Migration and Poverty: Toward Better Opportunities for the Poor, The International Bank for Reconstruction and Development / The World Bank, Washington DC.

Myrdal, Gunnar (1957). Rich Lands and Poor. The road to world prosperity, (Harper & Row, New York).

Natali, Claudia and Isaacs, Leon (2019): Remittances to and from Africa, in Africa Migration Report, International Organization for Migration, pp. 117–131.

Nussbaum, Martha (2000): Women and Human Development: The Capabilities Approach, Cambridge University Press: Cambridge.

ODI (2018): Africa's economic growth in a new global context, https://odi.org/en/events/africas-economic-growth-in-a-new-global-context/.

Piore, Michael J. (1979): Birds of passage. Migrant labour in industrial societies, (Cambridge University Press, Cambridge).

Polanyi, Karl (2001): The Great Transformation: The Political and Economic Origins of Our Time, Beacon Press.

Portes, A. (1997) 'Immigration theory for a new century: some problems and opportunities', *International Migration Review*, 31(4): 799–825.

Portes, A. and DeWind, J. (2004) 'A cross-Atlantic dialogue: the progress of research and theory in the study of international migration', *International Migration Review*, 38(3): 828–851.

Poster, Mark (1988): Baudrillard: Selected Writings, Stanford: Stanford University Press.

Rao, Smriti and Vakulabharanam, Vamsi (2019): Migration, Crisis, and Social Transformation in India, in The Oxford Handbook of Migration Crisis, Oxford University Press: Oxford, pp. 261–278.

Rigaud, Kanta Kumari et al. (2018): GROUNDSWELL: Preparing for Internal Climate Migration, International Bank for Reconstruction and Development / The World Bank.

Saleh, Mariam (2022): Internet usage in Africa – statistics & facts, https://www.statista.com/topics/9813/internet-usage-in-africa/.

Sanny, Josephine Appiah-Nyamekye and Selormey, Edem (2021): Africans welcome China's influence but maintain democratic aspirations, Afrobarometer Dispatch No. 489, 15 November 2021.

Sanny, Josephine Appiah-Nyamekye et al. (2019): In search of opportunity: Young and educated Africans most likely to consider moving abroad, Afrobarometer 2019, Dispatch No. 288 | 26 March 2019.

Schaub, Max Leonard (2012): Lines across the desert: mobile phone use and mobility in the context of trans-Saharan migration, *Information Technology for Development*, Vol. 18, No. 2, April 2012, 126–144.

Sen, Amartya (2000): Development as Freedom, Anchor Book, New York.

Sen, Amartya (2009): The idea of justice, Allen Lane: London.

Şentürk, Züliye Acar and Tarhan, Ahmet (2017): Life Styles and Consumption Culture Changing in Parallel with Globalization, Different aspects of globalization / Bünyamin Ayhan (ed.): Peter Lang, Frankfurt am Main, pp. 101–124.

Sintayehu, Firehwot (2021): The drivers of youth migration in Addis Ababa, in Asnake Kefale and Fana Gebresenbet, Youth on the move: Views from below on Ethiopian international migration, Hurst and Company: London, pp. 65–78.

Siochrú, Seán Ó. (2004): Social consequences of the globalization of the media and communication sector: some strategic considerations, *Policy Integration Department Word Commission on the Social Dimension of Globalization*, International Labour Organization, Geneva.

Somin, Ilya (2020): Free to Move: Foot Voting, Migration, and Political Freedom, Oxford University Press: Oxford.

Statista Research Department (2016): Number of TV households in Sub-Saharan Africa2010–2021,https://www.statista.com/statistics/287739/number-of-tv-households-in-sub-saharan-africa/.

Swai, Elinami Veraeli (2009): Trafficking of young women and girls: A case of "Au Pair" and domestic laborer in Tanzania, in, Toyin Falola and Niyi Afolabi (eds.), The Human Costs of African Migration, pp. 145–164.

The World Bank (2022): Worldwide Governance Indicators, https://databank.worldbank.org/source/worldwide-governance-indicators.

Todaro, Michael P. (1976): Internal migration in developing countries, (International Labor Office, Geneva).

Todd, Jennifer (2005): Social Transformation, Collective Categories, and Identity Change, *Theory and Society*, 2005, Vol. 34, No. 4, pp. 429–463.

Tufa, Fekadu Adugna et al. (2021): Say what the government wants and do what is good for your family: facilitation of irregular migration in Ethiopia, in Asnake Kefale and Fana Gebresenbet, *Youth on the move: Views from below on Ethiopian international migration*, Hurst and Company: London, pp. 43–64.

UN DESA (2022): World Population Prospects 2022, United Nations Department of Economic and Social Affairs, Population Division, New York.

UNCTAD (2018): Economic Development in Africa Report 2018, Migration for Structural Transformation, New York and Geneva.

UNDP (2009). 'Human Development Report 2009: Overcoming barriers: Human mobility and development'. New York: United Nations Development Programme, http://hdr.undp.org/en/reports/global/hdr2009/.

UNDP (2019): Scaling Fences: Voices of Irregular African Migrants to Europe, UNDP.

UNECA (2017): African Migration: Drivers of Migration in Africa, Draft Report Prepared for Africa Regional Consultative Meeting on the Global Compact on Safe, Orderly and Regular Migration, October 2017.

Walkey, Claire and Mbokazi, Sabelo (2019): The Migration–Development Nexus in Africa, in Africa Migration Report, International Organization for Migration, pp. 133–143.

World Bank Group (2022): A War in a Pandemic: Implications of the Ukraine crisis and COVID-19 on global governance of migration and remittance flows, Migration and Development Brief 36.

Wouter Pinxten and John Lievens (2014): The importance of economic, social and cultural capital in understanding health inequalities: using a Bourdieu-based approach in research on physical and mental health perceptions, *Sociology of Health & Illness*, Vol. 36 No. 7 2014 ISSN 0141-9889, pp. 1095–1110, https://doi.org/10.1111/1467-9566.12154.

Youngstedt, Scott (2013): Voluntary and involuntary homebodies: adaptations and lived experiences of Hausa "left behind" in Niamey, Niger, in Abdoulaye Kane and Todd Leedy, African Migration: patterns and perspectives, Indiana University Press, Bloomington, pp. 133–157.

Zelinsky, Wilbur (1971): The Hypothesis of the Mobility Transition, *Geographical Review*, Vol. 61, No. 2, pp. 219–249.

CHAPTER 9

# Summary and Conclusion

Besides the colonial legacies, there are also intra-African factors that keep Africans poor, violent, displaced, hungry, jobless, uneducated, beggars, desperate migrants to the West, etc. First, the patriarchal or authoritarian African cultural mindset keeps African women marginalized and oppressed and legitimizes the authority of the powerholders as something divine and unquestionable. Second, except in very few cases, we Africans find it hard to overcome our ethnicized politics and we are overwhelmingly incapable of constructing a common national identity. Thirdly, we adore the white West so much that in all that we do the West is our model. The more we deny this inner disposition and mindset of ours the more we confirm it. Our "desire for being" means the imitation of the white West. Fourth, for almost all our political or economic shortcomings we scapegoat outsiders: The West and the colonial time, Russia, China, IMF, the World Bank, etc.

Postcolonial African migration to the West is caused by the desire and hope for equality with the colonizer by imitating his lifestyle and living together with him in the West. This is the *migro ergo sum*, the desire for *being* through migration. Solving economic and political problems would not necessarily end African migration to the West. That is why I discussed in Chap. 2 the different migration policies of the European Union and various migration policy theories that examine the different causes of migration. As I argued in Chaps. 3 and 4, economic and political problems in Africa leading to African migration to the West are real but the main



B. Gebrewold, *Postcolonial African Migration to the West*, Politics of Citizenship and Migration,
https://doi.org/10.1007/978-3-031-58568-5_9

causes of the economic or political problems are Africans themselves, not the West or the East. I consider the postcolonial migration from Africa to the West as a special phenomenon essentially interconnected with the legacies of slavery and colonialism, as discussed in Chap. 5.

African desire for recognition through migration implies an African detrimental self-perception, as I discussed in Chap. 6. Migration to the West bestows subjectivity. To be recognized means to be. For the postcolonial African person, there is no being without becoming a bit European. Chapter 7 was the core of my argument: postcolonial African migration to the West is a desire for being which a quixotic undertaking is. This being is mimetic. Postcolonial African person hopes to gain identity and internal freedom through the imitation of the Europeans in all ways. The more Africans attempt to achieve this the more they lose their real being because they will be rejected and degraded rather than recognized as equals.

As discussed in Chap. 8, through interconnectedness and availability of information about the West and thanks to the decreasing liberty, resource, and knowledge gaps, postcolonial Africans believe that they can achieve their desired and mimetic being by imitating the West as much as possible. Migrants hope to change it economically by going where there are perceived opportunities. But, as discussed in this book, the causes of postcolonial African migration to the West are more than merely economic, political, or environmental. Therefore, I explored in this book postcolonial African migration from historical, social-psychological, and mimetic perspectives.

What postcolonial African migration to the West and the position of Africa in the international system show is that we Africans do not have self-respect or a sense of dignity. We either blindly flock to the West as the Serengeti Great Migration or slavishly beg for Russian wheat from an autocrat who completely disregards international law, invades an independent country, silences the opposition and democracy at home, disregards human rights, and whose government is guided by white supremacist ideologies. We have developed a pervasive beggar mentality and do not notice whom we are dealing with. We are physically free from slavery and colonization but we are still in mental slavery. We have the resources and human capital but we do not have the mentality and the determination to set ourselves free from the slavish adoration of the West, Russia, China, etc. We do not have the zeal to make our people prosperous and live in peace and security. When Russian President Putin organized the Russia-Africa summit in St. Petersburg, on 27–28 July 2023, 17 African Heads of State and Governments and many other government representatives from African countries flocked there despite the atrocities being committed against innocent Ukrainians by Russian troops. The African leaders' main worry was the grain export deal between Russia and Ukraine. According to the African Development Bank, Africa could feed more than 9 billion people but many African countries import almost half of their wheat from Russia and Ukraine and almost no African country is self-reliant for its food security. External powers manipulate Africa because of Africa's weaknesses: we don't have the ambition to be strong and prosperous as nations and as a continent, not only as individuals. We beg instead of utilizing our resources and managing our politics and economy better. We don't let our people work; we produce children but don't care for them adequately; we mistreat girls and women; we are not able to overcome our ethnicized politics as power holders as well as citizens. We don't care when our people are hungry, poor, thirsty, uneducated, or die in the Sahara Desert. For us, African lives don't count as European lives count for Europeans, for example. Instead of making a compromise we just kill our opponents. Instead of producing enough food for our people, we beg in the East or West as the African representatives did in the above-mentioned Russia-Africa.

Africa's beggar mentality and the lack of will for self-reliance and self-respect are shocking. Many illegal African migrants who come to Europe face severe difficulties in finding job permission. It is heartbreaking to see them more or less begging in front of many supermarkets such as in Austria where I live. They come from many resource-rich countries like the Democratic Republic of Congo, Guinea, etc. I had often opportunities to talk to them. What I noticed was how disillusioned they were, embarrassed, degraded, and humiliated but still unwilling to go back home empty-handed. In the cold Austrian winter, they stand the whole day in front of the supermarkets to sell some local news magazines freezing outside. Most people buy the magazines not because they read them but instead because they feel so sorry for these degraded African beings. Instead of buying food with the little money they earn, they send the biggest amount back home and beg for money to buy food. All of them who I talked to looked sad and humiliated. Their countries of origin are made poor by their own people, not by the West, Russia, China, or anyone else.

These Africans came here because for them to be meant to be in the West. As Fanon says, even in the twentieth century there is an attempt at total and collective cathartic identification with the white man (Fanon 2008: 124). The more we try to imitate the West, to be similar to the West, and as naïve allies of any external actor, the more we Africans will remain beggars, useful idiots, caressed like children, not treated like adults or equal human beings globally. What Africa needs is not isolation or detachment from the former colonizer, nor a slavish alliance or loyalty to anyone. As governments or as citizens we Africans need self-respect, which we lack currently.

Mbembe makes an interesting postcolonial analysis, which can be interpreted as a postcolonial-existentialist approach. "The transition from damaged consciousness to autonomous consciousness requires that slaves expose themselves and abolish the being-outside-of-self that is precisely their double ... The first abolition is not enough to achieve recognition and establish new relations of mutuality between former slaves and former masters. A second abolition is necessary, which is much more complex than the first, which fundamentally represents only an immediate negation. It is no longer simply a matter of abolishing the Other: it is a matter of abolishing oneself by ridding oneself of the part of oneself that is servile, and working to realize oneself as a singular figure of the universe" (Mbembe 2021: 51).

The mimetically violent world is a world of othering borders, fences, walls that divide and classify, and make hierarchies, and exclude, degrade, and look down on those who are considered inferior or dangerous. Humanity was stolen from these historical victims of slavery and colonialism. Psychological, economic, political, and cultural reparation is needed as a process of reassembling amputated parts, repairing broken links, relaunching reciprocity, and accepting and respecting. In the interest of common humanity what we need is a world freed from the burden of race, from resentment, and from the desire for vengeance that all racism calls into being (Mbembe 2017: 183). However, we will never get this from someone. We have to achieve our self-respect and dignity by ourselves. As Nietzsche says, "And you yourself should create what you have hitherto called the World: the world should be formed in your image by your reason, your will, and your love! And truly, it will be to your happiness, you enlightened men!" (Nietzsche 2003: 110).

> Colonization destroys relationships, destroys or petrifies institutions, and corrupts men, both colonizers and colonized. To live, the colonized needs to do away with colonization. To become a man, he must do away with the colonized being that he has become. If the European must annihilate the colonizer within himself, the colonized must rise above his colonized being. The liquidation of colonization is nothing but a prelude to complete libera-

tion, to self-recovery. In order to free himself from colonization, the colonized must start with his oppression, the deficiencies of his group. (Memmi 1967: 151)

Africans have overcome coercive, historical, and colonial slavery but the greatest challenge for postcolonial Africans is the will and determination to overcome the volitional, postcolonial, and emulative slavery. We postcolonial Africans struggle to live a meaningful and fulfilling life. However, we pursue it by way of a slavish imitation of whites. Migration is one type of such an attempt. The postcolonial migrant in the West will soon or later realize that he never achieves equality by imitation (Lukasik 2021: 297). Africans need, as Nietzsche would say, to become themselves. This is self-*becoming* through self-*overcoming*, the overcoming of the parts of yourself that are not, ultimately, of yourself or do not belong to your destiny, not wanting anything to be different from yourself by being someone else, white (Nietzsche 2007: xvii, 35). Postcolonial African imitation to be white is the flight from oneself, from one slavery to the next, as Dostoevsky in his *Grand Inquisitor* says. There are two types of slavery: *coercive slavery* (the historically imposed and dehumanizing of Africans) and *volitional slavery* or self-degradation in the attempt to be at least a bit white both literally and metaphorically.

Like Dostoevsky, I would say that the really liberating journey is the one that leads to the attainment of freedom from the self-destructive mimesis by finding the self within oneself (Dostoevsky 2003: 41). Dostoevsky brilliantly analyzes our inherent desire for *unfreedom* in his Grand Inquisitor: the *unbearability* of real freedom for human beings who are weak by nature. Dostoevsky designates this as the "unresolved historical contradictions of human nature throughout all the earth" (Dostoevsky 2003: 329). According to Dostoevsky, we look for freedom in slavery. We entrust ourselves and our freedom to those beings that enslave us. In the case of the first temptation to convert stones into loaves, Dostoevsky suggests that Jesus did not want to enslave people by just providing them with what they could eat. However, human beings prefer this slavery of begging. We do not care about our real freedom and still believe that we are free. We choose the one who gives us the loaves but who controls us and rules over us (Dostoevsky 2003: 331). The act of postcolonial Africans begging as individuals and as Heads of State and Government in the West is a kind of Dostoevskian self-enslavement.

Dostoevsky reminds us that "The secret of human existence does not consist in living, merely, but in what one lives for" (Dostoevsky 2003: 332). He says that peace of mind and even death are dearer to man than free choice and at the same time there is nothing more tormenting than this freedom from a mimetic self-enslavement. As Dostoevsky suggests, people need someone mighty who performs miracles and demonstrates his supernatural power, awakens servile ecstasies of these slaves, and takes away from them the restlessness, confusion, and unhappiness that come from real freedom, own responsibility, and independence (Dostoevsky 2003: 334). Postcolonial Africans slavishly prostrate before the West, China, or Russia on political and economic levels and before whites on an individual level. We renounce our freedom and become completely submissive. We look for someone who saves us from ourselves, someone who gives us loaves and keeps us the feeble creatures in reconciled happiness through slavery (Dostoevsky 2003: 334–337). On the one hand, we strive for freedom from the colonizer but on the other hand, paradoxically, we end up imitating him, instead of detaching ourselves from him and being free.

Postcolonial migration is by its very nature contradictory because it tries to heal one slavery through new slavery: subjugation. Postcolonial imitation of the West and the desire for being are self-induced slavery and colonialism out of free choice. It is not coerced, it is volitional. It is a search for identity and being through self-subjugation instead of freedom as Dostoevsky's Grand Inquisitor tells us. Postcolonial African migration to the West is a journey that is disguised as a journey of freedom but in reality, it is a journey of mimetic self-enslavement pushed by Dostoevskian self-delusion and the desire for being which however ends up in a new form of slavery or nothingness. We will be free if we are not afraid of being different as we are instead of trying to be a bit white.

Our blackness is still an issue. Racism against black Africans is globally widespread because we are doing miserably economically and politically in Africa. Of course, other countries in other regions do very poorly. However, these are exceptions, whereas good-performing countries are exceptions in Africa. To escape the nothingness inculcated in us during colonization and slavery, we hope to overcome it by imitating the whites instead of trying to do better by ourselves. We strive for freedom through imitation but end up in slavery because we search for it not in ourselves but in imitation of the Other whom we perceive as superior. The more we adore the white West, the more we will be degraded and rejected.

Cervantes instructs us the source of unhappiness and tragic ends are centered on a self-delusive desire for recognition, power, and an unscrupulous strive for social status. Sancho says to Don Quixote "… if you think I am going to sell my soul to the devil for the sake of being a governor let me tell you I am more interested in going as Sancho to heaven than as a governor to hell. 'By God, Sancho', said Don Quixote, I think you are worthy to be governor of a thousand islands just for those last few words you have spoken" (Cervantes 2003: 775).

Sancho talks to his donkey, the companion he left behind to become a governor, that "… ever since I climbed the towers of ambition and pride, my soul has been invaded by a thousand miseries, a thousand woes, and four thousand vexations. Now … let me go back to my old freedom" (Cervantes 2003: 848). Upon entering their village after their adventure, Sancho exclaims, "…my longed-for village, welcome your son Don Quixote, who has been conquered by other's arm but comes here as the conqueror of himself; and that, he has told me, is the best conqueror you can wish for." (Cervantes 2003: 970). Don Quixote recognizes his absurdity; he realizes that his mind has been restored to him, clear and free (Cervantes 2003: 976). And he dies as a real human being free from self-delusion guided by the desire for admiration, social status, and power. In the case of Cervantes's Don Quixote, however, we observe that, after having overcome his state of delusion, the state of pretending to be someone else, the world-famous Knight Errant not only became sane again but also ceased to call himself Don Quixote de la Mancha, instead Alonso Quixano, or even just the shepherd Quixotiz (Cervantes 2003, 973ff).

The unrelenting desire to be someone else leads not only to doubting and to denial of one's own existence but it even results in the ultimate destruction of the subject itself (Dostoevsky 2013, 180). Similarly, in Tolstoy's *Anna Karenina*, the heroine was not only striving for freedom from social constraints, but also her wish to achieve freedom through striving for power, recognition, and possession ended up in her demise.

Postcolonial African migrants are searching for their being to be free from the colonially imposed identity. Current African migration to the West is instigated by the traumas of slavery, colonization, and dehumanization of black Africans. Postcolonial Africans have been trying to overcome the legacy of this historical violence at least by two mechanisms: by scapegoating the colonial West for all their political and economic ills and at the same time by imitating it slavishly. The goal of the postcolonial African mimetic desire is to achieve equality with whites by becoming as white as possible. This endeavor has led even to more rejection and self-delusion. If postcolonial Africans strive for real equality, they have to free themselves from their Self-enslavement and self-degrading adoration of the white West, from black skin's relentless desire for white masks, from the imitation of Russian or Chinese dictatorship and authoritarianism for political power, and freedom from self-induced underage. As the Kenyan author, Wa Thiong'o says, Europeans are so successful in colonizing the African mind (Wa Thiong'o 1986) that black Africans need desperately white masks as Fanon would say. Postcolonial migration from Africa to the West is more than an economic undertaking and is a mindset of desire for being through imitation of the Western lifestyles and ideally by living in the West.

I do not agree with the vision for Africa by Mbembe and Fanon according to which, Africans will have to turn their gaze toward the new, especially if this new is meant the East instead of the West (Mbembe 2021: 230). According to me, what Africans need is to think about their people primarily, the historical sufferings of their people. They have to reshape their perception of the West. The rejection of the West and looking toward China or Russia is not a solution. They must choose their global partners only based on the criteria that benefit them. If they organize their partnerships wisely, any partner can be beneficial as long it promotes African interests without compromising human rights, democracy, good governance, and accountability. The choice should not be between the West and East but instead between the beneficial for the ordinary Africans.

Africans need self-transcendence: overcome their perpetual victim and beggar mentality and self-pitying. As long as Africans are not mentally free from the blind adoration of the West migration will continue even after poverty is overcome and people are not in danger of physical insecurity. The internalized African inferiority and slavish adoration of the West will perpetuate African mental enslavement. The self-degrading imitation of the West and blind alliance with Russia or China will keep Africa always the despised global beggar, poor house, and useful idiot for the likes of Putin. If we Africans individually and institutionally give freedom to our women, good education to our children, especially to girls, overcome our archaic mentality of ethnicity, let our economists, engineers, and scientists work, promote free speech and human rights, respect the inherent dignity of African lives, we will be secure, prosperous, and respected globally. There will be no need for flocking to the West like the Serengeti Great

Migration and no need for turning global summits into African begging agora.

Yes, postcolonial African migration to the West is also a journey of desire for being. However, we Africans will never find our liberty, equality, and our true being in flocking to the West, in slavish imitation of the West, or in our blind alliance with the East. Such attempts and mindsets will lead to quixotic delusion and Dostoevskian tragedies. As long as we seek our identity, dignity, and recognition through mimetic migration and imitation of the West our being will always remain a chimera, and we as beggars. As Dostoevsky would say, real freedom means accepting oneself (Dostoevsky 1991: 617). Raskolnikov discovered his freedom in prison where his desire for being is no longer dependent on dominating someone in the desire for recognition (Dostoevsky 1991: 648).

This book is not just about migration; it is about the postcolonial African mindset and self-consciousness to be really free and to be respected. In the name of tens of millions of enslaved and dehumanized Africans, we must make Africa peaceful, prosperous, and globally respected, and prove white racists wrong instead of flocking to the West like the Great Migration of the wildebeests in Masai Mara or Serengeti.

## Literature

Cervantes, Miguel de (2003): Don Quixote, Penguin Classics, London.
Dostoevsky, Fyodor (1991): Crime and Punishment: Penguin Classics, London.
Dostoevsky, Fyodor (2003): The Brothers Karamazov: Penguin Books: London.
Dostoevsky, Fyodor (2013): The Double, Alma Classics: Richmond.
Fanon, Frantz (2008): Black Skin, White Masks, Grove Press: New York.
Lukasik, Gail (2021): White like her: my family's story of race and racial passing, Skyhorse Publishing: New York.
Mbembe, Achille (2017): Critique of Black Reason, Duke University Press: Durham.
Mbembe, Achille (2021): Out of the dark night: Essays on decolonization, Columbia University Press, New York.
Memmi, Albert (1967): The colonizer and the colonized, Boston: Beacon Press.
Nietzsche, Frederick (2007): Ecce Homo, Oxford: Oxford University Press.
Nietzsche, Friedrich (2003): Thus spoke Zarathustra, Penguin Classics: London.
Wa Thiong'o, Ngugi (1986): Decolonizing the Mind: The Politics of Language in African Literature, James Currey: Woodbridge.

# Literature

Abdelmouman, Rachid (2020): The politics of recognition and cosmopolitanism: for an ethic of global citizenship, in Sbiri et al., Mobile identities: race, ethnicity, and borders in contemporary literature and culture, Cambridge Scholars Publishing: Newcastle, pp. 127–142.

Acemoglu, Daron and Robinson, James (2012): Why Nations Fail: The origins of power, prosperity, and poverty, Currency: New York.

Achebe, Chinua (2017): Things Fall Apart; Arrow of God; No Longer at Ease: The African Trilogy: Penguin Classics: New York.

Acker et al. (2020): Debt Relief with Chinese Characteristics. Working Paper No. 2020/39. China Africa Research Initiative, School of Advanced International Studies, Johns Hopkins University, Washington, DC. Retrieved from http://www.sais-cari.org/publications.

ACLED (2022): Wagner Group Operations in Africa: Civilian Targeting Trends in the Central African Republic and Mali, https://acleddata.com/2022/08/30/wagner-group-operations-in-africa-civilian-targeting-trends-in-the-central-african-republic-and-mali/#s4, 31.05.2023.

ACLED (2023): Fact Sheet: Conflict Surges in Sudan, https://acleddata.com/2023/04/28/fact-sheet-conflict-surges-in-sudan/, 30.05.2023.

ActionAid (2014): The Great Land Heist: How donors and governments are paving the way for corporate land grabs, ActionAid International.

Adam, Ilke and Trauner, Florian (2019): Ghana and EU Migration policy: studying an African response to the EU's externalization agenda, in Sergio Carrera



B. Gebrewold, *Postcolonial African Migration to the West*, Politics of Citizenship and Migration,
https://doi.org/10.1007/978-3-031-58568-5

et al. (eds.), Constitutionalizing the external dimensions of EU migration policies in times of crisis, Edward Elgar: Cheltenham, pp. 257–271.
Adegboye, Emmanuel(2022): Youth innovation can help shape the future of African cities, https://www.chathamhouse.org/2022/08/youth-innovation-can-help-shape-future-african-cities?gclid=CjwKCAiA7IGcBhA8EiwAFfUDsbPIUA0juJeEdAHTomGMS4cTCJZqCI0d15KtdZjsHFwqp3eXl6E-nxoC9nAQAvD_BwE.
Adem, Teferi Abate (2021): The moral economy of irregular migration and remittance distribution in south Wollo, in Asnake Kefale and Fana Gebresenbet, *Youth on the move: Views from below on Ethiopian international migration*, Hurst and Company: London, pp. 221–238.
Adepoju, Aderati (2010): The Politics of International Migration in Post-Colonial Africa, in Robin Cohen, *The Cambridge Survey of World Migration*, Cambridge University Press: Cambridge: pp. 166–171.
Adeyanju, Charles T. and Neverson, Nicole (2007): "There Will Be a Next Time": Media Discourse about an "Apocalyptic" Vision of Immigration, Racial Diversity, and Health Risks, *Canadian Ethnic Studies*, Volume 39, Number 1, 2007, pp. 79–105.
Adeyanju, Charles and Oriola, Temitope (2011): Colonialism and Contemporary African Migration: A Phenomelogical Approach, Journal of Black Studies, 42(6), 943–967.
Adichie, Chimamanda (2017): Americanah, 4th ESTATE: London.
Adusei, Aikins (2016): Hiding Africa's looted funds: The silence of Western media, *Third World Resurgence* No. 309, May 2016, pp. 21–23.
Aevarsdottir, Anna (2018): Electrifying Africa: The Oxford Institute for Energy Studies, September 2018, Issue 115.
African Development Bank (2014): African EconomicvOutlook 2013, Addis Abeba.
Africa Polling Institute (2020): Deconstructing the Canada Rush: A study on the motivations for Nigerians emigrating to Canada – Report March 2020, Abuja, Nigeria.
Africa research Bulletin (2019b): Migration: Boat disaster, Economic, Financial and Technical Series, November 16–December 15, V.56. No11, pp. 22779–814.
Africa Research Bulletin (2023): Sudan: Civil War Looms, *Political, Social and Cultural Series*, April 1st–30th 2023, p. 23977.
Africa Research Bulletin (2023a): Eritrea: Forced Conscription, *Political, Social and Cultural Series, February 1st–28th 2023*, p. 23920.
Africa Research Bulletin (2023b): Dakar 2 Summit: Food Sovereignty For Africa, *Economic, Financial and Technical Series*, January 16th – February 15th 2023, p. 24149.
Africa Research Bulletin (2023c): Ethiopia – China, *Economic, Financial and Technical Series*, December 16th–January 15th 2023, p. 24117.

Africa Research Bulletin (2023d): DR Congo: December Security Incidents, *Political, Social and Cultural Series*, December 1st – 31st 2022 p. 23845.
Africa Research Bulletin, Economic, Financial and Technical Series, Volume 56 Number 6, June 16th 2019a–July 15th 2019.
African Union (2022): Africa's Development Dynamics 2022: Regional Value Chains for a Sustainable Recovery, AUC/OECD.
African Union and IOM (2020): Africa Migration Report: Challenging The Narrative, International Organization for Migration (IOM).
Agamben, Giorgio (1998): Homo Sacer: Sovereign Power and Bare Life, Stanford: Stanford University Press.
Agamben, Giorgio (2005): State of Exception, Chicago University Press: Chicago.
Agblemagnon, N'sougan (1998): The existence and nature of African philsopphy in the past and in the present, In: Claude Sumner (ed.) African philosophy, Addis Abeba: Addis Abeba University Press, pp. 1–10.
Ahrendt, Hannah (2017): The Origins of Totalitarianism, Penguin Classics.
Akinbode, Ayomide (2021): The forgotten Arab slave trade, https://www.thehistoryville.com/arab-slave-trade/, 29.05.2023.
Akiwumi, Paul and Chiumya, Chiza Charles (2019): Migration and Trade in Africa, in Africa Migration Report, International Organization for Migration, pp. 103–115.
Alba, Davey, and Sheera Frenkel. 2019. "Russia Tests New Disinformation Tactics in Africa to Expand Influence." The New York Times, 30 October.
Albahari, Maurizio (2015): Crimes of Peace: Mediterranean Migration at the World's Deadliest Border, University of Pennsylvania Press: Philadelphia.
Alden, C. (2007), *China in Africa* (London: Zed Books).
Alden, Chris and Otele, Oscar M. (2022): Fitting China In: Local Elite Collusion And Contestation Along Kenya's Standard Gauge Railway, *African Affairs*, 121:484, pp. 443–466.
Alsop, Thomas (2021): Computer penetration rate among households in Africa 2005–2019, https://www.statista.com/statistics/748549/africa-households-with-computer/.
Anderson, Carol (2021): The Second: Race and guns in a fatally unequal America, Bloomsbury: London.
Andreas, Peter (2009): Border Games: Policing the US-Mexico Divide, Ithaca, NY: Cornell University Press.
Anette Hoffmann & Guido Lanfranchi (2023): Europe's re-engagement with Ethiopia Prioritising the prerequisites for lasting peace over geopolitical concerns, Clingendael – the Netherlands Institute of International Relations.
Ansprenger, Franz 1975. *Die Befreiungspolitik der Organisation der Afrikanischen Einheit (OAU) 1963–1975.* Kaiser: Grünewald.
Appiah, Anthony (1994): Identity, authenticity, survival: multicultural society and social reproduction, The politics of recognition, in Charles Taylor et al., Multiculturalism, Princeton University Press, Princeton, pp. 149–164.

Arana, Arantza Gomez and McArdle, Scarlett (2019): The EU and migration crisis: reinforcing a security-based approach to migration? in Sergio Carrera et al. (eds.), Constitutionalizing the external dimensions of EU migration policies in times of crisis, Edward Elgar: Cheltenham, pp. 272–289.

Arendt, Hannah (1998): The Human Condition, Chicago: The University of Chicago Press.

Arkhangelskaya, Alexandra and Shubin, Vladimir (2013): Russia's Africa Policy, Global Powers and Africa Programme, South African Institute of International Affairs, September 2013.

Arthur, J.K.M. (1998): African philosophers and African philosophy, In: Claude Sumner (ed.) African philosophy, Addis Abeba: Addis Abeba University Press, pp. 11–20.

Atanesian, Grigor (2023): Russia in Africa: How disinformation operations target the continent, https://www.bbc.com/news/world-africa-64451376.

Atlantic Council (2018a): *Turkey's Growing Presence in Africa, and Opportunities and Challenges To Watch in 2018,* http://www.atlanticcouncil.org/events/past-events/turkey-s-growing-presence-in-africa-and-opportunities-and-challenges-to-watch-in-2018, accessed 05.09.2018.

Atlantic Council (2018b): *Turkey's Growing Presence in Africa, and Opportunities and Challenges To Watch in* 2018, http://www.atlanticcouncil.org/events/past-events/turkey-s-growing-presence-in-africa-and-opportunities-and-challenges-to-watch-in-2018, accessed 05.09.2018.

Atuhaire, Patience (2023): Uganda Anti-Homosexuality bill: Life in prison for saying you're gay, https://www.bbc.com/news/world-africa-65034343.

Auth, Katie (2023): *How the U.S. Can Better Support Africa's Energy Transition*, Carnegie Endowment for International Peace.

Avrutin, Eugene M. (2022): Racism in Modern Russia: From the Romanovs to Putin, Bloomsbury: London.

Ayaz, Selin (2021): Globalization phenomenon and its effects. *KAUJEASF*, 12(24), 1199–1217.

Ayeni, Tofe (2022): reputational battle: China to overtake the EU as Africa's biggest trade partner by 2030, https://www.theafricareport.com/229297/china-to-overtake-the-eu-as-africas-biggest-trade-partner-by-2030/.

Babou, Cheikh (2013): Migration as a factor of cultural change abroad and at home: Senegalese female hair braiders in the United States, in Abdoulaye Kane and Todd Leedy, African Migration: patterns and perspectives, Indiana University Press, Bloomington, pp. 230–247.

Bajpai, Prableen (2019): The 5 Fastest Growing Economies in the World, https://www.nasdaq.com/articles/the-5-fastest-growing-economies-in-the-world-2019-06-27.

Bakewell, O. & Bonfiglio, A. (2013). Moving Beyond Conflict: Re-framing mobility in the African Great Lakes region. Working paper for the African Great Lakes Mobility Project (Vol. IMI Working paper 71). Oxford: International Migration Institutes, University of Oxford.

Bakewell, Oliver (2008): 'Keeping Them in Their Place': the ambivalent relationship between development and migration in Africa, *Third World Quarterly*, 29:7, 1341–1358.
Bakewell, Oliver (2021) Unsettling the boundaries between forced and voluntary migration: in Emma Carmel et al. Handbook on the Governance and Politics of Migration, Edward Elgar: Cheltenham, pp. 124–136.
Balibar, Ètienne (2002): Politics and the Other Scene, Verso: London.
Balytnikov, Vadim, et al. 209: Russia's Return to Africa: Strategy and Prospects, The Foundation for Development and Support of the Valdai Discussion Club, 2019, Moscow.
Banerjee, Abhijit and Duflo, Esther (2011): Poor Economics: A radical rethinking of the way to fight global poverty, Public Affairs: New York.
Banerjee, V. Abhijt and Duflo, Esther (2019): Good Economics for Hard Times, Public Affairs: New York.
Bartlett, Kate (2022): Are Rights Abuses Tarnishing China's Image in Africa? https://www.voanews.com/a/are-rights-abuses-tarnishing-china-s-image-in-africa-/6560353.html.
Basheska, Elena and Kochenov, Dimitry (2015): EuroMed, migration and frenemy-ship: pretending to deepen cooperation across the Mediterranean, in Francesca Ippolito and Seline Trevisanut, Migration in the Mediterranean: Mechanisms of International Cooperation, Cambridge: Cambridge University Press, pp. 43–67.
Bauböck, Rainer (2012): Migration and Citizenship: normative debates, in Marc Rosenblum and Daniel Tichenor, The Oxford Handbook of the politics of international migration, Oxford University Press: Oxford, pp. 594–613.
Bauman, Zygmunt (2016): Strangers at our door, Polity: Cambridge.
Baur, John (1998) *2000 years of Christianity in Africa: an African Church history*, Nairobi, Paulines Publications Africa.
Bayart, Jean-Francois (2009) The State in Africa: The Politics of the Belly, Polity.
BBC (2013): Japan's Shinzo Abe hails Africa as 'growth center', http://www.bbc.co.uk/news/world-africa-22758464, 26.05.2023.
BBC (2022): Ukraine war: Hungry Africans are victims of the conflict, Macky Sall tells Vladimir Putin, https://www.bbc.com/news/world-africa-61685383.
BBC (2023): 'Europe or death' - the teenage migrants risking it all to cross the Med, https://www.bbc.com/news/world-europe-66132875, 11.07.2023.
BBC (2005): Crossing deserts, scaling fences, http://news.bbc.co.uk/2/hi/africa/4332512.stm. accessed 26.03.2024.
Bedford, Nancy (2018): The most burning of Lavas: The Bible in Latin America, in Colonialism and the bible: contemporary reflections from the global south, Lexington books: Lanham, pp. 213–232.
Bell, David A. (2022): The Importance of Skin Colour in Central Eastern Europe: A Comparative Analysis of Racist Attitudes in Hungary, Poland and the Czech Republic, *Central and Eastern European Migration Review*, Vol. 11, No. 1, 2022, pp. 5–25, https://doi.org/10.54667/ceemr.2022.03.

Bell, Richard H. 2002. *Understanding African Philosophy: A Cross-cultural Approach to Classical and Contemporary Issues.* New York: Routledge.

Bello, Valeria (2017): International Migration and international security: why prejudice is a global security threat, Routledge: New York.

Benattia, Tahar et al. (2015): *Irregular Migration: Challenges and Solutions* Migration, Dialogue for West Africa 2015 Conference Research Paper, Altai Consulting: Abuja.

Bendandi, Barbara (2020): Migration induced by climate change and environmental degradation in the Central Mediterranean Route, in International Organization for Migration, Migration in West and North Africa and across the Mediterranean: Trends, risks, development and governance, Geneva, pp. 318–329.

Bentley, Thomas et al. (2015): African democracy update: Democratic satisfaction remains elusive for many, Afrobarometer Dispatch No. 45.

Berlin, Isaiah (2016): Liberty, Oxford University Press: Oxford.

Berthélemy, Jean-Claude, Beuran, Monica and Maurel, Mathilde (2009): Aid and Migration: Substitutes or Complements? World Development Vol. 37, No. 10, pp. 1589–1599, https://doi.org/10.1016/j.worlddev.2009.02.002.

Beth Elise Whitaker & Jason Giersch (2015) Political Competition and Attitudes towards Immigration in Africa, Journal of Ethnic and Migration Studies, 41:10, 1536–1557.

Bets, Alexander and Collier, Paul (2018): Refuge: Transforming a broken refugee system, Penguin Books: London.

Bhambra, G. K. (2007): Rethinking Modernity: Postcolonialism and the Sociological Immagination, Palgrave Macmillan: Basingstoke.

Birhanu, Kiros (2021): Heterogeneity in outcomes of migration: the case of irregular migration to Saudi Arabia in Harresaw, Eastern Tigray, in Asnake Kefale and Fana Gebresenbet, *Youth on the move: Views from below on Ethiopian international migration*, Hurst and Company: London, pp. 201–220.

Bisong, Amanda (2019): Trans-regional institutional cooperation as multilevel governance: ECOWAS migration policy and the EU, *Journal of Ethnic and Migration Studies*, 45:8, 1294–1309.

Bitterli, U. 1991. *Die Wilden und die Zivilisierten: Grundzüge einer Geistes- und Kulturgeschichte der Europäisch-Überseeischen Begegnung*. München: Beck.

Black, Richard (2011): Migration as adaptation, *Nature*, 478, 447–9.

Bleck, Aimie (2020): Migration aspirations from a youth perspective: focus groups with returnees and youth in Mali, Journal of Modern African Studies, 58:4, pp. 551–577.

Booty, Natasha (2023): Dragos Tigau: Romania recalls Kenya ambassador over racist monkey slur, https://www.bbc.com/news/world-africa-65867104, 07.07.2023.

Botwe-Asamoah, K. 2005. *Kwame Nkrumah's Politico-Cultural Thought and Policies: An African-Centred Paradigm for the Second Phase of the African Revolution.* New York: Routledge.

Bougroug, Anwar (2022): Anti-Blackness in the MENA-Region: time to reflect on racial bias, intersectionality, and paths to collective liberation, https://www.mykalimag.com/en/2020/07/23/anti-blackness-in-the-mena-region-time-to-reflect-on-racial-bias-intersectionality-and-paths-to-collective-liberation/, 25.05.2023.

Bourdieu, P. (2008) Esquisses algériennes. T. Yacine (Ed.). Paris: Seuil.

Bourdieu, P. (2014). On the State (1989–1992). Cambridge: Polity.

Bourdieu, P. (2016). Sociologie générale (Vol. II (1983–6)). Paris: Raisons d'Agir, Seuil.

Bratton, Michael and Bhoojedhur, Sadhiska (2019): Africans want open elections especially if they bring change, Afrobarometer Policy Paper No. 58 | June 2019.

Brautigam, D. (2019). A critical look at Chinese 'debt-trap diplomacy': The rise of a meme. Area Development and Policy, 5(1), 1–1430 October.

Bryant, Lisa (2022): Being Black in Tunisia, https://www.voanews.com/a/being-black-in-tunisia/6838272.html, 25.05.2023.

Burke, Peter and Stets, Jan (2009): Oxford: Oxford University Press.

Bürkle, Stefanie ed. (2016): *Migration* von Raum Architektur und Identität im Kontext türkischer Remigration, Vice Versa Verlag: Berlin.

Buzan, B. & Little, R. (2000), *International Systems in World History: Remaking the Study of International Relations,* (Oxford: Oxford University Press).

Campbell, John (2019): Conflicting perspectives on the "migrant crisis" in the Horn of Africa, in The Oxford Handbook of Migration Crisis, Oxford University Press: Oxford, pp. 229–243.

Camus, Albert (1991a): The myth of Sisyphus and other essays, New York: Vintage.

Camus, Albert (1991b): The Plague, Vintage: New York.

Camus, Albert (2013): The Fall, Penguin Classics, London: Penguin Books.

CARICOM (2023): 10-Point Reparation Plan, https://caricomreparations.org/caricom/caricoms-10-point-reparation-plan/, 07.07.2023.

Carling, Joergen (2001): Aspiration and ability in international migration: Cape Verdean experiences of mobility and immobility, Dissertations & Theses No. 5/2001, University of Oslo.

Carling, Jørgen & Collins, Francis (2018): Aspiration, desire and drivers of migration, *Journal of Ethnic and Migration Studies*, 44:6, pp. 909–926, https://doi.org/10.1080/1369183X.2017.1384134.

Carmel, Emma, Lenner, Katharina, and Paul, Regine (2021): Handbook on the Governance and Politics of Migration, Elgar Handbooks in Migration) Cheltenham.

Carrera, Sergio et al. (2019): The external dimension of EU migration and asylum policies in times of crisis, in Sergio Carrera et al. (eds.), Constitutionalizing the external dimensions of EU migration policies in times of crisis, Edward Elgar: Cheltenham, pp. 1–20.

Carruth, Lauren and Smith, Lahra (2022): Building one's own house: power and escape for Ethiopian women through international migration, Journal of Modern African Studies, 60 (1), pp. 85–109.

Carter, Donald (2013): Navigating diaspora: The precarious depths of the Italian immigration crisis, in Abdoulaye Kane and Todd Leedy, African Migration: patterns and perspectives, Indiana University Press, Bloomington, pp. 59–77.

Casas-Cortes, Maribel and Cobarrubias, Sebastian (2019): Genealogies of contention in concentric circles: remote migration control and its Eurocentric geographical imaginaries, in Critical geographies of migration, Edward Elgar, pp. 193–205.

Castles, S. (2008) Development and Migration*Migration and Development: What Comes First? New York: Social Science Research Council, http://essays.ssrc.org/developmentpapers/wp-content/uploads/2009/08/2Castles.pdf.

Castles, S. and Miller, M.J. (2009): The Age of Migration. Basingstoke: Palgrave Macmillan (4th edition).

Castles, Stephen (2001): Studying Social Transformation, International Political Science Review, Vol. 22, No. 1, pp. 13–32.

Castles, Stephen (2009): Development and Migration—Migration and Development: What Comes First? Global Perspective and African Experiences, *Theoria: A Journal of Social and Political Theory*, December 2009, Vol. 56, No. 121, pp. 1–31.

Castles, Stephen (2010) Understanding Global Migration: A Social Transformation Perspective, *Journal of Ethnic and Migration Studies*, 36:10, pp. 1565–1586.

Castles, Stephen (2012): Methodology and methods: conceptual issues, in Mohamed Berriane and Hein de Haas, African migration research: Innovative methods and methodologies, Africa World Press: Trenton, pp. 15–36.

Castles, et. al. (2014): The Age of Migration: International Population Movements in the Modern World. Fifth edition. Palgrave, Macmillan, London.

Cantat, Céline and Rajaram, Prem (2019): The politics of the Refugee Crisis, in The Oxford Handbook of Migration Crisis, Oxford University Press: Oxford, pp. 181–195.

Cattham, Martha et al. (2004): Introduction to political psychology, London: Lawrence Erlbaum Associates Publisher.

Caviedes, Alexander (2015): An Emerging 'European' News Portrayal of Immigration?, Journal of Ethnic and Migration Studies, 41:6, 897–917.

Center4s (2013): How Islamist militancy threatens Africa, http://www.centre4s.org/en/index.php?option=com_content&view=article&id=109:how-islamist-militancy-threatens-africa&catid=40:intervieuw&Itemid=58.

Cervantes, Miguel de (2003): Don Quixote, Penguin Classics, London.

Césaire, Aimé (2000): Discourse on Colonialism, Monthly Review Press: New York.

Cezne, Eric and Wethal, Ulrikke (2022): Reading Mozambique's mega-project developmentalism through the workplace: evidence from Chinese and Brazilian investments, *African Affairs*, 121: 484, pp. 343–370.

Chancel, Lucas et al. (2019): Income inequality in Africa, 1990–2017, WID.world Issue Brief 2019/6, September 2019.

Chang, Ha-Joon (2009): Economic History of the Developed World: Lessons for Africa, The lecture delivered in the Eminent Speakers Program of the African Development Bank, 26 February 2009.

Cherbib, Hamza (2018): Jihadism in the Sahel: Exploiting Local Disorders, *European Institute of the Mediterranean*, https://www.iemed.org/publication/jihadism-in-the-sahel-exploiting-local-disorders/, 31.05.2023.

China-Africa Research Initiative (2022): China-africa trade, http://www.sais-cari.org/data-china-africa-trade.

Christian Aid (2007): Human tide: the real migration crisis, London.

Clapham, Christopher 1996. *Africa and the International System: The Politics of State Survival*. Cambridge: Cambridge University Press.

Clemens, Michael A. (2014): *Does Development Reduce Migration? Center for Global Development*, NYU Financial Access Initiative and IZA, Discussion Paper No. 8592, October 2014.

Clemens, Michael A. (2020): The Emigration Life Cycle: How Development Shapes Emigration from Poor Countries, Center for Global Development, Working Paper 540 August 2020.

Clemens, Michael A. and Mendola, Mariapia (2020): Migration from Developing Countries: Selection, Income Elasticity, and Simpson's Paradox, Center for Global Development, Working Paper 539, August 2020.

Clemens, Michael A. and Postel, Hannah M. (2018): "Deterring Emigration with Foreign Aid: An Overview of Evidence from Low-Income Countries." CGD Policy Paper. Washington, DC: Center for Global Development. https://www.cgdev.org/publication/deterring-emigration-foreign-aid-overviewevidence-low-income-countries, 1–30.

Clement, Viviane et al. (2021): GROUNDSWELL: Acting on Internal Climate Migration, PART II, International Bank for Reconstruction and Development / The World Bank.

Clifford, Cayley and Gruzd, Steven (2022): Russian and African Media: Exercising Soft Power, South African Institute of International Affairs, African perspectives, Global insights.

Coen, A. and Henk, M. (2020): Bete für mich, in Die ZEIT, 7 May, 2020, N.20, p.14.

Cohen, Robin (ed.) (2010): *The Cambridge Survey of World Migration*, Cambridge University Press: Cambridge.

Cohen, Roger (2015): America's Bountiful Churn, https://www.nytimes.com/2015/12/31/opinion/americas-bountiful-churn.html, accessed, 21.04.2023.

Cohn, Theodore (2016): Global Political Economy: Theory and Practice, Routledge: New York.

Collier, Paul (2008): The Bottom Billion: Why the Poorest Countries are Failing and What Can Be Done About It, Oxford University Press: Oxford.

Collier, Paul (2014): Exodus: Immigration and multiculturalism in the 21st century, Penguin Books: London.

Collins, Francis L. (2018): Desire as a theory for migration studies: temporality, assemblage and becoming in the narratives of migrants, *Journal of Ethnic and Migration Studies*, 44:6, pp. 964–980, DOI: https://doi.org/10.1080/1369183X.2017.1384147.

Collyer, M. (2007). In-between places: Trans-Saharan transit migrants in Morocco and the fragmented journey to Europe. Antipode, 39(4), 668–690. https://doi.org/10.1111/j.1467-8330.2007.00546.x.

Collyer, Michael (2016): Geopolitics as a migration governance strategy: European Union bilateral relations with Southern Mediterranean countries, *Journal of Ethnic and Migration Studies*, 42:4, 606–624.

Collyer, Michael (2019): From Preventive to repressive: the changing use of development and humanitariansm to control migration, in Critical geographies of migration, Edward Elgar, pp. 170–181.

Conklin, Alice (1997): A mission to civilize: The republican idea of empire in France and West Africa, 1895–1930, Stanford University Press: Stanford.

Conrad Suso, Catherine (2019): Backway or bust: causes and consequences of Gambian irregular migration, *Journal of Modern African Studies*, 57:1, pp. 111–135.

Conrad, Joseph (2008): Heart of Darkness and Other Tales, Oxford University Press: Oxford.

Cotula, Lorenzo (2014): Addressing the human rights impacts of 'land grabbing', Directorate-General for External Policies of the Union Directorate B, Policy Department, European Parliament.

Cotula, Lorenzo et al. (2009): Land grab or development opportunity? Agricultural investment and international land deals in Africa, IIED/FAO/IFAD, London/Rome.

Council on Foreign Relations (2023): Conflict in the Democratic Republic of Congo, https://www.cfr.org/global-conflict-tracker/conflict/violence-democratic-republic-congo, 13.07.2023.

Crime and Punishment.

Crimmins, John (2023): The Psychological Impact of Skin Lightening, https://healthnews.com/mental-health/self-care-and-therapy/the-psychological-impact-of-skin-lightening/.

Cummings, Clare et al. (2015): *Why people move: understanding the drivers and trends of migration to Europe*, Overseas Development Institute, Working Paper 430, 2015s.

Czaika, Mathias (2015) Migration and Economic Prospects, Journal of Ethnic and Migration Studies, 41:1, 58–82.

Czaika, Mathias and De Haas, Hein (2017): Determinants of Migration to the UK, The Migration Observatory, Briefing, Oxford University, 2017.

Czaika, Mathias, and Mogens Hobolth. (2014): Deflection into irregularity? The (un)intended effects of restrictive asylum and visa policies. Working Paper, Oxford: University of Oxford: International Migration Institute.

Daftary, Karishma (2023): Uncovering the roots of skin bleaching: Colorism and its detrimental effects, *Journal of Cosmetic Dermatology*, 22: 1, pp. 337–338.

Danish Institute for International Studies (2021): Does information save migrant's lives? Knowledge and needs of West African migrants en route to Europe, Danish Institute for International Studies, Copenhagen.

Daras, Marina (2023): Why Turkey's election is being closely followed in Africa, https://www.bbc.com/news/world-africa-65671723, 26.05.2023.

Davies, Alys (2022): Equatorial Guinea seeks to block sale of confiscated Paris mansion, https://www.bbc.com/news/world-europe-63105426, 24.05.2023.

De Andrade, Pedro (2021): Globalization of Consumption, Lifestyles and 'Viral Society', E-*Revista de Estudos Interculturais do CEI–ISCAP*, N.° 9, maio de 2021.

De Genova, Nicholas (2013): Spectacles of migrant 'illegality': the scene of exclusion, the obscene of inclusion, *Ethnic and Racial Studies*, 36 (7), 1180–1198.

De Haan, Arjan and Yaqub, Shahin (2009): Migration and Poverty Linkages, Knowledge Gaps and Policy Implications, Social Policy and Development Programme Paper Number 40, United Nations Research Institute for Social Development (UNRISD).

De Haas, H. (2006) Turning the Tide? Why 'Development Instead of Migration' Policies are Bound to Fail. Oxford: International Migration Institute, IMI Working Paper, 2.

De Haas, H. (2011a). The Determinants of International Migration, DEMIG Working Paper 2. International Migration Institute, University of Oxford.

De Haas, H. (2011c): Mediterranean migration futures: Patterns, drivers and scenarios, Global *Environmental Change* 21S S59–S69.

De Haas, H. (2014). Migration theory: Quo vadis? DEMIG, Working Paper, International Migration Institute, University of Oxford.

De Haas, H. (2007a): Turning the Tide? Why Development Will Not Stop Migration. Dev. Change 38, 819–841. https://doi.org/10.1111/j.1467-7660.2007.00435.x.

De Haas, H., (2011d): Mediterranean migration futures: Patterns, drivers and scenarios. Glob. Environ. Change 21, S59–S69. https://doi.org/10.1016/j.gloenvcha.2011.09.003.

De Haas, H. (2010): The Internal Dynamics of Migration Processes: A Theoretical Inquiry, Journal of Ethnic and Migration Studies, 36:10, pp. 1587–1617.

De Haas, Hein (2011b): "The determinants of international migration: Conceiving and measuring origin, destination and policy effects". IMI Working Paper Series. Oxford: International Migration Institute, University of Oxford.

De Haas, Hein. 2007b. "Morocco's Migration Experience: A Transitional Migration 45 (4): 39–70. https://doi.org/10.1111/j.1468-2435.2007.00419.x.

Debalke, Mulugeta Gashaw (2021): Irregular migration among a Kambata community in Ethiopia: views from below and their implications, in Asnake Kefale and Fana Gebresenbet, *Youth on the move: Views from below on Ethiopian international migration*, Hurst and Company: London, pp. 137–158.

Degruy, Joy (2017): Post-Traumatic Slave Syndrome: America's Legacy of Enduring Injury and Healing, DeGruy.

Di Filippo, Marcello (2020): Fighting irregular forms of migration: the poisonous fruit of the securitarian approach to cooperation with Mediterranean countries, in Francesca Ippolito, et al., Bilateral Relations in the Mediterranean: Prospects for migration issues, Edward Elgar: Cheltenham, pp. 301–315.

Diop, Cheikh Anta (1998): Is there an African philosophy? In: Claude Sumner (ed.) African philosophy, Addis Abeba: Addis Abeba University Press, pp. 21–32.

Dom, Catherine (2021): Migration aspiration and *glocal* ideas of the good life in two rural communities in southern and north-eastern Ethiopia: a comparative perspective, in Asnake Kefale and Fana Gebresenbet, *Youth on the move: Views from below on Ethiopian international migration*, Hurst and Company: London, pp. 97–118.

Dopelfeld, Basilius 1994. *In der Mitte der Mensch.* Münsterschwarzach: Vier-Türme-Verlag.

Dostoevsky, Fyodor (1991): Crime and Punishment: Penguin Classics, London.

Dostoevsky, Fyodor (1993): Notes from Underground, Vintage Classics, New York: Vintage Books.

Dostoevsky, Fyodor (2003): The Brothers Karamazov: Penguin Books: London.

Dostoevsky, Fyodor (2013): The Double, Alma Classics: Richmond.

Dostoevsky, Fyodor (2021 [1863]): Winter Notes on Summer Impressions: Alma Classics, London.

Douglass, Frederick (1881): The Color Line, *The North American Review*, 132, 295, pp. 567–577.

Dougnon, Isaie (2013): Migration as coping with risk and state barriers: Malian migrants' conception of being far from home, in Abdoulaye Kane and Todd Leedy, African Migration: patterns and perspectives, Indiana University Press, Bloomington, pp. 35–58.

Doumbia, Djeneba(2020): The role of good governance in fostering pro-poor and inclusive growth, https://www.brookings.edu/blog/africa-in-focus/2020/07/01/the-role-of-good-governance-in-fostering-pro-poor-and-inclusive-growth/.

Driessen, Miriam (2022): Pidgin Play: Linguistic Subversion on Chinese-Run Construction Sites in Ethiopia, *African Affairs*,119: 476, pp. 432–451.

Du Bois, W.E.B (2009): Souls of Black Folk, Oxford University Press: Oxford.
Du Bois, W.E.B., (1903): The Souls of Black Folk, Gröls Classics.
Du Bois, W.E.B. (2007): Dusk of Dawn, Oxford University Press: Oxford.
Dugard, Martin (2003): Into Africa: the epic adventures of Stanley and Livingstone, Broadway Books: New York.
Duroy, Quentin (2011): North African Identity and Racial Discrimination in France: A Social Economic Analysis of Capability Deprivation, *Review of Social Economy*, 69: 3, pp. 307–332.
Easterly, William (2007): The White Man's Burden: Why the West's Efforts to Aid the Rest Have Done So Much Ill and So Little Good, Penguin Books: London.
Eatwell, Roger and Goodwin, Matthew (2018): National Populism: The Revolt Against Liberal Democracy, Pelican.
Eco, Umberto (2013): Inventing the enemy, Vintage: London.
Econfin Agency (2022): Africa imports seven times more Russian products than it exports to Moscow, https://www.ecofinagency.com/public-management/0903-43450-africa-imports-seven-times-more-russian-products-than-it-exports-to-moscow.
Economic Intelligence Unit (2022): Mining in Africa: The benefit of elevated prices amid inflationary pressures.
Economist Intelligence Unit (2022): A new Horizon for Africa-China Relations: Why Co-Operation will be Essential, The Economist Intelligence Unit Limited 2022.
Egypt Independent 2013: http://www.egyptindependent.com/news/former-egyptian-commander-striking-ethiopia-dam-impossible.
Eisele, Katharina (2019): The EU's readmission policy: of agreements and arrangements, in Sergio Carrera et al. (eds.), Constitutionalizing the external dimensions of EU migration policies in times of crisis, Edward Elgar: Cheltenham, pp. 135–154.
Eisenstadt, Shmuel N. (1954): *The Absorption of Immigrants*, London: Routledge and Kegan Paul.
EIU (2021): Democracy Index 2020: In sickness and in health? A report by The Economist Intelligence Unit.
EIU (2022a): Africa outlook 2023: the challenges ahead: Resilience amid disruption, The Economist Intelligence Unit Limited.
EIU (2022b): *Global Operational Risk Review: How War is Fueling Geopolitical Uncertainty*, The Economist Intelligence Unit Limited.
El Nour, Ashraf et al. (2019): Migration Futures for Africa, in *Africa Migration Report*, International Organization for Migration, pp. 199–211.
El-Badawy, Emman et al. (2022): Security, Soft Power and Regime Support: Spheres of Russian Influence in Africa, Tony Blair Institute for Global Change, March 23, 2022, https://institute.global/policy/security-soft-power-and-regime-support-spheres-russian-influence-africa.

Elder, Claire (2022): Logistics Contracts and the Political Economy of State Failure: Evidence From Somalia, *African Affairs,* 121: 484, pp. 395–417.
Elias, Norbert 1978. *The Civilising Process: The History of Manners.* Oxford: Blackwell.
Englisch Ausgabe von Gebrewold, Belachew and Bloom, Tendayi (eds.) (2018): Understanding Migrant Decisions: From Sub-Saharan Africa to the Mediterranean Region, Routledge: Abingdon.
Ennasri, Nabil (2013): Qatari-Ethiopian Relations: Prospects after the Emir's Visit, in http://studies.aljazeera.net/en/reports/2013/05/201352895218705530.html, accessed 04 July 2013.
Environmental Justice Foundation (2022): On the Precipice: Crime and corruption in Ghana's Chinese-owned trawler fleet, A report by the Environmental Justice Foundation.
Enzensberger, Hans Magnus (1994): Civil wars: From L.A. to Bosnia, New York: The New Press.
Erdal, Marta Bivand & Oeppen, Ceri (2018): Forced to leave? The discursive and analytical significance of describing migration as forced and voluntary, Journal of Ethnic and Migration Studies, 44:6, 981–998, https://doi.org/10.1080/1369183X.2017.1384149.
Esedebe, Olisanwuche P. 1982. *Pan-Africanism: The Idea and Movement 1776–1963.* Washington, DC: Howard University Press.
Estifanos, Yordanos (2021): Gender relations in a transnational space: Ethiopian irregular migration in South Africa, in Asnake Kefale and Fana Gebresenbet, *Youth on the move: Views from below on Ethiopian international migration*, Hurst and Company: London, pp. 159–178.
European Commission (2019): African Peace Facility: African Union Peace & Security Operations boosted by an additional €800 million from the European Union, https://ec.europa.eu/commission/presscorner/detail/it/ip_19_3432.
European Commission (2020): The European Union and Africa: Partners in Trade, https://trade.ec.europa.eu/doclib/docs/2022/february/tradoc_160053.pdf.
European Commission (2023): The European Union and Tunisia agreed to work together on a comprehensive partnership package, https://neighbourhood-enlargement.ec.europa.eu/news/european-union-and-tunisia-agreed-work-together-comprehensive-partnership-package-2023-06-11_en, 06.07.2023.
European Parliament (2019): Russia in Africa: A new arena for geopolitical competition, Briefing.
Eurostat (2022): Africa-EU trade in goods: €4 billion surplus, https://ec.europa.eu/eurostat/web/products-eurostat-news/-/edn-20220217-1.
EUTELSAT (2022): TV Viewing Trends in Sub-Sahara Africa, https://www.eutelsat.com/en/blog/6-TV-viewing-trends-in-sub-saharan-africa.html.

Fairhead, James et al. (2012): Green Grabbing: a new appropriation of nature?, Journal of Peasant Studies, 39:2, 237–261, https://doi.org/10.1080/03066150.2012.671770.

Falguere, X. Casademont (2019): Refugee and Romani immigrant populations in Barcelona, in The Oxford Handbook of Migration Crisis, Oxford University Press: Oxford, pp. 111–126.

Falola, Toyin (n.d.): Arab's race denial, https://www.laits.utexas.edu/africa/ads/900.html, 25.04.2023.

Fanon, Frantz (2001): The wretched of the earth, Penguin Books: London.

Fanon, Frantz (2008): Black Skin, White Masks, Grove Press: New York.

Fanon, Frantz 1966. *Die Verdammten dieser Erde*. Frankfurt am Main: Suhrkamp.

Fanon, Frantz 1991. *Black Skins, White Masks*, London: Pluto Press.

Farer, Tom (2020): Migration and Integration: The case for liberalism with borders, Cambridge University Press: Cambridge.

Fargues, P., Bonfanti, S., 2014. When the best option is a leaky boat: why migrants risk their lives crossing the Mediterranean and what Europe is doing about it.

Fasani, Francesco et al. (2019): Does Immigration increase crime? Migration policy and the creation of the criminal immigrant, Cambridge University Press: Cambridge.

Feagin, J. R., & Sikes, M. P. (1994). Living with racism. Boston: Beacon.

Feierstein, Gerald and Greathead, Craig (2017): The Fight for Africa: The New Focus of the Saudi-Iranian Rivalry, Middle East Institute, Policy Focus 2017–2, https://www.brookings.edu/blog/africa-in-focus/2022/03/09/figure-of-the-week-african-countries-votes-on-the-un-resolution-condemning-russias-invasion-of-ukraine/.

Ferguson, J. 2006. *Global Shadows: Africa in the Neoliberal World Order*. Durham, N.C.: Duke University Press.

Filindra, Alexandra and Junn, Jane (2012): Aliens and people of color: the multidimensional relationship of immigration policy abd radical classification in the United States, in Marc Rosenblum and Daniel Tichenor, The Oxford Handbook of the politics of international migration, Oxford University Press: Oxford, pp. 429–455.

Financial Times (2019): Putin seeks friends and influence at first Russia-Africa summit, https://www.ft.com/content/f20dbcc2-f17b-11e9-ad1e-4367d8281195, 20.06.2022.

Flahaux, Marie-Laurence and De Haas, Hein (2016a): African migration: trends, patterns, drivers; *Comparative Migration Studies* 4:1, pp. 1–25.

Flahaux, Marie-Laurence and De Haas, Hein (2016b): African migration: trends, patterns, drivers; *Comparative Migration Studies* (2016) 4:1, pp. 1–25.

Flam, Helena and Beauzamy, Brigitte (2011): Symbolic Violence, in Gerard Delanty et al. (eds.), Identity, Belonging and Migration, Liverpool University Press: Liverpool, p. 221–240.

Fleury, Anjali (2016): Understanding Women and Migration: A Literature Review, KNOMAD Working Paper Series, World Bank.
Flynn, Thomas (2006): Existentialism: A very short introduction, Oxford: Oxford University Press.
Forgiarini, Matteo et al. (2011): Racism and the empathy for pain on our skin, *Frontiers in Psychology*, 2, 108, pp. 1–7, https://doi.org/10.3389/fpsyg.2011.00108.
Fowler, Bridget (2020): Pierre Bourdieu on social transformation, with particular reference to political and symbolic revolutions, *Theory and Society* (2020) 49:439–463.
Fragile States Index (2022): Breaking the Cycle: Military Coups in West Africa, https://fragilestatesindex.org/2022/07/08/breaking-the-cycle-military-coups-in-west-africa/.
Francis, David J. 2006. *Uniting Africa: Building Regional Peace and Security Systems.* Aldershot: Ashgate.
Frazer, Garth (2006), 'Inequality and development across and within countries,' *World Development*, 34 (9), 1459–1481.
Freedman, Jane (2013): The feminization of asylum migration from Africa: problems and perspectives, in Abdoulaye Kane and Todd Leedy, African Migration: patterns and perspectives, Indiana University Press, Bloomington, pp. 211–229.
Freedom House (2023): Freedom in the World 2023, *Marking 50 Years in the Struggle for Democracy*, Freedom House.
Fuje, Habtamu and Yao, Jiaxiong (2022): Africa's Rapid Economic Growth Hasn't Fully Closed Income Gaps, https://www.imf.org/en/Blogs/Articles/2022/09/20/africas-rapid-economic-growth-hasnt-fully-closed-income-gaps.
Fukuyama, Francis (2018): Identity: Contemporary identity politics and the struggle for recognition, Profile Books: London.
Fumagalli, Corrado and Schaefer, Katja (2019): Migration and Urbanization in Africa, in Africa Migration Report, International Organization for Migration, pp. 41–51.
FurtherAfrica (2022): African countries with the highest number of mobile phones, https://furtherafrica.com/2022/07/19/african-countries-with-the-highest-number-of-mobile-phones/.
Fussell, Elisabeth (2012): Space, Time and Volition: Dimensions of migration theory, in Marc Rosenblum and Daniel Tichenor, The Oxford Handbook of the politics of international migration, Oxford University Press: Oxford, pp. 25–52.
Gakunzi, David (2018): The Arab-Muslim Slave Trade: Lifting the Taboo, *Jewish Political Studies Review*, 29: 3/4, pp. 40–42.
Gandhi, Leela (1998): Postcolonial Theory: A critical introduction, Columbia University Press: New York.

Garelli, Gelnda and Tazzioli, Martina (2019): Military-humanitarianism, in Critical geographies of migration, Edward Elgar, pp. 182–191.

Gebresenbet, Fana (2021): Hopelessness and future-making through irregular migration in Tigray, Ethiopia, in Asnake Kefale and Fana Gebresenbet, *Youth on the move: Views from below on Ethiopian international migration*, Hurst and Company: London, pp. 79–96.

Geisen, Thomas et al. (2008): (B)ordering and othering migrants by the European Union, Migration in a New Europe: People, Borders and Trajectories, Société Geografica Italiana: Rome.

Gershman, Sally (1976): Alexis de Tocqueville and Slavery, *French Historical Studies*, 9: 3, pp. 467–483.

Gerstter, Christiane (2011): An Assessment of the Effects on Land Ownership and Land Grab on Development, Policy Department DG External Policies.

Getachew, Adom (2019): Worldmaking after Empire: The rise and fall of self-determination, Princeton University Press: Princeton.

Gezahegne, Kiya (2021): Rituals of migration: socially entrenched practices among female migrants from Amhara national regional states, in Asnake Kefale and Fana Gebresenbet, *Youth on the move: Views from below on Ethiopian international migration*, Hurst and Company: London, pp. 119–136.

Gill Indermit S. and Karakülah, Kenan (2021): Is China Helping Africa? Growth and Public Debt Effects of the Subcontinent's Biggest Investor, Duke Global Working Paper, Paper 3, March 2019, Duke University Center for International and Global Studies.

Gilli, Aldo (1979) *Daniel Comboni: the man and his message; a selection from the writings of Bishop Daniel Comboni (1831–1881)*, Bologna, Editrice Missionaria Italiana.

Gilmartin, Mary and Kuusisto-Arponen, Anna-Kaisa (2019): Borders and Bodies: siting critical geographies of migration, in Critical geographies of migration, Edward Elgar, pp. 18–29.

Giménez-Gómez, José-Manuel et al. (2017): Trends in African Migration to Europe: Drivers Beyond Economic Motivations, Discussion Papers, Center for European, Governance and Economic Development Research, Georg-August-Universität Göttingen, ISSN: 1439–2305.

Girard, René (1965): Deceit, desire and the novel: self and other in literary structure, John Hopkins University: Baltimore.

Girard, René (1986): The Scapegoat, The Athlone Press: London.

Girard, René (2013): Violence and the Sacred, London: Bloomsbury.

Girard, René (2016): Violence and the Sacred, Bloomsbury Revelations: London.

Girard, René (2019): Things hidden since the foundation of the world, Bloomsbury Revelations: London.

Girard, René (2023): All desire is a desire for being, Penguin Classics: London.

Global Compliance News (2022): While challenges remain, continental free trade will further boost Africa's trade partners, https://www.globalcompliancenews.com/2022/06/14/africa-chinas-trade-ties-with-the-continent-continue-to-strengthen-31052022/.

GlobalR2P (2023): Democratic Republic of the Congo, *Global Centre for the Responsibility to Protect*, https://www.globalr2p.org/countries/democratic-republic-of-the-congo/, 01.06.2023.

Goedde, Lutz et al. (2019): Winning in Africa's agricultural market, February 2019, Agriculture Practice, McKinsey & Company.

Goethe, Johann W. (1989): The Sorrows of Young Werther, Penguin Classics, London: Penguin Books.

Gold, S.J. (2005) 'Migrant networks: a summary and critique of relational approaches to international migration', in Romero, M. and Magolis, E. (eds) *The Blackwell Companion to Social Inequalities*. Malden, MA: Blackwell, 257–85.

Goldin, I. et al. (2011): Exceptional People: How Migration shaped our world and will define our future, Princeton University Press: Princeton.

Gosetti-Ferencei, Jennifer Anna (2021): On Being and Becoming: An Existentialist Approach to Life, Oxford University Press: Oxford.

Goudsblom, J. 1998. *The Norbert Elias Reader: A Biographical Selection*. Oxford: Blackwell.

Gould, J. D. (1980): 'European Inter-Continental Emigration: The Role of "Diffusion" and "Feedback"', *Journal of European Economic History*, 9 (2), 267–315.

Graham, Alison et al. (2010): Land Grab study, CSO Monitoring 2009–2010 "Advancing African Agriculture" (AAA): The Impact of Europe's Policies and Practices on African Agriculture and Food Security, www.europafrica.info, 02.06.2014.

Green, M. (2019). China's debt diplomacy. Foreign Policy. 25 April.

Greenwood, Michael J. (1969): 'An Analysis of the Determinants of Geographic Labor Mobility in the United States,' *Review of Economics and Statistics*, 51 (2), 189–194.

Grieveson, Richard et al. (2021): Future Migration Flows to the EU: Adapting Policy to the New Reality in a Managed and Sustainable Way, Policy Notes and Reports 49, The Vienna Institute for International Economic Studies.

Gunpolicy 2008: Guns in North Africa, retrieved 12 March 2009, from http://www.gunpolicy.org/Topics/Guns_In_North_Africa.html.

Gurnah, Abdulrazak (2021): Memory of Departure, Bloomsbury, London.

Guttmann, A. (2022): Media in Africa – Statistics & Facts, https://www.statista.com/topics/5032/media-in-africa/#dossierKeyfigures.

Gyimah-Boadi, E. (2022): West Africa's Authoritarian Turn. Democratic Backsliding, Youth Resistance, and the Case for American Help, https://www.foreignaffairs.com/articles/west-africa/2022-07-11/west-africas-authoritarian-turn.

Gyimah-Boadi, E. and Asunka, Joseph (2021): Do Africans want democracy – and do they think they're getting it?, https://summit4democracy.org/do-africans-want-democracy-and-do-they-think-theyre-getting-it/.

Hagmann, Tobias and Reyntjens, Filip (2016): Aid and Authoriatarianism in Africa, Zed Books: London.

Hahonou, Eric (n.d.): Blackness, slavery and anti-racism activism in contemporary North Africa, Elliott School of International Affairs, Washington, DC, https://pomeps.org/blackness-slavery-and-anti-racism-activism-in-contemporary-north-africa, 25.05.2023.

Hainzl, Gerald (2021): The People's Republic of China's Presence in Africa, In Frank/Vogl (eds.): China's Footprint in Strategic Spaces of the European Union. New Challenges for a Multi-dimensional EU-China Strategy. Schriftenreihe der Landesverteidigungsakademie No. 11/2021.

Hallen, Barry (1998): Phenomenology and the exposition of African traditional thought, In: Claude Sumner (ed.) African philosophy, Addis Abeba: Addis Abeba University Press, pp. 47–66.

Hanlon, Joseph (2010): Mozambique: 'The war ended 17 years ago, but we are still poor', *Conflict, Security & Development*, 10:1, 77–102, https://doi.org/10.1080/14678800903553902.

Hardt, Michael & Negri, Antonio 2001. *Empire*. Cambridge: Harvard University Press.

Harouna, Samba et al. (2019): Environmental Degradation and Human Mobility Nexus in Africa, in Africa Migration Report, International Organization for Migration, pp. 65–75.

Hatton, Timothy (2009): The Rise and Fall of Asylum: What Happened and Why? *The Economic Journal*, Volume 119, Issue 535, 1 February 2009, Pages F183–F213.

Hegel, Georg W.F. (1975): Lectures on the philosophy of world history: Introduction, translated by H.B. Nisbet; Introduction by Duncan Forbes, Cambridge University Press: Cambridge.

Heidegger, Martin (2010): Being and Time, State University of New York.

Herder, Johann Gottfried von 2002. *Philosophical Writings*. Translated and ed. by Michael N. Forster. Cambridge: Cambridge University Press.

Herrmann, Richard (2003): Image Theory and strategic interaction in international relations, in: David O. Sear et al., Oxford Handbook of political psychology, Oxford University Press: Oxford, pp. 285–314.

Hertrich, Véronique and Lesclingand, Marie (2013): Adolescent Migration in Rural Africa as a Challenge to Gender and Intergenerational Relationships: Evidence from Mali, The Annals of the American Academy, AAPSS, 648, pp. 175–188.

Hetherwick, A. 1939. *The Romance of Blantyre: How Livingstone's dream came true*. Dunfermline: Lasodine Press.

Hilizah, Nisirine (2022): Racial Discrimination and Anti-Blackness in the Middle East and North Africa, Arab Barometer – Wave VII Racism Report, in https://www.arabbarometer.org/, accessed 02.05.2023.

Hofmann, Erin Trouth (2015): Choosing Your Country: Networks, Perceptions and Destination Selection among Georgian Labour Migrants, *Journal of Ethnic and Migration Studies*, 41:5, 813–834.

Hollifield, James (2012): Migration and International Relations, in Marc Rosenblum and Daniel Tichenor, The Oxford Handbook of the politics of international migration, Oxford University Press: Oxford, pp. 345–381.

Holtom, Paul 2008: International arms transfers. In: SIPRI Yearbook 2008: Armaments, Disarmament and International Security, 313–317.

Homer (1987): The Iliad, Penguin Classics, London: Penguin Books.

Hountondji, Paulin 1996. *African Philosophy: Myth and Reality.* Bloomington: Indiana University Press.

Houtum, Henk van & Naerssen, Ton van (2002): Bordering, Ordering and Othering, *Tijdschrift voor Economische en Sociale Geografie*, Vol. 93, No. 2, pp. 125–136.

Huete, R.; (2008) "Analysing the Social Perception of Residential Tourism Development" in Costa, C. and Cravo, P., Advances in Tourism Research. Aveiro: IASK. pp. 153–161.

Hugo, Victor (2006): The hunchback of Notre Dame, Dover: New York.

Human Rights Watch (2023): "They Fired on Us Like Rain": Saudi Arabian Mass Killings of Ethiopian Migrants at the Yemen-Saudi Border, AUGUST 2023, ISBN: 979-8-88708-065-9.

Hunter, Margaret (2007): The Persistent Problem of Colorism: Skin Tone, Status, and Inequality, Sociology Compass 1:1, pp. 237–254.

Ibhawoh, Bonny (2020): Seeking the political kingdom: universal human rights and the anti-colonial movement in Africa, in Decolonization, Self-determination and the rise of global human rights ed. A. Dirk Moses et al. Cambridge University Press: Cambridge, pp. 35–53.

Iliffe, John 1997. *Geschichte Afrikas.* München: Beck.

Indian Ministry of Commerce and Industry (2022): Foreign Trade (Africa), https://commerce.gov.in/about-us/divisions/foreign-trade-territorial-division/foreign-trade-africa/.

Internal Displacement Monitoring Center (2021): Global Report on Internal Displacement, Norwegian Refugee Council.

Internal Displacement Monitoring Center (2022): *Children and youth in internal displacement*, Internal Displacement Monitoring Center; Norwegian Refugee Council.

IOM (2020): The World Migration Report 2020, International Organization for Migration, Geneva.

IOM (2023): IOM Definition of "Migrant", https://www.iom.int/about-migration, 27.05.2023.

IPCC (2022): Climate Change 2022: Impacts, Adaptation and Vulnerability. Contribution of Working Group II to the Sixth Assessment Report of the Intergovernmental Panel on Climate Change, Cambridge University Press, Cambridge, UK.

Ippolito, Franseca (2020): The Rhetoric of human rights in EU external relations in the Mediterranean, in Francesca Ippolito, et al., Bilateral Relations in the Mediterranean: Prospects for migration issues, Edward Elgar: Cheltenham, pp. 270–300.

Jackson, Lauren (2023): *Tunisia's Influence in Europe*, https://www.nytimes.com/2023/04/05/world/africa/tunisia-europe-migration.html, 06.07.2023.

Jackson, Michael (2008): The Shock of the New: On Migrant Imaginaries and Critical Transitions, Ethnos: Journal of Anthropology, 73:1, 57–72, https://doi.org/10.1080/00141840801927533.

James, C.L.R. (2022): The black Jacobins: Toussaint Louverture and the San Domingo revolution, Penguin: Dublin.

James, William (1890): Principles of Psychology. New York: Holt Rinehart and Winston.

Jansiz, Ahmad (2014): The Ideology of Consumption: The Challenges Facing a Consumerist Society, *Journal of Politics and Law*; Vol. 7, No. 1.

Jeffers, Honoreé Fanonne (2022): The love songs of W. E. B. Du Bois, 4th State: London.

Jeffrey Sachs (2006): The End of Poverty: Economic Possibilities for Our Time: Penguin Books: London.

John Stuart Mill: https://www.gutenberg.org/cache/epub/34901/pg34901-images.html.

Johnson, Heather (2014): Borders, Asylum and Global Non-Citizenship: The other side of the fence, Cambridge University Press: Cambridge.

Jolly, Susie and Reeves, Hazel (2005): Gender and Migration, Overview Report, Institute of Development Studies, BRIDGE.

Jones, Paul and Krzyzanowski, Michal (2011): Identity, belonging and migration: beyond constructing 'Others', in Gerard Delanty et al. (eds.), Identity, Belonging and Migration, Liverpool University Press: Liverpool, pp. 38–53.

Jones, Robert P. et al. (2015): How Americans view Immigrants, and what they want from Immigration Reform, Public Religion Research Institute, Washington DC.

Jones, Bruce (2020): China and the Return of Great Power Strategic Competition, Global China: Assessing China's growing role in the world, The Brookings Institution, February 2020.

Kamer, Lars (2022a): African population without electricity 2000–2021, https://www.statista.com/statistics/1221698/population-without-access-to-electricity-in-africa/.

Kamer, Lars (2022b): Share of web traffic in Africa as of May 2022, by device, https://www.statista.com/statistics/1176693/web-traffic-by-device-in-africa/.

Kamer, Lars (2022c): Gender gap index in Sub-Saharan Africa 2022, by country, https://www.statista.com/statistics/1220485/gender-gap-index-in-sub-saharan-africa-by-country/.

Kane, Abdoulaye and Leedy, Todd (2013): African patterns of migration in a global era: new perspectives, in Abdoulaye Kane and Todd Leedy, African Migration: patterns and perspectives, Indiana University Press, Bloomington, pp. 1–18.

Kapchanga, Mark (2020): Africa may learn from china's political path to revitalize its own development plans, https://www.globaltimes.cn/content/1189460.shtml, 2020/5/25.

Kasparek, Bernd and Schmidt-Sembdner, Matthias (2019): Renationalization and spaces of migration: the European border regime after 2015, in Critical geographies of migration, Edward Elgar, pp. 206–219.

Kaya, Hülya (2020): The EU-Turkey Statement on Refugees: Assessing its impact on Fundamental Rights, Cheltenham: Edward Edgar.

Kefale, Asnake and Gebresenbet, Fana (2021): Youth on the Move: Views from below on Ethiopian International Migration, Hurst and Company: London.

Keulder, Christiaan (2021): Africans see growing corruption, poor government response, but fear retaliation if they speak out, Afrobarometer Dispatch No. 421.

Kibreab, G., 2013. The national service/Warsai-Yikealo Development Campaign and forced migration in post-independence Eritrea. J. East. Afr. Stud. 7, 630–649. https://doi.org/10.1080/17531055.2013.843965.

King, Stephen (n.d.): Anti-Black Racism and Slavery in Desert and Non-Desert Zones of North Africa, https://pomeps.org/blackness-slavery-and-anti-racism-activism-in-contemporary-north-africa, 25.05.2023.

King, Stephen J. (2020): Ending Denial: Anti-Black Racism in Morocco, Arab Reform Initiative, Bawader, 28 August 2020.

Kingsley, Patrick (2017): The new Odyssey: The story of Europe's refugee crisis, The Guardian: London.

Kinyongo, J. (1998): From discursivity to philosophical discourse in Africa, In: Claude Sumner (ed.) African philosophy, Addis Abeba: Addis Abeba University Press, pp. 123–138.

Kirk, G. S. (1975): Heraclitus: The cosmic fragments, Cambridge University Press: Cambridge.

Kleemann, Linda et al. (2013): Economic and ethical challenges of "land grabs" in sub-Saharan Africa, Kiel Policy Brief, No. 67, Kiel Institute for the World Economy (IfW), Kiel.

Klobucista, Calire (2021): Africa's 'Leaders for Life', https://www.cfr.org/backgrounder/africas-leaders-life.

Knaus, Gerald (2020): Welche Grenzen brauchen wir: zwischen Empathie und Angst – Flucht, Migration und die Zukunft von Asyl, Piper: München.

Koffi, Niamkey and Abdou, Toure (1998): Controversy on the existence of an African philosophy, In: Claude Sumner (ed.) African philosophy, Addis Abeba: Addis Abeba University Press, pp. 157–174.

Kohr, Leopold 1995. *Small is Beautiful: Ausgewählte Schriften aus dem Gesamtwerk.* Wien : Deuticke.

Kortunov, Andrey et al. (2020): Africa-Russia+: Achievements, Problems, Prospects: Report № 53/2020, Russian International Affairs Council (RIAC), Africa Business Initiative Union.

Koser, K., & McAuliffe, M. (2013). Est*ablishing an Evidence-Base for Future Policy Development on Irregular Migration to Australia*, Irregular Migration Research Program. Irregular Migration Research Programme, Occasional Paper Series.

Koslowski, Rey (2012): Immigration, crime and terrorism, in Marc Rosenblum and Daniel Tichenor, The Oxford Handbook of the politics of international migration, Oxford University Press: Oxford, pp. 511–531.

Kritz, M., Lim, L. L. and Zlotnik, H. (eds.) (1992): *International migration systems: A global approach* (Oxford: Clarendon Press).

Kuschminder, Katie et al. (2015): *Irregular Migration Routes to Europe and Factors Influencing Migrants' Destination Choices*, Maastricht Graduate School of Governance.

Kuusisto-Arponen and Gilmartin, Mary (2019): Embodied immigration and the geographies of care: the worlds of unaccompanied refugee minors, in Critical geographies of migration, Edward Elgar, pp. 80–91.

Kymlicka, Will (2015): Solidarity in diverse societies: beyond neoliberal multiculturalism and welfare chauvinism, Comparative Migration Studies, 3, 17, pp. 1–19.

Lakmeeharan, Kannan (2020): Solving Africa's infrastructure paradox, McKinsey & Company.

Lanati, Mauro and Thiele, Rainer (2018): The impact of foreign aid on migration revisited, *World Development 111 (2018)*, 59–74.

Landau, Loren (2013): Belonging amid shifting sands: insertion, self-exclusion, and the remaking of African urbanism, in Abdoulaye Kane and Todd Leedy, African Migration: patterns and perspectives, Indiana University Press, Bloomington, pp. 93–112.

Landes, David (1999): The Wealth and poverty of nations: why some are so rich and some so poor, Norton and Company, New York.

Langan, Mark (2018): Neo-Colonialism and the Poverty of 'Development' in Africa (Contemporary African Political Economy), Palgrave Macmillan: New Castle.

Leanne Weber and Claudia Tazreiter, eds. (2021): Handbook of Migration and Global Justice, Edward Elgar, Cheltenham.

Ljubas, Zdravko (2019): Russia Enters Africa with Soft Power, https://www.occrp.org/en/daily/10959-russia-enters-africa-with-soft-power, 22.06.2020.

Loprete, Giuseppe (2016): MIGRO ERGO SUM – I Migrate, Therefore I Am – Social Pressure as a Driver of Economic Migration from West Africa, https://blogs.lse.ac.uk/africaatlse/2016/01/14/migro-ergo-sum-i-migratethere-fore-i-am-social-pressure-as-a-driver-of-economic-migration-from-west-africa/, accessed 26.03.2024.

Loschmann, C., and Siegel, M. (2014): The influence of vulnerability on migration intentions in Afghanistan. Migr. Dev. 3, 142–162. https://doi.org/10.1080/21632324.2014.885259.

Lukasik, Gail (2021): White like her: my family's story of race and racial passing, Skyhorse Publishing: New York.

Lunt, Peter Kenneth and Livingstone, Sonia (1992): Mass consumption and personal identity: everyday economic experience, Buckingham : Open University Press.

Lyotard, Jean-Francois 1999. *The Postmodern Condition: A Report on Knowledge.* Manchester University Press.

Mabogunje, A. L. (1970): Systems Approach to a Theory of Rural-Urban Migration. Geographical Analysis, 2(1): 1–18.

Madrid-Morales, Dani et al. (2021): It is about their story: How China, Turkey and Russia influence the media in Africa, Konrad-Adenauer-Stiftung Regional Media Program, Sub-Sahara Africa.

Maeymaekers, Timothy (2019): The law of impermanence: displacement, sovereignty, subjectivity, in in Critical geographies of migration, Edward Elgar, pp. 58–68.

Mahler, Sarah (1995): American Dreaming, Princeton: Princeton University Press.

Mangalu, Agbada (2012): Congolese migrants towards Africa, Europe, America, and Asia, in Mohamed Berriane and Hein de Haas, African migration research, Africa World Press: Trenton, pp. 147–176.

Mannoni, O. (1990): Prospero and Caliban: The psychology of colonization, Ann Arbor: Michigan University Press.

Maquet, J. 1971. *Power and Society in Africa.* London: World University Library.

Martin, Susan (2012): War, natural disasters, and forced migration, in Marc Rosenblum and Daniel Tichenor, The Oxford Handbook of the politics of international migration, Oxford University Press: Oxford, pp. 53–73.

Martin-Prével, Alice (2016): The unholy Alliance: Five Western Donors Shape a Pro-Corporate Agenda for African Agriculture, The Oakland Institute, Oakland.

Martin-Prével, Alice et al. (2016): The Unholy Alliance: Five Western Donors Shape a Pro-Corporate Agenda for African Agriculture, The Oakland Institute, Oakland.

Massey, D.S. (1990) 'Social structure, household strategies, and the cumulative causation of migration', Population Index, 56(1): 3–26.

Massey, D.S., Arango, J., Hugo, G., Kouaouci, A., Pellegrino, A. and Taylor, J.E. (1998a) *Worlds in Motion, Understanding International Migration at the End of the Millenium*. Oxford: Clarendon Press.

Massey, D.S., Arango, J., Hugo, G., Kouaouci, A., Pellegrino, A. and Taylor, J.E. (1993) 'Theories of international migration: a review and appraisal', Population and Development Review, 19(3): 431–66.

Massey, D.S., et al. (1998b) Worlds in Motion: Understanding International Migration at the End of the Millennium. Oxford: Clarendon Press.

Massey, Douglas S., Joaquín Arango, Graeme Hugo, Ali Kouaouci, Adela Pellegrino and J Edward Taylor (2006) Theories of International Migration: A Review and Appraisal. In: Messina Anthony and Lahav Gallya (Eds.) The Migration Reader. Exploring Politics and Policies (Lynne Rienner, Boulder), 34–62.

Mattes, Robert and Bratton, Michael (2016): Do Africans still want democracy? Afrobarometer Policy Paper No. 36 | November 2016.

Mayblin, Lucy (2021): Postcolonial perspectives on migration governance, in Emma Carmel et al. Handbook on the Governance and Politics of Migration, Edward Elgar: Cheltenham, pp. 25–35.

Mazrui, A. A. 1967. *On the Concept of 'We are All Africans', Towards a Pax-Africana: A Study of Ideology and Ambition*. London: Weidenfeld and Nicolson.

Mbaye, Linguère Mously (2013): Understanding Illegal Migration from Senegal, IZA DP No. 7728, Discussion Paper Series, Forschungsinstitut zur Zukunft der Arbeit Institute for the Study of Labor.

Mbembe, Achille (2001): On the Postcolony, University of California Press: Berkeley.

Mbembe, Achille (2003): Necropolitics, *Public Culture*, 15, 1, pp. 11–40.

Mbembe, Achille (2017): Critique of Black Reason, Duke University Press: Durham.

Mbembe, Achille (2021): Out of the dark night: Essays on decolonization, Columbia University Press, New York.

Mbuwayesango, Dora Rudo (2018): The bible as a tool of colonization: The Zimbabwean context, in Colonialism and the bible: contemporary reflections from the global south, Lexington books: Lanham, pp. 31–42.

McAuliffe, M. (2013). *Seeking the views of irregular migrants: Decision-making, drivers and migration journeys*. Irregular Migration Research Programme Occasional Paper Series.

Melachrinos, Constantinos et al. (2020): Using big data to estimate migration "push factors" from Africa, in International Organization for Migration, Migration in West and North Africa and across the Mediterranean: Trends, risks, development and governance, Geneva, pp. 98–116.

Memmi, Albert (1967): The colonizer and the colonized, Boston: Beacon Press.

Memmi, Albert (1968): Dominated Man: notes toward a portrait, Orion Press: London.

Menin, Laura (2018): Being 'black' in North Africa and the Middle East, https://www.opendemocracy.net/en/beyond-trafficking-and-slavery/being-black-in-north-africa-and-middle-east/, 25.05.2023.

Menjivar, Cecilia et al. (2019): Migration Crisis: Definitions, Critiques, and Global Contexts, in The Oxford Handbook of Migration Crisis, Oxford University Press: Oxford, pp. 1–20.

Meredith, M. 2005. The Fate of Africa: a History of Fifty years of Independence. New York: PublicAffairs.

Mill, John Stuart (1859 [2016]): On Liberty, Enhanced Media: Los Angels.

Mills, Greg (2010): Why Africa is Poor: And What Africans Can Do About It?, Penguin.

Mitchell, Jason (2019): IMF: African economies are the world's fastest growing, https://www.fdiintelligence.com/content/news/imf-african-economies-are-the-worlds-fastest-growing-75841.

Mitsilegas, Valsamis (2019): Extraterritorial immigration control, preventive justice and the rule of law in turbulent times: lessons from the anti-smuggling crusade, in Sergio Carrera et al. (eds.), Constitutionalizing the external dimensions of EU migration policies in times of crisis, Edward Elgar: Cheltenham, pp. 290–308.

Mixed Migration Centre (2019): Navigating borderlands in the Sahel: border security governance and mixed migration in Liptako-Gourma. Available at www.mixedmigration.org, 12.06.2022.

Moyo, Dambisa (2010): Dead Aid: Why aid is not working and how there is another way for Africa, Penguin : London.

Mugisha, Ivan (2019): Rwanda approves nuclear power deal with Russia, https://www.theeastafrican.co.ke/business/Rwanda-approves-nuclear-power-deal-with-Russia/2560-5318000-view-asAMP-815wfl/index.html?__twitter_impression=true, 20.06.2022.

Murrugarra, Edmundo et al. (2011): Migration and Poverty: Toward Better Opportunities for the Poor, The International Bank for Reconstruction and Development / The World Bank, Washington DC.

Myrdal, Gunnar (1957). Rich Lands and Poor. The road to world prosperity, (Harper & Row, New York).

Nandy, Ashis (1983): The Intimate Enemy: Loss and Recovery of Self Under Colonialism: Oxford University Press: New Delhi.

Natali, Claudia and Isaacs, Leon (2019): Remittances to and from Africa, in Africa Migration Report, International Organization for Migration, pp. 117–131.

Natanasabapathy, Puvanambihai & Maathuis-Smith, Sandra (2019): Philosophy of being and becoming: A transformative learning approach using threshold concepts, *Educational Philosophy and Theory*, 51:4, pp. 369–379, https://doi.org/10.1080/00131857.2018.1464439.

Natter, Katharina (2014): Fifty years of Maghreb emigration : How states shaped Algerian, Moroccan and Tunisian emigration, Working Papers, Paper 95, July 2014, DEMIG project paper 21.
Nawyn, Stephanie (2019): Refugees in the United States and Politics of Crisis, in The Oxford Handbook of Migration Crisis, Oxford University Press: Oxford, pp. 163–179.
Ndi, Frankline A. (2017): Land Grabbing, Local Contestation, and the Struggle for Economic Gain: Insights From Nguti Village, South West Cameroon, *SAGE Open*, 1–14.
Ndi, Frankline A. and Batterbury, Simon (2017): Land Grabbing and the Axis of Political Conflicts: Insights from Southwest Cameroon, in *Africa Spectrum*, 52, 1, 33–63.
Ngugi (1986): Decolonising the Mind: The Politics of Language, in African Literature: James Currey: Nairobi.
Ngwa, Kenneth (2018): Postwar hermeneutics: bible and colony-related necropolitics, in Colonialism and the bible: contemporary reflections from the global south, Lexington books: Lanham, pp. 43–74.
Nietzsche, Frederick (2007): Ecce Homo, Oxford: Oxford University Press.
Nietzsche, Friedrich (2003): Thus spoke Zarathustra, Penguin Classics: London.
Nigerian National Bureau of Statistics (2022): *Nigeria Launches its Most Extensive National Measure of Multidimensional Poverty*, https://nigerianstat.gov.ng/news/78, 06.07.2023.
Nkrumah, Kwame 1970. *Class Struggle in Africa*. London: PANAF.
Nkrumah, Kwame 1978. *Consciencism: Philsophy and Ideology for De-colonization*. London: PANAF.
Nogueira, Simone Gibran (2013): Ideology of White Racial Supremacy: Colonization and De-Colonization Processes, *Psicologia & Sociedade*, 25, pp. 23–32.
Norris, John and Bruton, Bronwyn (2011): Twenty Years of Collapse and Counting the Cost of Failure in Somalia, A Joint Report from the Center for American Progress and One Earth Future Foundation.
Norwegian Refugee Council (2022): Global Report on Internal Displacement 2022, Children and youth in internal displacement, Internal Displacement Monitoring Center.
Nussbaum, Martha (2000): Women and Human Development: The Capabilities Approach, Cambridge University Press: Cambridge.
Nyerere, Julius K. 1968. *Ujamaa: Essays on African Socialism*. Dar es Salam.
Nyman, Jopi (2020): Narratives of contemporary Im/mobility: writing forced migration in the borderscape, in Sbiri et al., Mobile identities: race, ethnicity, and borders in contemporary literature and culture, Cambridge Scholars Publishing: Newcastle, pp. 15–34.
O'Rourke, Dennis (2014): Why do we migrate A retrospective, in Michael Crawford and Benjamin Campbell, *Causes and Consequences of Human*

*Migration: An evolutionary perspective,* Cambridge University Press: Cambridge, pp. 527–536.
Oatley, Thomas (2019): International Political Economy, Routledge: New York.
Ochonu, Moses E. (2020): South African Afrophobia in local and continental contexts, *Journal of Modern African Studies*, 58: 4, pp. 499–519.
ODI (2018): Africa's economic growth in a new global context, https://odi.org/en/events/africas-economic-growth-in-a-new-global-context/.
Omuziligbo, Maya (n.d.): About Physical and Mental Health Effects of Racism and Discrimination on African Americans, https://smhp.psych.ucla.edu/qf/ptsd.htm.
Online etymology dictionary (2023): Colonize, Online Etymology Dictionary, https://www.etymonline.com/search?q=colonize.
Onyewuenyi, Innocent Chilaka 1998. *Is There an African philosophy*, in: Claude Sumner (ed.) African philosophy. Addis Abeba: Addis Abeba University Press, pp. 243–260.
Oram, Julian (2014): The Great Land Heist: How donors and governments are paving the way for corporate land grabs, ActionAid International.
Ostebo, Terje (2012): Islamic militancy in Africa, Africa Security Brief, The Africa Center for Strategic Studies, No 23, November 2012.
OXFAM (2013): Poor Governance, Good Business : How land investors target countries with weak governance, OXFAM MEDIA BRIEFING, 7 February 2013 Ref: 03/2013.
Painter, Nell Irvin (2010): The History of white people, Norton and Company: New York.
Palaver, Wolfgang (2003): René Girards mimetische Theorie, LIT: Münster.
Palaver, Wolfgang (2019): Populism and religion: On the politics of fear, Dialog, 58: pp. 22–29, https://doi.org/10.1111/dial.12450
Papastavridis, Efthymios (2020): Search and rescue at sea: shared responsibilities in the Mediterranean See, in Francesca Ippolito, et al., Bilateral Relations in the Mediterranean: Prospects for migration issues, Edward Elgar: Cheltenham, pp. 229–249.
Pattanayak, Banikinkar (2022): India-Africa trade ties offer huge potential but challenges remain, https://www.financialexpress.com/economy/india-africa-trade-ties-offer-huge-potential-but-challenges-remain/2603269/.
Paul, Anju Mary (2011): "Stepwise International Migration: A Multistage Migration Pattern for the Aspiring Migrant." American Journal of Sociology 116 (6): 1842–1886. https://doi.org/10.1086/659641.
Pedersen, Anne and Hartley L. (2015): Can We Make a Difference? Prejudice Towards Asylum Seekers in Australia and the Effectiveness of Antiprejudice Interventions, *Journal of Pacific Rim Psychology*, 9, 1–14.
Petit, Véronique and Charbit, Yves (2018): Economic Migrations, Households and Patriarchy In Africa And Southern Europe, HAL Id: halshs-02113339, https://halshs.archives-ouvertes.fr/halshs-02113339 (accessed 04.05.2023).

Phillips, Jon (2018): Who's in Charge of Sino-African Resource Politics? Situating African State Agency in Ghana, *African Affairs*, 118/470, 101–124.
Pildegovičs, Tomass et al. (2021): Russia's Activities in Africa's Information Environment: Case Studies: Mali and Central African Republic, NATO Strategic Communications Centre of Excellence.
Pilling, David and Schipani, Andres (2023) War in Tigray may have killed 600,000 people, peace mediator says, https://www.ft.com/content/2f385e95-0899-403a-9e3b-ed8c24adf4e7, 30.05.2023.
Piore, Michael J. (1979). Birds of passage. Migrant labour in industrial societies, (Cambridge University Press, Cambridge).
Pitts, Jennifer (2005): A Turn to empire: the rise of imperial liberalism in Britain and France, Princeton University Press: Princeton.
Polanyi, Karl (2001): The Great Transformation: The Political and Economic Origins of Our Time, Beacon Press.
Pombo, Maria Dolores (2019): Violence at the US-Mexican border, in The Oxford Handbook of Migration Crisis, Oxford University Press: Oxford, pp. 485–500.
Pope Alexander VI. (1493): Demarcation Bull Granting Spain Possession of Lands Discovered by Columbus, Rome, May 4, 1493.
Portes, A. (1997) 'Immigration theory for a new century: some problems and opportunities', *International Migration Review*, 31(4): 799–825.
Portes, A. and DeWind, J. (2004) 'A cross-Atlantic dialogue: the progress of research and theory in the study of international migration', *International Migration Review*, 38(3): 828–851.
Poster, Mark (1988): Baudrillard: Selected Writings, Stanford: Stanford University Press.
Quartey, Peter et al. (2020): Migration across West Africa: development-related aspects, in International Organization for Migration, Migration in West and North Africa and across the Mediterranean: Trends, risks, development and governance, Geneva, pp. 270–278.
Rabanka, Reiland (2015): The Negritude Movement, Lexington Books: Lanham.
Rachman, Gideon (2022): The Age of the Strong-Man, The Bodley Head, London.
Rae, H. 2002. *State Identities and the Homogenisation of Peoples.* Cambridge: Cambridge University Press.
Rahner, K. 1979. *Theological Investigations.* Vol. 16, translated by David Morland. London: Darton, Longman & Todd.
Rahner, Karl 1976. *Theological Investigations.* Vol. 14, translated by David Bourke. London: Darton: Longman & Todd.
Ramani, Samuel (2021): Russia and China in Africa: Prospective Partners or Asymmetric Rivals? South African Institute of International Affairs, Policy Insights 120, December 2021.

Ramani, Samuel (2022): Russia Has Big Plans for Africa. America Must Push Back—Without Getting Dragged In, https://www.foreignaffairs.com/print/node/1128391.

Rao, Smriti and Vakulabharanam, Vamsi (2019): Migration, Crisis, and Social Transformation in India, in The Oxford Handbook of Migration Crisis, Oxford University Press: Oxford, pp. 261–278.

Ray, Rashwan (2022): The Russian invasion of Ukraine shows racism has no boundaries, https://www.brookings.edu/articles/the-russian-invasion-of-ukraine-shows-racism-has-no-boundaries/, 07.07.2023.

Reslow, Natasja (2019): Transformation or continuity? EU external migration policy in the aftermath of the migration crisis, in Sergio Carrera et al. (eds.), Constitutionalizing the external dimensions of EU migration policies in times of crisis, Edward Elgar: Cheltenham, pp. 95–115.

Reuters (2018): Factbox: Russian military cooperation deals with African countries, https://www.reuters.com/article/us-africa-russia-factbox-idUSKCN1MR0KH, 20.06.2022.

Reyntjens, Filip (2020): Respecting and Circumventing Presidential Term Limits in Sub-Saharan Africa: A Comparative Survey, *African Affairs*,119: 475, pp. 275–295.

Richards, Eric (2019): Migrants in Crisis in Nineteenth-Century Britain, *in The Oxford Handbook of Migration Crisis, Oxford University Press: Oxford*, pp. 37–56.

Rigaud, Kanta Kumari et al. (2018): GROUNDSWELL: Preparing for Internal Climate Migration, International Bank for Reconstruction and Development / The World Bank.

Robilliard, Anne-Sophie (2022): What's New About Income Inequality in Africa? World Inequality Lab Issue Brief 2022-09, November 2022.

Rockefeller, Steven (1994): Comment, The politics of recognition, in Charles Taylor et al., multiculturalism, Princeton University Press, Princeton, pp. 87–98.

Roos, Christof and Zaun, Natascha (2016): The global economic crisis as a critical juncture? The crisis's impact on migration movements and policies in Europe and the U.S., *Journal of Ethnic and Migration Studies*, 2016, Vol. 42, No. 10, 1579–1589.

Rosnick, Danielle (2022): What does the war in Ukraine mean for Africa? https://www.brookings.edu/blog/africa-in-focus/2022/02/25/what-does-the-war-in-ukraine-mean-for-africa/?utm_campaign=Global%20Economy%20and%20Development&utm_medium=email&utm_content=205330541&utm_source=hs_email, 20.06.2022.

RSF (2022): Reporters Without Borders Index, https://rsf.org/en/index.

Ruch, E.A. 1998. *African Philosophy.* Regressive or Progressive? In: Claude Sumner (ed.) African philosophy, Addis Abeba: Addis Abeba University Press, pp. 261–276.

Ruth Hall (2011) Land grabbing in Southern Africa: the many faces of the investor rush, *Review of African Political Economy*, 38:128, 193–214, https://doi.org/10.1080/03056244.2011.582753.

Sager, Alex (2019): Ethics and Migration Crisis, in The Oxford Handbook of Migration Crisis, Oxford University Press: Oxford, pp. 589–602.

Said, Edward (1989): Representing the Colonized: Anthropology's Interlocutors, *Critical Inquiry*, 15, 2, pp. 205–225.

Saleh, Mariam (2022): Internet usage in Africa – statistics & facts, https://www.statista.com/topics/9813/internet-usage-in-africa/.

Sanny, Josephine Appiah-Nyamekye and Selormey, Edem (2021): Africans welcome China's influence but maintain democratic aspirations, Afrobarometer Dispatch No. 489, 15 November 2021.

Sanny, Josephine Appiah-Nyamekye et al. (2019): In search of opportunity: Young and educated Africans most likely to consider moving abroad, Afrobarometer 2019, Dispatch No. 288 | 26 March 2019.

Sartre, Jean-Paul (1993a): Being and Nothingness, Washington Square Press: Washington.

Sartre, Jean-Paul (1993b): No Exit, Vintage: New York.

Sartre, Jean-Paul (2007): Existentialism is a Humanism, New Haven: Yale University Press.

Saunders, H.W. (1956): 'Human Migration and Social Equilibrium', in: J.J. Spengler & O.D. Duncan (Eds), Population Theory and Policy, New York: Free Press.

Sbiri, Kamal et al. (2020): Introduction, in Sbiri et al., Mobile identities: race, ethnicity, and borders in contemporary literature and culture, Cambridge Scholars Publishing: Newcastle, pp. 1–14.

Sbri, Kamal (2020): Black transits: Transition, recognition, and bordering in Marie Ndiaye's *Three strong women*, in Sbiri et al., Mobile identities: race, ethnicity, and borders in contemporary literature and culture, Cambridge Scholars Publishing: Newcastle, pp. 35–56.

Schapendonk, J., van Moppes, D., (2007). Migration and Information: Images of Europe, migration encouraging factors and en route information sharing (No. 16), Working Papers Migration and Development Series. Radboud University, Nijmegen.

Schaub, Max Leonard (2012): Lines across the desert: mobile phone use and mobility in the context of trans-Saharan migration, *Information Technology for Development*, Vol. 18, No. 2, April 2012, 126–144.

Schneidman, Witney (2022): Will Biden deliver on his commitment to Africa in 2022?, https://www.brookings.edu/blog/africa-in-focus/2022/01/10/will-biden-deliver-on-his-commitment-to-africa-in-2022/.

Schöfberger, Irene (2020): Migration aspirations in West and North Africa: what do we know about how they translate into migration flows to Europe? In International Organization for Migration, Migration in West and North Africa and across the Mediterranean: Trends, risks, development and governance, Geneva, pp. 87–97.

Schöfelberger, Irene et al. (2020): Migration aspirations in West and North Africa: what do we know about how they translate into migration flows to Europe? Migration in West and North Africa and across the Mediterranean Trends, risks, development and governance International Organization for Migration (IOM), Geneva, pp. 87–97.

Scott, C. D. (2021). Does China's involvement in African elections and politics hurt democracy? Democracy in Africa. 27 September.

Sen, Amartya (2000): Development as Freedom, Anchor Book, New York.

Sen, Amartya (2006): Identity and violence: the illusion of destiny, Norton: New York.

Sen, Amartya (2009): The idea of justice, Allen Lane: London.

Senghor, Léopold S.1967. *Négritude und Humanismus*. Düsseldorf.

Senghor, Leopold Sedar (1998): Negritude and African socialism, in Pieter H. Coetzee and Abraham Roux (eds.), The African Philosophy reader, Routledge: New York, pp. 438–448.

Şentürk, Züliye Acar and Tarhan, Ahmet (2017): Life Styles and Consumption Culture Changing in Parallel with Globalization, Different aspects of globalization / Bünyamin Ayhan (ed.): Peter Lang, Frankfurt am Main, pp. 101–124.

Shakespeare, William (2006): Othello, Oxford World Classics, Oxford: Oxford University Press.

Shakespeare, William (2015): A Midsummer Night's Dream, Penguin Classics, London: Penguin Books.

Sharma, Ruchir (2016): The rise and fall of nations: forces of change in the post-crisis world, Norton and Company: New York.

Shillington, K. 2005. *History of Africa*. 2nd rev. ed., Oxford: Macmillan Education.

Shinn, David (2015): Turkey's Engagement in Sub-Saharan Africa: Shifting Alliances and Strategic Diversification, Chatham House, the Royal Institute of International Affairs, Research Paper.

Siegle, Joseph (2021): Russia and Africa: Expanding, Influence and Instability, in Graeme P. Herd, ed., Russia's Global Reach: A Security and Statecraft Assessment (Garmisch-Partenkirchen: George C. Marshall European Center for Security Studies, 2021).

Siegle, Joseph (2022b): The future of Russia-Africa relations, in *Foresight Africa*: top priorities for the continent in 2022, pp. 117–118.

Siegle, Joseph and Smith, Jeffrey (2022): Putin's World Order Would Be Devastating for Africa. Moscow is already deeply involved in destabilizing wars, https://foreignpolicy.com/2022/05/30/russia-war-africa/.

Signé, Landry (2017): Innovating Development in Africa: The role of international, regional and national actors, Cambridge University Press: Cambridge.

Signé, Landry (2019): Vladimir Putin is resetting Russia's Africa agenda to counter the US and China, https://www.brookings.edu/opinions/vladimir-putin-is-resetting-russias-africa-agenda-to-counter-the-us-and-china/, 20.06.2022.

Signé, Landry (2022): US Secretary of State Blinken to visit Africa as tension with China and Russia intensifies, https://www.brookings.edu/blog/africa-in-focus/2022/08/05/us-secretary-of-state-blinken-to-visit-africa-as-tension-with-china-and-russia-intensifies/?utm_campaign=Global%20Economy%20and%20Development&utm_medium=email&utm_content=222851075&utm_source=hs_email.

Sikainga, Ahmad (2009): 'The World's Worst Humanitarian Crisis': Understanding the Darfur Conflict, https://origins.osu.edu/article/worlds-worst-humanitarian-crisis-understanding-darfur-conflict?language_content_entity=en, 30.05.2023.

Silverman, Max 1999: *Facing Postmodernity.* London: Routledge.

Sindima, H. J. 1998. *Religious and Political Ethics in Africa: A moral inquiry.* Westport: Conn., Greenwood Press.

Singer, J. D. (1960), International conflict: Three levels of analysis, *World Politics*, 12:3, 453–461.

Sintayehu, Firehwot (2021): The drivers of youth migration in Addis Ababa, in Asnake Kefale and Fana Gebresenbet, Youth on the move: Views from below on Ethiopian international migration, Hurst and Company: London, pp.65–78.

Siochrú, Seán Ó. (2004): Social consequences of the globalization of the media and communication sector: some strategic considerations, *Policy Integration Department Word Commission on the Social Dimension of Globalization*, International Labour Organization, Geneva.

Sjaastad, L. A. (1962): The costs and returns of human migration. Journal of Political Economy, 70(5): 80–93.

Snyder, Timothy (2018): The Road to Unfreedom: Russia, Europe, and America, Tim Duggan Books: New York.

Somin, Ilya (2020): Free to Move: Foot Voting, Migration, and Political Freedom, Oxford University Press: Oxford.

Sophocles (1984): The Three Theban Plays: Antigone, Oedipus the King, Oedipus at Colonos, Penguin Classics, London: Penguin Books.

Soulé, Folashadé (2020): 'Africa+1'summit Diplomacy And The 'New Scramble'narrative: Recentering African Agency, *African Affairs*,119:477, pp. 633–646.

Soyinka, Wole (2012): Of Africa, Yale University Press: New Haven.

Speke, John Hanning (1864): THE DISCOVERY OF THE SOURCE OF THE NILE, https://www.gutenberg.org/files/3284/3284-h/3284-h.htm#link2HCH0014, 04.04.2023S.

Stapleton, Sarah Opitz et al. (2017): Climate change, migration and displacement: The need for a risk-informed and coherent approach, UNDP.

Stark, O. (1991): The migration of labor. Basil Blackwell, Cambridge, MA.
Statista (2022a): General government gross debt in relation to Gross Domestic Product (GDP) in Sub-Saharan Africa as of 2021, by country, https://www.statista.com/statistics/1223393/national-debt-in-relation-to-gdp-in-sub-saharan-africa-by-country/.
Statista (2022b): Number of smartphone subscriptions in Sub-Saharan Africa from 2011 to 2027, https://www.statista.com/statistics/1133777/sub-saharan-africa-smartphone-subscriptions/.
Statista Research Department (2016): Number of TV households in Sub-Saharan Africa 2010–2021, https://www.statista.com/statistics/287739/number-of-tv-households-in-sub-saharan-africa/.
Stearns, Jason K. (2022): Rebels without a Cause: The New Face of African War, *Foreign Affairs,* May/June 2022, pp. 143–156.
Stendhal (2002): The Red and the Black, Penguin Classics: London: Penguin Books.
Stoller, Paul (2013): Strangers are like the mist: language in the push and pull of the African diaspora, in Abdoulaye Kane and Todd Leedy, African Migration: patterns and perspectives, Indiana University Press, Bloomington, pp. 158–172.
Stouffer, Samuel A. (1962): Social Research to Test Ideas Free Press: New York.
Stronski, Paul (2019): Late to the Party: Russia's Return to Africa, Carnegie Endowment for International Peace, October 2019.
Stronski, Paul (2023): Russia's Growing Footprint in Africa's Sahel Region, February 28, 2023, https://carnegieendowment.org/2023/02/28/russia-s-growing-footprint-in-africa-s-sahel-region-pub-89135.
Sudan Tribune (2013): In unusual rebuke, Saudi Arabia accuses Ethiopia of posing threats to Sudan & Egypt, http://www.sudantribune.com/spip.php?article45666, accessed August 12, 2013.
Sun, Irene Yuan et al. (2017): Dance of the lions and dragons: how are Africa and China engaging, and how will the partnership evolve? JUNE 2017, McKinsey & Company.
Sun, Yun (2021): FOCAC 2021: China's retrenchment from Africa? https://www.brookings.edu/blog/africa-in-focus/2021/12/06/focac-2021-chinas-retrenchment-from-africa/, 29.06.2022.
Swai, Elinami Veraeli (2009): Trafficking of young women and girls: A case of "Au Pair" and domestic laborer in Tanzania, in, Toyin Falola and Niyi Afolabi (eds.), The Human Costs of African Migration, pp. 145–164.
Sweet, Rachel (2020): Bureaucrats at War: The Resilient State in the Congo, *African Affairs,*119: 475, pp. 224–250.
Tadesse Abebe, Tsion and Mugabo, John (2019): Migration and Security in Africa: Implications for the Free Movement of Persons Agenda, in Africa Migration Report, International Organization for Migration, pp. 145–12.
Taylor, Charles (1994): The politics of recognition, in Charles Taylor et al., Multiculturalism, Princeton University Press, Princeton, pp. 25–74.

Taylor, Ian (2006): China and Africa: engagement and compromise, London: Routledge.
Tazreiter, Claudia (2019): Narratives of Crisis Migration and Power of Visual Culture, in The Oxford Handbook of Migration Crisis, Oxford University Press: Oxford, pp. 619–614.
Tertullianus, Q. S. F. 1952. *Apologeticum*, München: Kösel.
The Economist (2019a): The new scramble for Africa: This time, the winners could be Africans themselves, 7 March 2019.
The Economist (2019b): A sub-Saharan seduction. Africa is attracting ever more interest from powers elsewhere: They are following where China led, 7 March 2019.
The Economist (2021a): EU border policy: New kings of the wild frontier, & March, pp. 19–21.
The Economist (2021b): Fighting Jihadists: Macron's African mission, 20 February, p. 10.
The Economist (2021c): African Odyssey: Many more Africans are migrating within Africa than to Europe, 30 October, pp. 25–27.
The Economist (2021d): China and Africa: Pomp and Circumspection, 4 December 2021, p. 33.
The Economist (2021e): Democracy in Africa: Time out, 9 January, 2021, p. 11.
The Economist (2021f): Too many African countries are letting presidential term limits slip: Few presidencies improve with time, https://www.economist.com/leaders/2021/01/07/too-many-african-countries-are-letting-presidential-term-limits-slip, 23.07.2023.
The Economist (2022a): Business in Africa: Ottomanpower, May 7th 2022, p. 58.
The Economist (2022b): Turkey and Africa: The call of the South, April 23rd, p. 31.
The Economist (2022d): Africa's debt crisis: Debt and denial, April 30th April, p. 29.
The Economist (2022e): Geopolitics: Friends like these, April 16th April, p. 50.
The Economist (2022g): Mercenaries: Vladimir's Army, April 9 April, p. 48.
The Economist (2022h): The ripples of Putin's war: Bread and Oil, 12 March, pp. 26–28.
The Economist (2022j): China and Africa: Chasing the dragon, 19 February, pp. 29–31.
The Economist (2022k): Jihadists in the Sahel: French Leave, 19 April, p. 31.
The Economist (2022l): African Economies: When you are in a hole…, 8 January 2022, pp. 24–25.
The Economist (2022m): Crossing the Mediterranean: An EU-funded horror story, 15 January, pp. 31–32.
The Economist (2022n): Russia and Africa: Wagner, worse than it sounds, 15 January, pp. 31–32.

The Economist (2022o): Unequal Partnership, Special Report: China in Africa, pp. 1–12.

The Economist (2022q): Echoes of war, A resurgence of regional rivalries imperils eastern Congo: Meddling neighbors add to the mayhem, 30 June 2022, https://www.economist.com/middle-east-and-africa/2022/06/30/a-resurgence-of-regional-rivalries-imperils-eastern-congo.

The Economist (2022r): How oil-rich Nigeria failed to profit from an oil boom, https://www.economist.com/middle-east-and-africa/2022/09/11/how-oil-rich-nigeria-failed-to-profit-from-an-oil-boom, 24.05.2023.

The Economist (2022s): The son also rises. Is Uganda heading for a dynastic succession? It would set the country on a dangerous course, https://www.economist.com/middle-east-and-africa/2022/06/16/is-uganda-heading-for-a-dynastic-succession.

The Economist (2022t): A dark state. Vladimir Putin is in thrall to a distinctive brand of Russian fascism: That is why his country is such a threat to Ukraine, the West and his own people, https://www.economist.com/briefing/2022/07/28/vladimir-putin-is-in-thrall-to-a-distinctive-brand-of-russian-fascism.

The Economist (2022u): Hunger in the Horn. Somalia is on the brink of starvation: Drought, and the war in Ukraine, are causing the first famine of the global food crisis, https://www.economist.com/middle-east-and-africa/2022/07/25/somalia-is-on-the-brink-of-starvation.

The Economist (2022v): How al-Qaeda and Islamic State are digging into Africa: The terrorist group's African franchises are now punchier than those in the Middle East, 13 August 2022, pp. 27–18, https://www.economist.com/middle-east-and-africa/2022/08/11/how-al-qaeda-and-islamic-state-are-digging-into-africa.

The Economist (2022w): The ones who sweep: In the Gulf 99% of Kenyan migrant workers are abused, a poll finds, https://www.economist.com/middle-east-and-africa/2022/09/15/in-the-gulf-99-of-kenyan-migrant-workers-are-abused-a-poll-finds, 30.06.2023.

The Economist (2023): Myths and mummies: A new docu-drama about Cleopatra has riled officials in Cairo, https://www.economist.com/culture/2023/05/10/a-new-docu-drama-about-cleopatra-has-riled-officials-in-cairo, 07.07.2023.

The Guardian (2013): Chinese firm steps up investment in Ethiopia with 'shoe city', http://www.theguardian.com/global-development/2013/apr/30/chinese-investment-ethiopia-shoe-city, accessed August 12, 2013.

The World Bank (2022a): Net official development assistance received (constant 2020 US$) – Sub-Saharan Africa, https://data.worldbank.org/indicator/DT.ODA.ODAT.KD?locations=ZG.

The World Bank (2022b): Worldwide Governance Indicators, https://databank.worldbank.org/source/worldwide-governance-indicators.

Thielemann, E. (2003): 'Does Policy Matter? On Governments' Attempts to Control Unwanted Migration', *European Institute Working Paper*, May 2003, available at: http://www.lse.ac.uk/collections/europeaninstitute/pdfs/Eiworkingpaper2003-02.pdf, accessed May 05, 2005.
Thiong'o, Ngugi wa (2005): Decolonizing the mind: the politics of language in African literature, James Curry: Nairobi.
Thompson-Miller, Ruth and Feagin, Joe R. (2007): Continuing Injuries of Racism: Counseling in a Racist Context, *The Counseling Psychologist*, 35, 1, pp. 106–115.
Thornton, John (2001), 'The Kuznets inverted-U hypothesis: panel data evidence from 96 countries.' *Applied Economics Letters* 8 (1): 15–16.
Thussu, Daya (2016): The Scramble for Asian Soft Power in Africa, La ruée vers l'Afrique du Soft power asiatique, pp. 224–237.
Todaro, Michael P. (1976). Internal migration in developing countries, (International Labor Office, Geneva).
Todd, Jennifer (2005): Social Transformation, Collective Categories, and Identity Change, *Theory and Society*, 2005, Vol. 34, No. 4, pp. 429–463.
Tolstoy, Leo (2007): War and Peace, Penguin Books: London.
Tolstoy, Leo (2008): The Death of Ivan Ilyich and Other Stories, Penguin Classics, London: Penguin Books.
Tom Wilson, David Blood, and David Pilling (2019): Congo voting data reveal huge fraud in poll to replace Kabila, https://www.ft.com/content/2b97f6e6-189d-11e9-b93e-f4351a53f1c3, accessed 15.06.2022.
Transparency International (2019): Where are Africa's billions? https://www.transparency.org/en/news/where-are-africas-billions.
Transparency International (2020): A $100 million question for Nigeria's asset recovery efforts, https://www.transparency.org/en/blog/a-100-million-question-for-nigerias-asset-recovery-efforts.
Tufa, Fekadu Adugna et al. (2021): Say what the government wants and do what is good for your family: facilitation of irregular migration in Ethiopia, in Asnake Kefale and Fana Gebresenbet, *Youth on the move: Views from below on Ethiopian international migration*, Hurst and Company: London, pp. 43–64.
Turner, Bryan S. (2010): Enclosures, Enclaves, and Entrapment, *European Journal of Social* Theory 10(2):287–303.
Twagiramungu, Noel (2019): Re-describing transnational conflict in Africa, Journal of Modern African Studies, 57:3, pp. 377–391.
TWN (2016): Hiding Africa's looted funds: The silence of Western media, https://www.twn.my/title2/resurgence/2016/309/cover04.htm, 21.07.2023.
Twomey, Hannah (2014): Displacement and dispossession through land grabbing in Mozambique: The limits of international and national legal instruments, The Refugee Studies Centre (RSC) Working Paper Series.

UN (2015): Slave Trade: International Day of Remembrance of the Victims of Slavery and the Transatlantic Slave Trade, https://www.un.org/en/observances/decade-people-african-descent/slave-trade, 29.05.2023.

UN DESA (2022): World Population Prospects 2022, United Nations Department of Economic and Social Affairs, Population Division, New York.

UNCTAD (2018): Economic Development in Africa Report 2018, Migration for Structural Transformation, New York and Geneva.

UNCTAD (2020): *Tackling Illicit Financial Flows for Sustainable Development in Africa*, Economic Development in Report 2020, United Nations: Geneva T.

UNECA (2017): African Migration: Drivers of Migration in Africa, Draft Report Prepared forAfrica Regional Consultative Meeting on the Global Compact on Safe, Orderly and Regular Migration, October 2017.

UNDP (2009b): HUMAN DEVELOPMENT REPORT 2009: Overcoming barriers: Human mobility and development.

UNDP 2009. 'Human Development Report 2009: Overcoming barriers: Human mobility and development'. New York: United Nations Development Programme, http://hdr.undp.org/en/reports/global/hdr2009/.

UNDP (2019): Human Development Report: Scaling Fences: Voices of Irregular African Migrants to Europe, UNDP.

United States Census Bureau (2022): Trade in Goods with Africa, https://www.census.gov/foreign-trade/balance/c0013.html.

Usman, Zainab (2021): How Biden Can Build U.S.-Africa Relations Back Better, https://carnegieendowment.org/2021/04/27/how-biden-can-build-u.s.-africa-relations-back-better-pub-84399.

Usman, Zainab and Abayo, Aline (2022): Will U.S.-China Competition Shape Kenya's Trade Trajectory? https://carnegieendowment.org/2022/09/15/will-u.s.-china-competition-shape-kenya-s-trade-trajectory-pub-87919?mkt_tok=ODEzLVhZVS00MjIAAAGG4YA0W9biEIfzLVVwEhsPL1bXJrDH0qqXzxqwgyeb4_DrZSaVTjZDq3PoOpIPxJwEbMKSIj0K-4fjIR-ayvrw6W7Ziri86MB9ez7AGsyC, 24.05.2023.

Utsey, Shawn O. and Payne, Yasser (2000): Psychological Impacts of Racism In a Clinical Versus Normal Sample of African American Men, *Journal of African American Men*, 5, 3, pp. 57–72.

Van de Walle, Nicolas (2001): African Economies and the Politics of Permanent Crisis, Cambridge University Press: Cambridge.

Van Hear, Nicholas et al. (2018): Push-pull plus: reconsidering the drivers of migration, Journal of Ethnic and Migration Studies, 44:6, 927–944, https://doi.org/10.1080/1369183X.2017.1384135.

Wa Thiong'o, Ngugi (1986): Decolonizing the Mind: The Politics of Language in African Literature, James Currey: Woodbridge.

Walia, Harsha (2021): Border and Rule: Global migration, capitalism, and the rise of racist nationalism, Haymarket Books: Chicago.

Walkey, Claire and Mbokazi, Sabelo (2019): The Migration–Development Nexus in Africa, in Africa Migration Report, International Organization for Migration, pp. 133–143.
Waltz, K. N. (1979), *Theory of international politics* (New York: McGraw-Hill).
Washington Post (2013): Egyptians up in arms as Ethiopia builds giant hydro dam on Nile River; minister rules out war, http://www.washingtonpost.com/world/africa/egyptians-up-in-arms-as-ethiopia-builds-giant-hydro-dam-on-nile-river-minister-rules-out-war/2013/05/30/bb284bb4-c95a-11e2-9cd9-3b9a22a4000a_story_1.html#, accessed 13 June 2013.
Weima, Yolanda and Hindman, Jennifer (2019): Managing displacement: negotiating transnationalism, encampment and return, in Critical geographies of migration, Edward Elgar, pp. 30–44.
Wendt, A. (1987): The agent-structure problem in international relations theory, *International Organization*, 41, 3, pp. 335–370, https://doi.org/10.1017/S002081830002751X.
Wendt, A. (1992), Anarchy is what the states make of it: The social construction of power politics, *International Organization*, 46:2, 391–425.
Werbner, Pnina (2012): Migration and culture, in Marc Rosenblum and Daniel Tichenor, The Oxford Handbook of the politics of international migration, Oxford University Press: Oxford, pp. 215–242.
Wezeman, Pieter D. 2009: Arms transfers to Central, North and West Africa. In: SIPRI Background Paper, retrieved 21 July 2009, from http://books.sipri.org/files/misc/SIPRIBP0904b.pdf.
Wezeman, Pieter D. et al. 2019: *Trends in international arms transfers*, in SIPRI Factsheet, March 2020.
Wezeman, Pieter et al. (2022): Trends in International Arms Transfers, SIPRI Fact Sheet, March 2021.
Whitehouse, Bruce (2013): Overcoming the economistic fallacy: social determinants of voluntary migration from the Sahel to the Congo basin, in Abdoulaye Kane and Todd Leedy, African Migration: patterns and perspectives, Indiana University Press, Bloomington, pp. 19–34.
Wodak, Ruth (2011): 'Us' and 'Them': Inclusion and Exclusion – Discrimination via Discourse, in Gerard Delanty et al. (eds.), Identity, Belonging and Migration, Liverpool University Press: Liverpool, p. 54–77.
Wodak, Ruth (2015): The politics of fear: what right-wing populist discourses mean, Sage: Los Angeles.
Wonders, Nancy and Jones, Lynn (2021): Challenging the borders of difference and inequality: power in migration as a social movement for global justice, in Handbook of Migration and Global Justice, Leanne Weber and Claudia Tazreiter, eds., Edward Elgar, Cheltenham, pp. 296–313.

World Bank Group (2022): A War in a Pandemic: Implications of the Ukraine crisis and COVID-19 on global governance of migration and remittance flows, Migration and Development Brief 36.

World Economic Forum (2022): The Horn of Africa is facing an unprecedented drought. What is the world doing to help solve it? https://www.weforum.org/agenda/2022/07/africa-drought-food-starvation/.

World Peace Foundation (2015): Mass Atrocity Endings: Angola: civil war, https://sites.tufts.edu/atrocityendings/2015/08/07/angola-civil-war/, 13.07.2023.

Wouter Pinxten and John Lievens (2014): The importance of economic, social and cultural capital in understanding health inequalities: using a Bourdieu-based approach in research on physical and mental health perceptions, *Sociology of Health & Illness*, Vol. 36 No. 7 2014 ISSN 0141-9889, pp. 1095–1110, https://doi.org/10.1111/1467-9566.12154.

Youngstedt, Scott (2013): Voluntary and involuntary homebodies: adaptations and lived experiences of Hausa "left behind" in Niamey, Niger, in Abdoulaye Kane and Todd Leedy, African Migration: patterns and perspectives, Indiana University Press, Bloomington, pp. 133–157.

Zajontz, Tim (2022): 'Win-win' contested: negotiatingthe privatisation of Africa's Freedom Railway with the 'Chinese of today', *Journal of Modern African Studies*, 60: 1, pp. 111–134.

Zane, Damian (2022): How ex-Gambia President Yahya Jammeh's US mansion was seized, https://www.bbc.com/news/world-africa-58924630.

Zatari, Amalia (2020): Racism in Russia: Stories of prejudice, https://www.bbc.com/news/world-europe-53055857, 07.07.2020.

Zelinsky, Wilbur (1971): The Hypothesis of the Mobility Transition, *Geographical Review*, Vol. 61, No. 2, pp. 219–249.

Zhou, Hang (2022): Western and Chinese Development Engagements In Uganda's Roads Sector: An Implicit Division of Labor, *African Affairs*, 121:482, pp. 29–59.

Žižek, Slavoj 2002. *Welcome to the Desert of the Real*! London: Verso.

Zlotnik, H. (1998): The theories of international migration. Paper for the Conference on International Migration: Challenges for European Populations, Bari, Italy, 25–27 June 1998.

# Index




B. Gebrewold, *Postcolonial African Migration to the West*, Politics of Citizenship and Migration,
https://doi.org/10.1007/978-3-031-58568-5

## C



## D

R



S